HAWKING

ON

THE BIG BANG
AND BLACK
HOLES

ADVANCED SERIES IN ASTROPHYSICS AND COSMOLOGY

Series Editors: Fang Li Zhi and Remo Ruffini

Advanced Series in Astrophysics and Cosmology – Vol. 8

HAWKING

ON
THE BIG BANG
AND BLACK
HOLES

STEPHEN HAWKING

Lucasian Professor of Mathematics
Department of Applied Mathematics and
Theoretical Physics
University of Cambridge
England

World Scientific
Singapore • New Jersey • London • Hong Kong

Published by

World Scientific Publishing Co. Pte. Ltd.

P O Box 128, Farrer Road, Singapore 9128

USA office: Suite 1B, 1060 Main Street, River Edge, NJ 07661

UK office: 73 Lynton Mead, Totteridge, London N20 8DH

We are grateful to the following publishers for their permission to reproduce the articles found in this volume:

The American Physical Society (*Phys. Rev. D* and *Phys. Rev. Lett.*)
Elsevier Science Publishers (*Nucl. Phys. B* and *Phys. Lett. B*)
Springer-Verlag (*Commun. Math. Phys.*)
Cambridge University Press
Gordon and Breach
The Royal Society of London

HAWKING ON THE BIG BANG AND BLACK HOLES

Cover picture from the Image Bank.

ISBN   981-02-1078-7
      981-02-1079-5 (pbk)

Printed in Singapore.

CONTENTS

INTRODUCTION

This collection of papers reflects the problems that I have worked on over the years. With hindsight, it might appear that there had been a grand and premeditated design to address the outstanding problems concerning the origin and evolution of the universe. But it was not really like that. I did not have a master plan; rather I followed my nose and did whatever looked interesting and possible at the time.

There has been a great change in the status of general relativity and cosmology in the last thirty years. When I began research in the Department of Applied Mathematics and Theoretical Physics (DAMTP) at Cambridge in 1962, general relativity was regarded as a beautiful but impossibly complicated theory that had practically no contact with the real world. Cosmology was thought of as a pseudo-science where wild speculation was unconstrained by any possible observations. That their standing today is very different is partly due to the great expansion in the range of observations made possible by modern technology. But it is also because we have made tremendous progress on the theoretical side, and this is where I can claim to have made a modest contribution.

Before 1960, nearly all work on general relativity had been concerned with solving the Einstein equations in particular coordinate systems. One imposed enough symmetry assumptions to reduce the field equations either to ordinary differential equations or to the Laplacian in three dimensions. It was regarded as a great achievement to find *any* closed form solution of the Einstein equations. Whether it had any physical significance was a secondary consideration. However a more geometric approach began to appear in the early 1960s in the work of Roger Penrose and others. Penrose introduced global concepts and showed how they could be used to establish results about spacetime singularities that did not depend on any exact symmetries or details of the matter content of the universe. I extended Penrose's methods and applied them to cosmology. This phase of work on global properties came to an end in about 1972 when we had solved most of the qualitative problems in classical general relativity. The major problem that remains outstanding is the Cosmic Censorship Conjecture. This is very difficult to prove, but all attempts to find genuine counter-examples have failed, so it is probably true.

This global classical phase of my work is represented by the first three papers in this volume. They deal with the classical properties of the two themes that recur throughout my work: the Big Bang and black holes. Nowadays everyone accepts it as natural that the universe had a beginning about 15 billion years ago and that, before that, time simply was not defined. But opinions were very different in the early 1960s. The Steady State school believed that the universe had existed forever more or less as we see it today. Even among those who thought the universe was evolving with time, there was a general feeling that one could not extrapolate back to the extreme conditions near the initial singularity of the Friedmann models and that it was probably just an artifact of the high degree of symmetry of these solutions. Indeed in 1963 Lifshitz and Khalatnikov claimed to have shown that singularities would not occur in fully general solutions of the Einstein equations without exact symmetries. Presumably this would have implied that the universe had a contracting phase and some sort of bounce before the present expansion.

The discovery of the microwave background in 1964 ruled out the Steady State Theory and showed that the universe must have been very hot and dense at some time in the past. But the observations themselves did not exclude the possibility that the universe bounced at some fairly large but not extremely high density. This was ruled out on theoretical grounds by the singularity theorems that Penrose and I proved. The first singularity theorems involved the assumption that the universe had a Cauchy surface. Thus they proved *either* that a singularity would occur *or* that a Cauchy horizon would develop. But in 1970 Penrose and I published "The Singularities of Gravitational Collapse and Cosmology" [1]. This was an all purpose singularity theorem that did not assume the existence of a Cauchy horizon. It showed that the classical concept of time must have a beginning at a singularity in the past (the Big Bang) and that time would come to an end for at least part of spacetime when a star collapsed. Most of my work since then has been concerned with the consequences and implications of these results.

Up to 1970, my work had been concerned with cosmology and in particular with the question of whether the universe had a beginning at a singularity in the past. But in that year I realized that one could also apply the global methods that Penrose and I had developed for the singularity theorems to study the black holes that formed around the singularities that the theorems predicted would occur in gravitational collapse. This was what Kip Thorne has called "The Golden Age of Black Holes", two or three years in which the concept of a black hole as an entity distinct from the collapsing star was established and its main classical properties were deduced. This was a case where theory definitely had the lead over observation. Black holes were predicted theoretically some time before possible black hole candidates were detected observationally.

My two most important contributions to the classical theory of black holes were probably the Area Theorem, which stated that the total area of black hole event horizons can never decrease, and the part I played in proving the No Hair Theorem, which states that black holes settle down to a stationary state that depends only on the mass, angular momentum and charge of the black hole. Most of my work on classical black holes was described in "The Event Horizon" [2], my lectures given at the 1972 Les Houches Summer School on black holes, which was the culmination of the Golden Age. One important part that was not in these lecture notes because it was work carried out actually at Les Houches was a paper on "The Four Laws of Black Hole Mechanics" [3] with J. Bardeen and B. Carter. In it we pointed out that the area of the event horizon and a quantity we called the surface gravity behaved very much like entropy and temperature in thermodynamics. However, they could not be regarded as the actual physical entropy and temperature as Bekenstein had suggested. This was because a black hole could not be in equilibrium with thermal radiation since it would absorb radiation but, as everyone thought at that time, a black hole could not emit anything.

The situation was completely changed however when I discovered that quantum mechanics would cause a black hole to emit thermal radiation with a temperature proportional to the surface gravity. I announced this first in a letter in *Nature* and then wrote a longer paper, "Particle Creation by Black Holes" [4], which I submitted to *Communications in Mathematical Physics* in March 1974. I did not hear anything from them for a year, so I wrote to enquire what was happening. They confessed they had lost the paper and asked me to send another copy. They then added insult to injury by publishing it with a submission date of April 1975, which would have made it later than some of the great flood of papers my discovery led to on the quantum mechanics of black holes. I myself have written a number of further papers on the subject, the most significant of which are "Action Integrals and Partition Functions in Quantum Gravity" [5] with G. W. Gibbons in which we derived the temperature and entropy of a black hole from a Euclidean path integral, and "Breakdown of Predicitability in Gravitational Collapse" [6] in which I showed that the evaporation of black holes seemed to introduce a loss of quantum coherence in that an initial pure quantum state would appear to decay into a mixed state. Interest in this possibility of a non-unitary evolution from initial to final quantum states has recently been reinvigorated by the study of gravitational collapse in two-dimensional field theories in which one can consistently take into account the back reaction to the particle creation. I have therefore included a recent paper of mine, "Evaporation of Two Dimensional Black Holes" [7], as an example.

Event horizons occur in exponentially expanding universes as well as in black holes. G. W. Gibbons and I used Euclidean methods in "Cosmological Event Horizons, Thermodynamics and Particle Creation" [8] to show that de Sitter space had a temperature and entropy like a black hole. The physical significance of this temperature was realized a few years later when the inflationary model of the universe was introduced. It led to the prediction that small density perturbations would be generated in the expanding universe; see "The Development of Irregularities in a Single Bubble Inflationary Universe" [9]. This was the first paper on the subject but it was soon followed by a number of others, all predicting an almost-scale-free spectrum of density perturbations. The detection of fluctuations in the cosmic microwave background by the COBE satellite has confirmed these predictions and can claim to be the first observation of a quantum gravitational process.

In "Zeta Function Regularization of Path Integrals in Curved Spacetime" [10], I introduced to physics what was then a new technique for regularizing determinants of differential operators on a curved background. This was used in "The Path Integral Approach to Quantum Gravity" [11] to develop a Euclidean approach to quantum gravity. This in turn led to a possible answer to the problem that my early work on singularities had raised: How can physics predict how the universe will begin because all the laws will break down in the Big Bang? In "Wave Function of the Universe" [12], J. B. Hartle and I put forward the No Boundary Proposal: The quantum state of the universe is determined by a path integral over all compact positive definite (Euclidean) metrics. In other words, even though spacetime has boundaries at singularities in real Lorentzian time, it has no boundaries in the imaginary direction of time. The action of spacetime is therefore well defined, so the path integral can predict the expectation values of physical quantities without any assumption about initial conditions. In "Quantum Cosmology" [13], another set of Les Houches lectures, I showed that the No Boundary Proposal would imply that the universe would expand in an inflationary manner and in "Origin of Structure in the Universe" [14], J. J. Halliwell and I showed that it would imply that the universe would contain gravitational and density perturbations with an almost-scale-free spectrum. These density perturbations are just what is required to explain the formation of galaxies and other structures in the universe and they agree with the COBE observations. Thus the No Boundary Proposal can explain why the universe is the way it is.

In "Arrow of Time in Cosmology" [15], I pointed out that the results of the "Origin of Structure" paper implied that the universe would have started out in a smooth and ordered state, and would have evolved to a more irregular and disordered state as it expanded. Thus the No Boundary Proposal would explain the existence of a Thermodynamic Arrow of Time that pointed in the direction in which the universe was expanding. However I also claimed that if the universe were to reach a point of maximum size and start to recontract, the Thermodynamic Arrow would reverse. Shortly after writing this paper, I realized that the Thermodynamic Arrow would not in fact reverse in a contracting phase. I added a note to the proofs of the "Arrow of Time" paper but did not get round to writing a fuller explanation until "The No Boundary Proposal and the Arrow of Time" [16].

Another important outcome of the Euclidean approach to quantum gravity was "The Cosmological Constant is Probably Zero" [17]. In it I showed that if the cosmological constant could take a range of values, then zero would be overwhelmingly the most probable. In my opinion this is the only plausible mechanism that has been advanced to account for the extremely low observational upper limits on the cosmological constant. This explanation received fresh impetus when, in "Wormholes in Spacetime" [18], I put forward the idea that there might be thin tubes or wormholes connecting different regions of spacetime. Sydney Coleman showed that such wormholes would change the values of physical constants and could therefore implement this mechanism to make the cosmological constant zero. Coleman went on to suggest that it might determine all the other constants of physics as well. My doubts on this latter claim were expressed in "Do Wormholes Fix the Constants of Nature?" [19].

Recently my interest in the global structure of the universe led me to consider whether the macroscopic topology of spacetime could change. In "Selection Rules for Topology Change" [20], G. W. Gibbons and I showed that there was an important restriction if there was to be a Lorentz metric which allowed spinors to be defined consistently. Roughly speaking, wormholes or handles could be added to the topology of spatial sections only in pairs. However, any topology change necessarily requires the existence of closed time-like curves which in turn implies that one might be able to go back into the past and change it with all the paradoxes that this could lead to. In "Chronology Protection Conjecture" [21], I examined how closed time-like curves might appear in spacetimes that did not contain them initially and

4

I presented evidence that the laws of physics would conspire to prevent them. This would seem to rule out time machines.

I can claim that my work so far has shed light (maybe an unfortunate metaphor) on the Big Bang and black holes. But there are many problems remaining, like the formulation of a consistent theory of quantum gravity and understanding what happens in black hole evaporation. Still, that is all to the good: the really satisfying feeling is when you find the answer to part of Nature's puzzle. There is plenty left to be discovered.

Stephen Hawking
14 January 1993

REFERENCES

[1] The Singularities of Gravitational Collapse and Cosmology (with R. Penrose), *Proc. Roy. Soc.* **A314**, 529 (1970).

[2] The Event Horizon, in *Black Holes* (eds. DeWitt & DeWitt), Gordon and Breach (1973).

[3] The Four Laws of Black Hole Mechanics (with J. M. Bardeen & B. Carter), *Commun. Math. Phys.* **31**, 161 (1973).

[4] Particle Creation by Black Holes, *Commun. Math. Phys.* **33**, 323 (1973).

[5] Action Integrals and Partition Functions in Quantum Gravity (with G. Gibbons), *Phys. Rev.* **D15**, 2725 (1977).

[6] Breakdown of Predictability in Gravitational Collapse, *Phys. Rev.* **D14**, 2460 (1976).

[7] Evaporation of Two-Dimensional Black Holes, *Phys. Rev. Lett.* **69**, 406–409 (1992).

[8] Cosmological Event Horizons, Thermodynamics, and Particle Creation (with G. Gibbons), *Phys. Rev.* **D15**, 2738 (1977).

[9] The Development of Irregularities in a Single Bubble Inflationary Universe, *Phys. Lett.* **B115**, 295–297 (1982).

[10] Zeta Function Regularization of Path Integrals in Curved Spacetime, *Commun. Math. Phys.* **56**, 133 (1977).

[11] The Path-Integral Approach to Quantum Gravity, in *General Relativity: An Einstein Centenary Survey* (ed. with W. Israel), Cambridge University Press (1979).

[12] Wave Function of the Universe (with J. B. Hartle), *Phys. Rev.* **D28**, 2960–2975 (1983).

[13] Quantum Cosmology, Les Houches Lectures, in *Relativity Groups and Topology* (eds. B. Dewitt & R. Stora), North-Holland (1984).

[14] Origin of Structure in the Universe (with J. J. Halliwell), *Phys. Rev.* **D31**, 8 (1985).

[15] Arrow of Time in Cosmology, *Phys. Rev.* **D32**, 2489 (1985).

[16] The No-Boundary Proposal and the Arrow of Time, in *Physical Origins of Time Asymmetry* (eds. J. J. Halliwell, J. Perez-Mercader & W. H. Zurek) Cambridge Univ. Press (1992).

[17] The Cosmological Constant is Probably Zero, *Phys. Lett.* **B134**, 403 (1984).

[18] Wormholes in Spacetime, *Phys. Rev.* **D37**, 904 (1988).

[19] Do Wormholes Fix the Constants of Nature? *Nucl. Phys.* **B335**, 155–165 (1990).

[20] Selection Rules for Topology Change (with G. Gibbons), *Commun. Math. Phys.* **148**, 345–352 (1992).

[21] Chronology Protection Conjecture, *Phys. Rev.* **D46**, 603–611 (1992).

Proc. Roy. Soc. Lond. A. **314**, 529–548 (1970)

Printed in Great Britain

The singularities of gravitational collapse and cosmology

By S. W. Hawking

Institute of Theoretical Astronomy, University of Cambridge

and R. Penrose

Department of Mathematics, Birkbeck College, London

(*Communicated by H. Bondi, F.R.S.—Received* 30 *April* 1969)

A new theorem on space-time singularities is presented which largely incorporates and generalizes the previously known results. The theorem implies that space-time singularities are to be expected if *either* the universe is spatially closed *or* there is an 'object' undergoing relativistic gravitational collapse (existence of a trapped surface) *or* there is a point p whose past null cone encounters sufficient matter that the divergence of the null rays through p changes sign somewhere to the past of p (i.e. there is a minimum apparent solid angle, as viewed from p for small objects of given size). The theorem applies if the following four physical assumptions are made: (i) Einstein's equations hold (with zero or negative cosmological constant), (ii) the energy density is nowhere less than minus each principal pressure nor less than minus the sum of the three principal pressures (the 'energy condition'), (iii) there are no closed timelike curves, (iv) every timelike or null geodesic enters a region where the curvature is not specially alined with the geodesic. (This last condition would hold in any sufficiently general physically realistic model.) In common with earlier results, timelike or null geodesic incompleteness is used here as the indication of the presence of space-time singularities. No assumption concerning existence of a global Cauchy hypersurface is required for the present theorem.

1. Introduction

An important feature of gravitation, for very large concentrations of mass, is that it is essentially *unstable*. This is due, in the first instance, to its r^{-2} attractive character. But, in addition, when general relativity begins to play a significant role, other instabilities may also arise (cf. Chandrasekhar 1964). The instability of gravitation is not manifest under normal conditions owing to the extreme smallness of the gravitational constant. The pull of gravity is readily counteracted by other forces. However, this instability *does* play an important dynamical role when large enough concentrations of mass are present. In particular, as the work of Chandrasekhar (1935) showed, a star of mass greater than about 1.3 times that of the Sun, which has exhausted its resources of thermal and nuclear energy, cannot sustain itself against its own gravitational pull, so a *gravitational collapse* ensues. It has sometimes been suggested also that, on a somewhat larger scale, some form of gravitational collapse may be taking place in quasars, or perhaps in the centres of (some?) galaxies. Finally, on the scale of the universe as a whole, this instability shows up again in those models for which the expansion eventually reverses, and the entire universe becomes involved in a gravitational collapse. In the reverse direction in time there is also the 'big bang' initial phase which is common to most relativistic expanding models. This again may be regarded as a manifestation of the instability of gravitation (in reverse).

But what is the ultimate fate of a system in gravitational collapse? Is the picture that is presented by symmetrical exact models accurate, according to which a *singularity in space-time* would ensue? Or may it not be that any asymmetries present might cause the different parts of the collapsing material to miss each other, so possibly to lead to some form of *bounce*? It seems that until comparatively recently many people had believed that such an asymmetrical bounce might indeed be possible to achieve, in a manner consistent with general relativity (cf. particularly, Lindquist & Wheeler 1957; Lifshitz & Khalatnikov 1963). However, some recent theorems† (Penrose 1965a; Hawking 1966a, b; H; Geroch 1966) have ruled out a large number of possibilities of this kind. The present paper carries these results further, and considerably strengthens the implication that a singularity-free bounce (of the type required) does not seem to be realizable within the framework of general relativity.

In the first theorem (referred to as I; see Penrose 1965a; cf. also Penrose 1966; P; Hawking 1966c) the concept of the existence of a *trapped surface*‡ was used as a characterization of a gravitational collapse which has passed a 'point of no return'. On the basis of a *weak energy condition*,‡ the intention was to establish the existence of space-time singularities from the existence of a trapped surface. Unfortunately, however, theorem I required, as an additional hypothesis, the existence of a non-compact global Cauchy hypersurface. Although 'reasonable' from the point of view of classical Laplacian determinism, the assumption of the existence of a global Cauchy hypersurface is hard to justify from the standpoint of general relativity. Also, it is violated in a number of exact models. Furthermore, the non-compactness assumption used in theorem I applies only if the universe is 'open'.

The second theorem (Hawking 1966a), and its improved version (referred to as II, see H; cf. also Hawking (1966c) and P), required the existence of a compact spacelike hypersurface with everywhere diverging normals. Thus it applies to 'closed', everywhere expanding, universe models. For such models II implies the existence of an initial (e.g. 'big bang' type) singularity. However, this condition on the normals may well not be applicable to the actual universe (particularly if there are local collapsing regions), even if the universe *is* 'closed'. Also, the condition is virtually unverifiable by observation.

The third and fourth results (referred to as III and IV; see Geroch (1966) and Hawking (1966b), respectively) again apply to 'closed' universe models (i.e. containing a compact, spacelike hypersurface), but which do not have to be assumed to be everywhere expanding. However, III required the somewhat unnatural assumption of the non-existence of 'horizons', while IV required that the given compact hypersurface be a global Cauchy hypersurface. Thus, III and IV could be objected to on grounds similar to those of I.

† We use H for referring to Hawking (1967) and P for referring to Penrose (1968).
‡ The precise meanings of these terms will be given in §3.

The fifth theorem (referred to as V; see H, also Hawking (1966c) and P) does not suffer from objections of this kind, but the requirement on which it was based—namely that the divergence of all timelike and null geodesics through some point p changes sign somewhere to the past of p—is somewhat stronger than one would wish. Theorem V would be considerably more useful in application if the above requirement referred only to *null* geodesics.

In this paper we establish a new theorem, which, with two reservations, effectively incorporates all of I, II, III, IV and V while avoiding each of the above objections. In its physical implications, our theorem falls short of completely superseding these previous results only in the following two main respects. In the first instance we shall require the non-existence of closed time like curves. Theorem II (and II alone) did not require such an assumption. Secondly, in common with II, III, IV and V, we shall require the slightly stronger energy condition given in (3.4), than that used in I. This means that our theorem cannot be directly applied when a *positive cosmological constant* λ is present. However, in a collapse, or 'big bang', situation we expect large curvatures to occur, and the larger the curvatures present the smaller is the significance of the value of λ. Thus, it is hard to imagine that the value of λ should qualitively affect the singularity discussion, except in regions where curvatures are still small enough to be comparable with λ. We may take I as a further indication (though not a proof) of this. In a similar way, II may be taken as a strong indication that the development of closed timelike curves is not the 'answer' to the singularity problem. Of course, such causality violation would carry with it other very serious problems, in any case.

The energy condition (3.4) used here (and in II, III, IV and V) has a very direct physical interpretation. It states, in effect, that 'gravitation is always attractive' (in the sense that neighbouring geodesics near any one point accelerate, on the average, towards each other). Our theorem will apply, in fact, in theories other than classical general relativity provided gravitation remains attractive. In particular, we can apply our results in the theory of Brans & Dicke (1961), using the metric for which the field equations resemble Einstein's (cf. Dicke 1962). The gravitational constant could, in principle, change sign in this theory, but only via a region at which it becomes infinite. Such a region could reasonably be called a 'singularity' in any case. On the other hand, gravitation does not always remain attractive in the theory of Hoyle & Narliker (1963) (owing to the effective negative energy of the C-field) so our theorem is not directly applicable in this theory. We note, finally, that in Einstein's theory (with 'reasonable' sources) it is only $\lambda > 0$ which can prevent gravitation from being always attractive, the λ term representing a 'cosmic repulsion'.

In common with all the previous results I,..., V, our theorem will not give very much information as to the nature of the space-time singularities that are to be inferred on the basis of Einstein's theory. If we accept that 'causality breakdown' is unlikely to occur (because of philosophical difficulties encountered with closed timelike curves and because theorem II suggests that such curves probably do not help in the singularity problem in any case), then we are led to the view that the instability of gravitation presumably† results in regions of enormously large curvature occurring in our universe. These curvatures would have to be so large that our present concepts of local physics would become drastically modified. While the quantum effects of gravitation are normally thought to be significant only when curvatures approach 10^{33} cm^{-1}, all our local physics is based on the Poincaré group being a good approximation of a local symmetry group at dimensions greater than 10^{-13} cm. Thus, if curvatures ever even approach 10^{13} cm^{-1}, there can be little doubt but that extraordinary local effects are likely to take place.

When a singularity results from a collapse situation in which a trapped surface has developed, then any such local effects would not be observable outside the collapse region. It is an open question whether physically realistic collapse situations, resulting in singularities, will sometimes arise *without* trapped surfaces developing (cf. Penrose 1969). If they do, it is likely that such singularities could (in principle) be observed from outside. Of course, the initial 'big bang' singularity of the Robertson–Walker models is an example of a singularity of the observable type. However, our theorem yields no information as to the observability of singularities in general. We cannot even rigorously infer whether the implied singularities are to be expected in the 'past' or the 'future'. (In this respect our present theorem yields somewhat less information than I, II, or V.)

Our theorem will be directly applicable to *any one* of the following three situations. First, to the existence of a trapped surface; secondly, to the existence of of a compact space-like hypersurface; thirdly, to the existence of a point whose null-cone begins to 'converge again' somewhere to the past of the point. We assume the energy condition and the non-existence of closed timelike curves. On the basis of this (and another very minor assumption which merely rules out some highly special models) we deduce that singularities will develop in fully general situations involving a collapsing star, *or* in a spatially closed universe, *or* (taking the point in question in the third case to be the earth at the present time) if the apparent solid angle subtended by an object of a given intrinsic size reaches some minimum when the object is at a certain distance from us. We show, in an appendix, that this last condition is indeed likely to be satisfied in our universe, assuming the correctness of the normal interpretation of the 2.7 K background radiation. A similar discussion was given earlier by Hawking & Ellis (1968) in connexion with theorem V. Since we now have a stronger theorem, we can use somewhat weaker physical assumptions concerning the radiation.

In §2 we give a number of lemmas and definitions that will be needed for our theorem. The precise statement of the theorem will be given in § 3. This statement

---

† We must always bear in mind that a local 'energy-condition' (cf. (3.4)) is being assumed here, which might be violated not only in a modified Einstein theory (e.g. 'C-field'), but also in the standard theory if we were allowed to have very 'peculiar' matter under extreme conditions. The quantum field-theoretic requirement of positive-definiteness of energy (in order that the vacuum remain stable) is of great relevance here, but its status is perhaps not completely clear (cf. Sexl & Urbantke 1967 for example).

is presented in a rather general form, which is somewhat removed from the actual applications. The main applications are given in a corollary to the theorem. One slight advantage of the form of statement that we have chosen will be that it enables a small amount of information to be extracted about the actual nature of the singularities. This is that (at least) one timelike or null geodesic must enter (or leave) the singularity not only in a finite proper (or affine) time, but also in such a way that none of the neighbouring initially parallel geodesics has time to be focused towards it before the singularity is encountered.

2. Definitions and lemmas

A four-dimensional differentiable (Hausdorff and paracompact†) manifold M will be called a *space-time* if it possesses a pseudo-Riemannian metric of hyperbolic normal signature $(+, -, -, -)$ and a time-orientation. (In fact the following arguments will apply equally well if M has any dimension $\geqslant 3$; also, the time-orientability of M need not really be assumed if we are prepared to apply the arguments to a twofold covering of M.) There will be no real loss of generality in physical applications if we assume that M and its metric are both C^∞. However, the arguments we use actually only require the metric to be C^2.

We shall be concerned with *timelike curves* and *causal curves* on M. (When we speak of a 'curve', we shall, according to context, mean either a continuous map into M of a connected closed portion of the real line, or else the image in M of such a map.) For definiteness we choose our timelike curves to be *smooth*, with future-directed tangent vectors everywhere strictly timelike, including at its end-points. A causal curve is a curve obtainable as a limiting case of timelike curves‡ (cf. Siefert 1967; Carter 1967); it is continuous but not necessarily everywhere smooth; where smooth, its tangent vectors are either timelike or null. A timelike or causal curve will require end-points if it can be extended as a causal curve either into the past or the future (cf. P, p. 187). If it continues indefinitely into the past [*resp.* future] it will be called *past-inextendible* [*resp. future-inextendible*]. If both past- and future-inextendible it is called *inextendible*.

If $p, q \in M$, we write $p \ll q$ if there is a timelike curve with past end-point p and future end-point q; we write $p \prec q$ if either $p = q$ or there is a causal curve from p to q (cf. Kronheimer & Penrose 1967). If $p \prec q$ but not $p \ll q$, then there is a null geodesic from p to q, or else $p = q$. If $p \ll q$ and $q \prec r$, or if $p \prec q$ and $q \ll r$, then $p \ll r$. We do not have $p \ll p$ unless M contains *closed timelike curves*. A subset of M is called *achronal* if it contains no pair of points p, q with $p \ll q$.

† Geroch (1968b) has shown that the assumption of paracompactness is not actually necessary for a space-time, being a consequence of the other assumptions for a space-time manifold.

‡ Except for very minor parts of our discussion, the fact that we are allowing our causal curves not to be smooth plays no significant role in this paper, but it is useful for the general theory. A continuous map of the connected closed interval $\Gamma \subset \mathfrak{R}$, into M, can be characterized as a causal curve by the fact that if $[a, b] \in \Gamma$ and if A, B and C are neighbourhoods in M of the images of a, b and $[a, b]$, respectively, then there exists a timelike curve lying in C with one end-point in A and another end-point in B.

We shall, for the most part, use terminology, definitions and some basic results as given in **P**. (However we use 'causal' for curves referred to in **P** as 'nonspace-like' and 'achronal' for sets referred to in **P** as 'semispacelike'; cf. Carter 1967.) As in Kronheimer & Penrose (1967), we write $I^+(p)$ for the open future of a point $p \in M$, i.e. $I^+(p) = \{x : p \ll x\}$ and $I^+[S]$ for the open future of a set $S \subset M$, i.e. $I^+[S] = \bigcup_{p \in S} I^+(p)$. (The sets $I^+[S]$ are open in the manifold topology for M.) Similarly, $J^+(p) = \{x : p \prec x\}$; $J^+[S] = \bigcup_{p \in S} J^+(p)$. These are not always closed sets.) We define

$$E^+(S) = J^+[S] - I^+[S]. \tag{2.1}$$

Then $E^+(S)$ is *part* of the *boundary* $\dot{I}^+[S]$ of $I^+[S]$ but not necessarily all of it. The sets $I^-(p)$, $I^-[S]$, $J^-(p)$, $J^-[S]$ and $E^-(S)$ are defined similarly, but with future and past interchanged.

For any set $S \in M$ we can define the (*future*) *domain of dependence* $D^+(S)$ and *Cauchy horizon* $H^+(S)$ by

$$D^+(S) = \{x : \text{every past-inextendible timelike curve through } x \text{ meets } S\} \tag{2.2}$$

and

$$H^+(S) = \{x : x \in D^+(S), \ I^+(x) \cap D^+(S) = \varnothing\}$$
$$= D^+(S) - I^-[D^+(S)]. \tag{2.3}$$

The sets $D^-(S)$ and $H^-(S)$ are correspondingly defined. (These definitions are chosen to agree with **P**; they differ somewhat from those of **H**.) We shall be concerned only with the cases when S is an *achronal closed set*. Then $D^+(S)$ is a closed set and $H^+(S)$ is an *achronal closed set*. One easily verifies:

$$I^+[H^+(S)] = I^+[S] - D^+(S). \tag{2.4}$$

Define the *edge* of an achronal closed set S to be the set of points $p \in S$ such that† if $r \ll p \ll q$, with γ a timelike curve from r to q, containing p, then every neighbourhood of γ contains a timelike curve from r to q not meeting S. It follows that edge (S) is in fact the set of points in whose vicinity S fails to be a C°—manifold (S achronal and closed). We have (cf. **P**, p. 191) edge $(S) \subset H^+(S)$. (In fact edge$(S) = $ edge $(H^+(S))$.) Furthermore:

LEMMA (2.5). *Every point of $H^+(S) - $ edge(S) is the future end-point of a null geodesic on $H^+(S)$ which can be extended into the past on $H^+(S)$ either indefinitely, or until it meets edge(S).*

For the proof, see **P**, p. 217 (compare **H**).

A similar result (which follows at once from **P**, p. 216; **H**) is (with S closed and achronal).

LEMMA (2.6). *Every point $p \in \dot{I}^+[S] - S$ is the future end-point of a null geodesic on $\dot{I}^+[S]$ which can be extended into the past on $\dot{I}^+[S]$ either indefinitely (if $p \in \dot{I}^+[S] - E^+(S)$) or until it meets edge$(S)$ (whence $p \in E^+(S)$).*

We say that *strong causality* holds at p if arbitrarily small neighbourhoods of p exist, each intersecting no timelike curve in a disconnected set. (Roughly speaking,

---

† This replaces the definition of edge (S) given in **P**, which was not quite correctly stated.

this means that timelike curves cannot leave the vicinity of p and then return to it; i.e. M does not 'almost' contain closed timelike curves.) We must say 'arbitrarily small', rather than 'every', in the above definition because of the existence of 'hour-glass shaped' (or even 'ball shaped') neighbourhoods of any point in *any* space-time, which are left and re-entered by a timelike curve. To avoid this feature, let us call an open set Q *causally convex* (**P**, p. 224) if Q intersects no timelike curve in a disconnected set. Thus, strong causality holds at p if and only if p possesses arbitrarily small causally convex neighbourhoods (in which case, the 'Alexandrov neighbourhoods' $I^+(q) \cap I^-(r)$ will suffice, with $q \ll p \ll r$). A causally convex open set which lies inside a convex normal coordinate ball with compact closure† will be called a *local causality neighbourhood* (**H**, p. 192). Strong causality holds at every point of a local causality neighbourhood. The only properties of a local causality neighbourhood that we shall in fact use, are that it is open and causally convex, that it contains no past- (or future) -inextendible null geodesic and that any point at which strong causality holds possesses such a neighbourhood.

A property of $D^+(S)$ we shall require is the following. Again, S is to be achronal and closed.

LEMMA (2.7). *If $p \in \mathrm{int}\, D^+(S)$, then $J^-(p) \cap J^+[S]$ is compact.*

This follows from **H**. (See also **P**, p. 227: if $\mathrm{edge}(S) = \varnothing$, and strong causality holds at each point‡ of S, we have the stronger result that $\mathrm{int}\, D^+(S)$ is precisely the set of $p \in I^+[S]$ for which $J^-(p) \cap J^+[S]$ is both compact and contains no point at which strong causality fails. Lemma (2.7) follows by similar reasoning.)

We shall require the concept of *conjugate points* on a causal (i.e. timelike or null) geodesic. Two points p and q on a causal geodesic γ are said to be *conjugate* if a geodesic 'neighbouring' to γ 'meets' γ at p and at q. Somewhat more precisely, the congruence of geodesics through p in the neighbourhood of γ has q as a *focal* point, that is, a point where the divergence of the congruence becomes infinite. (This focal point will in general be an 'astigmatic' focal point. It is a point of the 'caustic' of the congruence. Precise definitions of conjugate points will be found in Milnor (1963), Hicks (1965), Hawking (1966a).) The relation of conjugacy is symmetrical in p and q. The above definition still holds if the roles of p and q are reversed. The property of conjugate points that we shall require is the following (for the timelike case, see Boyer (1964), Hawking (1966a, c), cf. Milnor (1963); for the null case see Hawking (1966c) and also **P**, p. 215, for an equivalent result).

LEMMA (2.8). *If a causal geodesic γ from p to q contains a pair of conjugate points between p and q, then there exists a timelike curve from p to q whose length exceeds that of γ.*

We use the term 'length' for a causal curve to denote its proper time integral. A timelike geodesic is *locally* a curve of maximum length. As a corollary of lemma (2.8) we have:

† This condition was not explicitly included in the definition given in **H**.
‡ This condition should have been included in the conditions on $\mathscr{H}$ in lemma V of **P**.

LEMMA (2.9). *If γ is a null geodesic lying on $I^+[S]$ or on $H^+(S)$ for some $S \subset M$, then γ cannot contain a pair of conjugate points except possibly at its end-points.*

Another consequence of lemma (2.8) is the following result:

LEMMA (2.10). *If M contains no closed timelike curves and if every inextendible null geodesic in M possesses a pair of conjugate points, then strong causality holds throughout M.*

Proof. The result has been given in Hawking (1966c). We repeat the argument here since this reference is not readily available. Suppose strong causality fails at p. Let B be a normal coordinate neighbourhood of p and Q_i a nested sequence of neighbourhoods of p converging on p. Now there is a timelike curve originating in Q_i which leaves B at a point $q_i \in \dot{B}$, re-enters B and returns to Q_i. As $i \to \infty$ the q_i have an accumulation point q on $\dot{B}$ ($\dot{B}$ being compact). The geodesic pq in B cannot be timelike (since otherwise $I^-(q)$ would contain some Q_i, so closed timelike curves would result), nor spacelike. It must therefore be null. Furthermore, strong causality must also fail at q. Repeating the argument with q in place of p, we obtain a new null geodesic qr. In fact this must be the continuation of pq, since otherwise closed timelike curves would result. Continuing the process indefinitely both into the future and into the past we get an inextendible null geodesic γ at every point of which strong causality must fail. By hypothesis γ contains a pair of conjugate points. Thus by lemma (2.8) two of its points can be connected by a timelike curve. It follows that each point of some neighbourhood of one of these point can be joined by a timelike curve to each point of some neighbourhood of the other. This leads at once to the existence of closed timelike curves (because of strong causality violation), contrary to hypothesis. This establishes the lemma.

An important consequence of strong causality is the following result.

LEMMA (2.11). *Let $p \ll q$ be such that the set $J^+(p) \cap J^-(q)$ is compact and contains no points at which strong causality fails. Then there is a timelike geodesic from p to q which attains the maximum length for timelike curves connecting p to q.*

This result was proved by Siefert (1967). The result is, in effect, also contained in the earlier work of Avez (1963). (Unfortunately Avez's analysis contains some errors owing to the fact that the possibility of strong causality breakdown is not duly taken into account.) Lemma (2.11) follows also from lemma V in **P** (p. 227) in conjunction with VI of **P** (p. 228), as applied to the closed achronal set $\dot{I}^-(q)$. In fact, lemma (2.11) can be generalized: if C is a compact subset of M containing no points at which strong causality fails, then the maximum length for all timelike curves contained in C is *attained* (though not necessarily by a geodesic). The essential feature of this situation is that the *space of causal curves* contained in C is compact, the length of a causal curve being an upper semi-continuous function of the curve. For this, we need the appropriate topology on the space of causal curves. (See Seifert (1967); cf. also Avez (1963)). But it will not be necessary to enter into the general discussion here, as lemma (21.1) is all we shall need.

We define a *future-trapped* [*resp. past-trapped*] set to be a non-empty achronal closed† set $S \subset M$ for which $E^+(S)$ [*resp.* $E^-(S)$] is compact. (Note that $E^+(S)$ [*resp.* $E^-(S)$] must then be a closed achronal set.) Any future-trapped set S must itself be compact, since $S \subset E^+(S)$.) An example of a future-trapped set is illustrated in figure 1. We now come to our main lemma.

LEMMA (2.12). *If S is a future-trapped set for which strong causality holds at every point of $\bar{I}^+[S]$, then there exists a future-inextendible timelike curve $\gamma \subset \mathrm{int}\, D^+(E^+(S))$.*

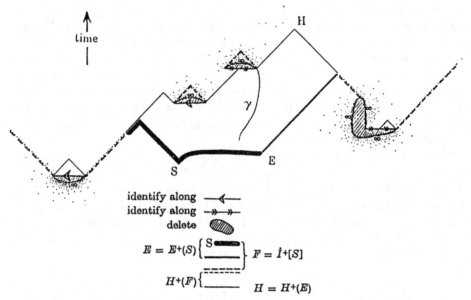

FIGURE 1. A future-trapped set S, together with the associated achronal sets $E = E^+(S)$, $F = \dot{I}^+[S]$, $H^+(F)$, $H = H^+(E)$. (For the proof of lemma (2.12).) The figure is drawn according to the conventions whereby null lines are inclined at 45°. The diagonally shaded portions are excluded from the space-time and some identifications are made. The symbol ∞ indicates regions 'at infinity' with respect to the metric. A future-inextendible timelike curve $\gamma \in D^+(E)$ is depicted, in agreement with the conclusion of lemma (2.12).

Proof.‡ We first make some remarks concerning the relation between $E = E^+(S)$ and $F = \dot{I}^+[S] = \dot{I}^+[E]$, and between their domains of dependence and their Cauchy horizons. We have $E \subset F$, whence $D^+(E) \subset D^+(F)$. We have $\mathrm{edge}(F) = \varnothing$, so it follows from lemma (2.5) that each point of $F-E$ lies on a past-inextendible null geodesic on $F-E$. (These null geodesics extend into the future, while remaining

† The condition that S be closed could be omitted from this definition if desired. For, if S is achronal with $E^+(S)$ compact, then $E^+(S) = E^+(\bar{S})$. Another apparent weakening of the definition of 'future-trapped' for a closed achronal non-empty set S would be to say that $E^+(S)$ has compact closure. ($E^+(S)$ is not always a closed set, for general S.) This definition would be equivalent to the one we use, provided strong causality holds.

‡ This argument follows, to some extent, one given in H (pp. 198–9). It may also serve as a replacement for the final argument given in P (on p. 230) which was not stated correctly.

on $F-E$, perhaps reaching a future end-point on edge(E). We readily obtain $D^+(F) - D^+(E) = H^+(F) - H^+(E) = F - E$, so int $D^+(E) =$ int $D^+(F)$.)

We shall show that $H = H^+(E)$ is non-compact or empty. For, suppose H is compact. Then we can cover H with a finite number of local causality neighbourhoods B_i. If H is non-empty, then $D^+(E) \not\supset I^+[S]$. Let $p \in I^+[S] - D^+(E)$ with p near H and suppose $p \in B_k$. Since $p \in I^+[S]$, a timelike curve η exists connecting S to p. Since $p \notin D^+(E)$, it follows that η meets H at a point p_0, say. We wish to construct a point $q \in I^+[S] - D^+(E)$ with $q \ll p$, $q \notin B_k$ and $q \in B_l$, say. If $p_0 \notin \bar{B}_k$ we can achieve this by taking q just to the future of p_0 on η. If $p_0 \in \bar{B}_k$ we follow the past-inextendible null geodesic ζ through p_0 on $H^+(F)$ (cf. (2.5)). Now ζ must leave $\bar{B}_k$ (since $\bar{B}_k$ is compact) and so contains a point $p_1 \notin \bar{B}_k$ on $H^+(F)$. We have $p_1 \prec p_0 \ll p$, so $p_1 \ll p$. Choosing q near p_1, with $p_1 \ll q \ll p$, we have $q \notin B_k$ and $q \in B_l$, say, where $q \in I^+(p_1) \subset I^+[H^+(F)] = I^+[S] - D^+(E)$ as required (cf. (2.4)). Repeating the procedure, we can find $r \in I^+[S] - D^+(E)$ with $r \ll q$, $r \notin B_l$ and $r \in B_m$, say, etc. Since the B_i are finite in number, there must be two of $p, q, r, \ldots$, in the same B_i, hence violating causal convexity. Thus, H if non-empty, must be non-compact, as required.

Now by a well known theorem (cf. Steenrod 1951, p. 201) we can choose a smooth (future-directed) timelike vector field on M. Form the integral curves $\{\mu\}$ of this vector field. Then each μ which meets H must also meet E (since $H \subset D^+(E)$), but there must be *some* $\mu = \mu_0$ which meets E but *not* H. Otherwise the μ's would establish a homeomorphism between E and H, which is impossible since E is compact and non-empty, while H is non-compact or empty. Choose $\gamma = \mu_0 \cap I^+[E]$. Then $\gamma \subset$ int $D^+(E)$ and is future-inextendible as required.

3. The theorem

We shall begin by giving a precise statement of our theorem. The form of statement we adopt is made primarily for the sake of generality and for certain mathematical advantages. But in order that the theorem may be directly applied to physical situations, we single out the main special cases of interest in a corollary. This recasts our main result in a much more suggestive and immediately usable form. However, the generality of the statement given in the theorem will also yield some advantages as regards applications. It will enable a small amount of information to be extracted as to the actual nature of the space-time singularities. Also, it is by no means impossible that the theorem, as stated, may have relevance in physical situations other than precisely those which we have considered here. We shall follow the statement of the theorem with some explanations and interpretations.

THEOREM. *No space-time M can satisfy all of the following three requirements together:*

(3.1)   *M contains no closed timelike curves,*

(3.2)   *every inextendible causal geodesic in M contains a pair of conjugate points,*

(3.3)   *there exists a future- (or past-) trapped set $S \subset M$.*

Let us examine each of these three conditions in turn. With regard to (3.1), the existence of closed timelike curves in any space-time model leads to very severe interpretative difficulties. It might perhaps be argued that the presence of a closed timelike world-line could be admissible, provided the world-line entered a region of such extreme physical conditions, or involved such large accelerations, that no physical observer could 'survive' making this trip into his own past, so that any 'memory' of events would necessarily be destroyed in the course of the trip. However, it seems highly unlikely that the physical consequences of closed timelike curves can be eliminated by considerations of this kind. The existence of such curves can imply serious global consistency conditions on the solutions of hyperbolic differential equations.† We are reassured by the theorem referred to as II in § 1 (cf. H) that the singularity problem of general relativity is not forcing us into consideration of closed timelike curves.

Condition (3.2) of the theorem—namely that for any timelike or null geodesic, there is a 'neighbouring geodesic' which meets it at two distinct points—may, at *first sight* appear to be a strong one. However, this is not so. The condition is in fact one that could be expected to hold in *any* physically realistic non-singular space-time. It is a consequence of three requirements: *causal geodesic completeness*, the *energy condition* and a *generality* assumption.

The requirement of *causal geodesic completeness* is simply that every timelike and null geodesic can be extended to arbitrarily large affine parameter value both into the future and into the past. (In the case of timelike geodesics we can use the proper time as such a parameter.) In crude terms we could interpret this condition as saying: 'photons and freely moving particles cannot just appear or disappear off the edge of the universe'. A completeness condition of this kind is sometimes used as virtually a *definition* of what is meant by a non-singular space-time (cf. Geroch 1968a). Since one must normally 'delete' any actual singular points from consideration as part of the space-time manifold, it is by some criterion such as 'incompleteness' that the 'holes' left by the removal of the singularities may be detected.

The *energy condition* may be expressed as

$$t^a t_a = 1 \; implies \; R_{ab} t^a t^b \leqslant 0. \tag{3.4}$$

(We use a $+ - - -$ signature, with Riemann and Ricci tensor signs fixed by $2\nabla_{[a}\nabla_{b]}k_c = k_d R^d_{aab}$, $R_{ab} = R^c_{acb}$.) With Einstein's equations

$$R_{ab} - \tfrac{1}{2} R g_{ab} = -K T_{ab}, \tag{3.5}$$

(3.4) becomes
$$t^a t_a = 1 \; implies \; T_{ab} t^a t^b \geqslant \tfrac{1}{2} T^c_c. \tag{3.6}$$

(We have $K > 0$. To incorporate a cosmological constant λ, we would have to replace T_{ab} in the above by $T_{ab} + \lambda K^{-1} g_{ab}$. Thus, (3.6), as it stands, would still

† For example, $\phi = $ const. is the only solution of $\partial^2\phi/\partial t^2 - \partial^2\phi/\partial x^2 = 0$, on the (x, t)-torus, for which (t, x) is identified with $(t+n, x+m\pi)$ for each pair of integers n, m.

imply (3.4) so long as $\lambda \leqslant O$.) [If, in an eigentetrad of T_{ab}, E denotes the *energy density* and p_1, p_2, p_3 denote the three *principal pressures*, then (3.6) can be written as

$$E + \Sigma p_i \geqslant 0, \tag{3.7}$$

together with

$$E + p_i \geqslant 0, \tag{3.8}$$

where $i = 1, 2, 3$.

The *weak* energy condition is

$$l^a l_a = 0 \text{ implies } R_{ab} l^a l^b \leqslant 0, \tag{3.9}$$

which is a consequence of (3.4) (as follows by a limiting argument). This is equivalent, assuming Einstein's equations, to (3.8) (*without* (3.7)) and follows from the *positive-definiteness* of the energy expression $T_{ab} t^a t^b$, for $t^a t_a = 1$. (This is now irrespective of the value of λ.)

The assumption of *generality* we require (compare Hawking 1966 b) is that every causal geodesic γ contains some point for which

$$k_{[a} R_{b]cd[e} k_{f]} k^c k^d \neq 0, \tag{3.10}$$

where k_a is tangent to γ. If γ is timelike, we can rewrite (3.10) as

$$R_{abcd} k^b k^c \neq 0. \tag{3.11}$$

(To see this, transvect (3.10) with $k^a k^f$.)

In any physically realistic 'generic' model, we would expect (3.10) to hold for each γ. For example, the condition can fail for a timelike geodesic γ *only* if $R_{ab} k^a k^b$ vanishes at every point on γ, and then only if the Weyl tensor is related in a very particular way to γ (i.e. $C_{abcd} k^b k^c = 0$) at *every* point on γ. (For a generic space-time this would not even occur at *any* point of any γ!) The condition can fail for a null geodesic γ only if $R_{ab} k^a k^b$ vanishes at every point of γ and the Weyl tensor has the tangent direction to γ as a principal null direction at *every* point of γ (cf. **P**, p. 162). (In a generic space-time, there would not be *any* null geodesic γ which is directed along a principal null direction at *six* or more of its points. This is because null geodesics form a five-dimensional system. It is n conditions on a null geodesic that it be directed along a principal null direction at n of its points, so such null geodesics form a $(5-n)$-dimensional system in a generic space-time.) We can thus reasonably say that it is only in very 'special' (and therefore physically unrealistic) models that the condition will fail.

We must now show why these three conditions together imply (3.2). The fact that they do is essentially a consequence of the *Raychaudhuri* effect (1955, cf. also **P**, p. 169; compare also Myers 1941). The idea here is to proceed so far along the causal geodesic γ that we get beyond the focal length of the effective 'lens system' due to the curvature along γ (compare Penrose 1965 b). Consider a causal geodesic γ belonging to a hypersurface orthogonal congruence Γ of causal geodesics. We are interested in the members of Γ only in the immediate neighbourhood of γ.

When γ is a null geodesic, we shall, for convenience, specify that all the other members of Γ shall also be null. In this case we shall, in fact, be interested only in those members of Γ, near γ, which generate a null hypersurface containing γ. When γ is time-like we define the vector field t^a to be the unit future-directed tangents to the curves of Γ. When γ is null, we choose a vector field l^a to be smoothly varying future-directed tangents to the curves of Γ, where l^a is parallelly propagated along each curve. We have

$$\nabla_a t_b = \nabla_b t_a, \quad t^a t_a = 1, \quad D t^a = 0, \quad \text{with} \quad D = t^a \nabla_a \qquad (3.12)$$

and

$$l_{[c} \nabla_a l_{b]} = 0, \quad l^a l_a = 0, \quad D l^a = 0, \quad \text{with} \quad D = l^a \nabla_a \qquad (3.13)$$

respectively.

Let us first consider the timelike case. Ricci identities give, with (3.12),

$$R_{abcd} t^b t^d = D(\nabla_c t_a) + (\nabla_c t^d)(\nabla_d t_a). \qquad (3.14)$$

Now $R_{abcd} t^b t^d$ and $\nabla_c t_d$ each annihilate t^a when transvected with it on any free index. Introduce an orthonormal basis frame, with t^a as one of the basis elements. Let $Q_{\alpha\beta}$ and $U_{\alpha\beta}$ denote the symmetric (3×3) matrices of spatial components of $R_{abcd} t^d t^d$ and $\nabla_a t_b$, respectively. Then (3.14) becomes

$$Q_{\alpha\beta} = D U_{\alpha\beta} - U_{\alpha\gamma} U_{\gamma\beta}. \qquad (3.15)$$

The matrix $Q_{\alpha\beta}$ defines the geodesic deviation (relative acceleration) of Γ; the trace-free part of $U_{\alpha\beta}$ defines the shear of Γ. We define the *divergence* θ of Γ to be

$$\theta = \nabla_a t^a = - U_{\alpha\alpha}. \qquad (3.16)$$

Taking the trace of (3.15), we get

$$D\theta + \tfrac{1}{3}\theta^2 = \tfrac{1}{3}(U_{\alpha\beta} U_{\alpha\beta} \delta_{\rho\sigma} \delta_{\rho\sigma} - U_{\alpha\beta} \delta_{\alpha\beta} U_{\rho\sigma} \delta_{\rho\sigma}) - Q_{\gamma\gamma} \leqslant 0 \qquad (3.17)$$

by Schwarz's inequality and the energy condition (3.4) (which asserts $Q_{\gamma\gamma} \geqslant 0$). Equality holds only when $Q_{\gamma\gamma} = 0$ *and* $U_{\alpha\beta}$ is proportional to $\delta_{\alpha\beta}$ (so that the shear would have to vanish).

Suppose $R_{abcd} t^b t^d \neq 0$ at some point x of γ, in accordance with (3.11). Then $Q_{\alpha\beta} \neq 0$ at x. We shall show, first, that this implies that the *strict* inequality holds in (3.17) at some point y on γ with $x \prec y$. For if it turns out that $Q_{\alpha\beta} = \mu \delta_{\alpha\beta}$ at x (for some μ), then clearly $Q_{\alpha\beta} \neq 0$ at x implies $Q_{\gamma\gamma} \neq 0$ at x, so that strict inequality holds at $y = x$. On the other hand, suppose $Q_{\alpha\beta}$ is *not* of this form at x. Then by (3.15) $U_{\alpha\beta}$ cannot be proportional to $\delta_{\alpha\beta}$ throughout any open segment of γ whose closure includes x. Thus, the expression in parentheses in (3.17) must fail to vanish at some point $y \in \delta$ with $x \prec y$, so the strict inequality in (3.17) must hold at y.

Let the real quantity W be defined along γ as a non-zero solution of

$$DW = \tfrac{1}{3}\theta W \qquad (3.18)$$

(so that W^3 measures a spacelike 3-volume element orthogonal to γ and Lie transported along the curves of Γ). Then (3.17) gives†

$$D^2 W \leqslant 0 \qquad (3.19)$$

along γ, provided W remains positive. Furthermore the strict inequality holds at y. Choosing $W > 0$ at x, we see from (3.18) and (3.19) that if $\theta \leqslant 0$ at x, then W becomes zero at some point q on γ with $x \ll q$. Furthermore, if $\theta > 0$ at x, then W becomes zero at some $p \in \gamma$ with $p \ll x$. This is *provided* we assume that γ is a *complete* geodesic. (By (3.12), we can interpret the 'D' in (3.17), (3.18), (3.19) as d/ds, where s is a proper time parameter on γ. The completeness condition ensures that the range of s is unbounded.) When W becomes zero, we have a *focal point* of Γ (point of the caustic) at which θ becomes infinite (since $\theta = 3D \ln W$).

Now fix the causal geodesic γ and fix a point x on it at which (3.11) holds: then allow the congruence Γ to vary. Thus, we consider solutions of (3.15), where the matrix $Q_{\alpha\beta}$ is a given function of s. We shall be interested, in the first instance, in solutions for which $\theta \geqslant 0$ at x. Then by the above discussion there will be a first focal point q_Γ on γ, for each Γ (with $x \ll q_\Gamma$). Each solution of (3.15) is fixed once the value of $U_{\alpha\beta} = \overset{\circ}{U}_{\alpha\beta}$ is fixed at x (with $\overset{\circ}{U}_{\alpha\alpha} \geqslant 0$). Thus, q_Γ is a function of the nine $\overset{\circ}{U}_{\alpha\beta}$. Furthermore, it must be a continuous function. We note that if any component of $\overset{\circ}{U}_{\alpha\beta}$ is very large, then q_Γ is very near x (since, in the limit $Q_{\alpha\beta}$ becomes irrelevant and the solution resembles the flat space-time case). It follows that the q_Γ's must lie in a bounded portion ζ of γ. (The one-point compactification of the space of $\overset{\circ}{U}_{\alpha\beta}$, with $\overset{\circ}{U}_{\alpha\alpha} \geqslant 0$ is mapped continuously into γ, with the point at infinity being mapped to x itself. Thus, the image must be compact.) Choose a point $q \in \gamma$, to the future of ζ and let Γ consist of the timelike geodesics (near γ) through q. If there were no conjugate point to q on γ, then the Γ congruence would be non-singular to the past of q. We cannot have $\theta \leqslant 0$ at x, since this would imply $q \in \zeta$. But we have seen that $\theta > 0$ implies another focal point to the past of x. This establishes the existence of a pair of conjugate points on γ in the timelike case.

When γ is null, the argument is essentially similar. In place of (3.14) we can use the Sachs equations (cf. **P**, p. 167) which have a matrix form similar to (3.15). The components of the curvature tensor which enter into these equations are just the four independent real (or two independent complex) components of $l_{[a}R_{b]cd[e}l_{f]}l^c l^d$. The analogue of θ is $-2\rho = \nabla_a l^a$. In place of W we have a 'luminosity parameter' L, satisfying $DL = -\rho L$ and $D^2 L \geqslant 0$. The conclusion is the same: If (3.10) holds at some point on γ, if γ is complete and if the energy condition holds (in this case the weak energy condition (3.9) will suffice), then γ contains a pair of conjugate points.

† Equation (3.19), which follows from $R_{ab}k^a k^b \leqslant 0$, is essentially the statement that 'gravitation is always attractive' (cf. §1). It tells us that the geodesics of Γ, neighbouring to γ, have a tendency to accelerate towards γ—in the sense that freely falling 3-volumes accelerate inwards.

We now come to (3.3), the final condition of the theorem. A drawback of this condition, when it comes to applications, is that we may require considerable information of a global character concerning the space-time M, in order to decide whether or not a given set S is future-trapped. However, in certain special cases, we can invoke the weak energy condition and null-completeness, to enable us to infer, on the *basis of these two properties*, that a certain set should be future-trapped. An example of such a set S is a *trapped surface* (Penrose 1965a; P, p. 211), defined as a compact spacelike 2-surface with the property that *both* systems of null geodesics which intersect S orthogonally *converge* at S, as we proceed into the future. (For simplicity, suppose S to be achronal.) We expect trapped surfaces to arise when a gravitational collapse of a localized body (e.g. a star) to within its Schwarzschild radius takes place, which does not deviate too much from spherical symmetry. The significant feature of a trapped surface arises from the fact that the null geodesics meeting it orthogonally are the generators of $E^+(S)$. If these null geodesics start out by converging ($\rho > 0$) then by the earlier discussion (Raychaudhuri effect in the null case—weak energy condition and null completeness assumed), they must continue to converge until they encounter a focal point. Either then, or before then, they must leave $E^+(S)$ (cf. P, p. 218). Since S is compact and since the focal points must move continuously with the geodesic (being obtainable via integration of curvature), it follows that the geodesic segments joining S to the focal points must sweep out a compact set. Thus $E^+(S)$, being the intersection of this compact set with the closed set $\dot{I}^+[S]$, must also be compact— so S is future-trapped and the theorem applies.

Precisely the same argument will apply in more general situations. For example, if S is any compact achronal set whose edge is smooth and at which the null geodesics which form the local boundary of its future (these will be orthogonal to edge(S)) *converge* at edge(S) as we proceed into the future, then (again *assuming* null completeness and the weak energy condition) S will be future-trapped. More generally still, we need not require that the null geodesics which form the local boundary of the future of S actually converge *at* edge(S). It is only necessary that we should have some reason for believing that they converge *somewhere* to the future of S. In particular, S might contain but a single point p, located somewhere near the centre of a collapsing body, but at a time before the collapse has drastically affected the geometry at p. Then, under suitable circumstances the future null cone of p can encounter sufficient collapsing matter that it (locally) starts converging again. Thus every null geodesic through p will encounter a point conjugate to p in the future (assuming null completeness and the weak energy condition), so again these null geodesic segments sweep out a compact set. Its intersection with $\dot{I}^+(p)$ is $E^+(\{p\})$, implying that $E^+(\{p\})$ is compact, so $\{p\}$ is future-trapped and the theorem applies.

In its time-reversed form, this last example has relevance to cosmology. If the point p refers to the earth at the present epoch, the null geodesics into the past, through p sweep out a region which can be taken to represent that portion of the universe which is visible to us now. If sufficient matter (or curvature in general) encounters these null geodesics, then the divergence $(-\rho)$ of the geodesics may be expected to change sign somewhere to the past of p. This sign change occurs where an object of given size intercepting the null ray subtends its *maximum solid angle* at p. Thus, the *existence* of such a maximum solid angle for objects in each direction, may be taken as the physical interpretation of this type of past-trapped set $\{p\}$. Again the theorem applies. In an appendix we give an argument to show that the required condition on p seems indeed to be satisfied in our universe.

Another example of a future- (or past-) trapped set is any achronal set which is a *compact spacelike hypersurface*. (If we do not assume that the hypersurface is achronal, we can produce a 'copy' of it which *is* achronal by taking a suitable covering manifold of the entire space-time, cf. H. Thus, we actually lose no generality by assuming that S is achronal.) In this case, since edge$(S) = \emptyset$, we have $E^+(S) = S$, so $E^+(S)$ is compact. Hence the theorem applies to 'closed universe' models. It is possible that still other situations of physical interest might arise in which a future- (or past-) trapped set S would be inferred as existing (perhaps on the basis of completeness or energy assumptions).

We are now in a position to state the corollary to our theorem.

COROLLARY. A *space-time M cannot satisfy causal geodesic completeness if, together with Einstein's equations (3.5), the following four conditions hold:*

(3.20) *M contains no closed timelike curves.*

(3.21) *the energy condition (3.6) is satisfied at every point,*

(3.22) *the generality condition (3.10) is satisfied for every causal geodesic,*

(3.23) *M contains either*

     (i) *a trapped surface,*

or   (ii) *a point p for which the convergence of all the null geodesics through p changes sign somewhere to the past of p,*

or   (iii) *a compact spacelike hypersurface.*

We may interpret failure of the causal geodesic completeness condition in our corollary as virtually a statement that any space-time satisfying (3.20)–(3.23) 'possesses a singularity' (cf. Geroch 1968a and our earlier remarks). However, one cannot conclude, on the basis of the corollary, that such a singularity need necessarily be of the 'infinite curvature' type. Although one might infer that in *some sense* a 'maximally extended' space-time satisfying (3.20)–(3.23) should obtain arbitrarily large curvatures, there are, nevertheless, other possibilities to consider (cf. H). In fact, very little is known about the nature of the space-time singularities arising in general relativity other than in highly symmetrical situations. For this reason, it is worth pointing out the minor inference that can be made about the nature of these singularities if we revert back to our original statement of the theorem. The implication is, virtually, that a space-time satisfying

(3.20)–(3.23) must contain a causal geodesic which possesses no pair of conjugate points. At a first guess, one might have imagined that causal geodesics entering very large curvature regions would be inclined to possess many pairs of conjugate points. Instead, we see that our theorem implies that *some* causal geodesic 'enters a singularity' (i.e. is compelled to be geodesically incomplete) before any repeated focusing has time to take place.

Proof of the theorem:

Take S as future-trapped. Then, by lemma (2.12), there is a future-inextendible timelike curve $\gamma \subset \operatorname{int} D^+(E^+(S))$. (That strong causality holds for M follows from lemma (2.10).) Define $T = \bar{I}^-[\gamma] \cap E^+(S)$. We shall show that T is past-trapped. (That T is closed and achronal follows at once since $\bar{I}^-[\gamma]$ is closed and $E^+(S)$ is closed and achronal.) Now, since $\gamma \subset D^+(E(S))$, every past-inextendible timelike curve with future end-point on γ must cross $E^+(S)$. More particularly, it must cross T. Also, $\bar{I}^-[T] \subset I^-[\gamma]$. Thus $I^-[T]$ is simply a portion of $I^-[\gamma]$ 'cut off' by T. Examining the boundaries of these sets, we see $\dot{I}^-[T] \subset T \cup \dot{I}^-[\gamma]$. We are interested in $E^-(T) - T$. This is generated by null geodesics $\{\beta\}$ on $\dot{I}^-[T]$ with future end-point on T (at edge(T)). These null geodesics can be continued on $\dot{I}^-[\gamma]$ inextendibly into the future. (For, by lemma (2.6), each point of $\dot{I}^-[\gamma]$ is the past end-point of a null geodesic on $\dot{I}^-[\gamma]$ which continues future-inextendibly unless it meets γ. But it clearly cannot meet γ, since γ is timelike and future-inextendible.) But, by (3.2), every generator β of $\dot{I}^-[T]$ must, when maximally extended, contain a pair of conjugate points p, q, with $p \prec q$, say. By lemma (2.9), p cannot lie on $\dot{I}^-[\gamma]$ (so $p \in I^-[\gamma]$). Thus β must contain a past end-point either at p, or to the future of p. Now T and edge(T) are compact (being closed subsets of the compact set $E^+(S)$). Since β meets edge(T) and since conjugate points vary continuously, (being obtainable as integrals of curvature, cf. Hicks 1964, H) we can choose p and q, for each β, so that the segment of the extension of β from p to q sweeps out a compact region. Thus, the segment of the extension of β from p to edge(T) also sweeps out some compact region C of M. We have $E^-(T) = \dot{I}^-[T] \cap (C \cup T)$, showing that $E^-(T)$ is a closed subset of the compact set $C \cup T$ and is therefore itself compact. Thus, T is past-trapped, as required.

By lemma (2.12) there exists a past-inextendible timelike curve $\alpha \in \operatorname{int} D^-(E^-(T))$. Choose a point $a_0 \in \alpha$. We have $a_0 \in I^-[\gamma]$, so we find $c_0 \in \gamma$ with $a_0 \ll c_0$. Choose the sequence $a_0, a_1, a_2, \ldots, \in \gamma$, receding into the past indefinitely (i.e. with no limit point). Similarly choose $c_0, c_1, c_2, \ldots \in \gamma$ proceeding into the future indefinitely. We have $a_i \ll c_i$ for all i. Now $a_i \in \operatorname{int} D^-(E^-(T))$ and $c_i \in \operatorname{int} D^+(E^+(S))$. Thus by lemma (2.7) $J^+(a_i) \cap J^-[T]$ is compact (with strong causality holding throughout) and so is $J^-(c_i) \cap J^+[S]$. It is easily seen that $J^+(a_i) \cap J^-(c_i)$, is a closed subset of $\{J^-(c_i) \cap J^+[S]\} \cup \{J^+(a_i) \cap J^-[T]\}$ and so is also compact with strong causality holding throughout. Thus, by lemma (2.11) there is a maximal causal geodesic μ_i from a_i to c_i. Now μ_i must meet T, which is compact, at q_i, say. As $i \to \infty$, there will be an accumulation point q in T and an accumulation causal direction at q.

Choose the causal geodesic μ, through T, in this direction, so μ is approached by μ_i. By (3.2), μ contains a pair of conjugate points, u and v, say, with $u \prec v$. Since conjugate points vary continuously, we must have u as a limit point of some $\{u_j\}$ and v as a limit point of some $\{v_j\}$ where u_j and v_i are conjugate points on the maximal extension of μ_j, the $\{\mu_j\}$ being chosen to converge on μ. But $\{a_i\}$ and $\{c_i\}$ cannot accumulate at any point of the segment uv of μ. Hence, for some large enough j, a_j will lie to the past of u_j in μ_j and c_j to the future of v_j on μ_j. This contradicts lemma (2.8) and the maximality of μ_j. The theorem is thus established.

The authors are grateful to C. W. Misner and to R. P. Geroch for valuable discussions.

References

Avez, A. 1963 *Inst. Fourier* **105**, 1.

Boyer, R. H. 1964 *Nuovo Cim.* **33**, 345.

Brans, C. & Dicke, R. H. 1961 *Phys. Rev.* **124**, 925.

Carter, B. 1967 Stationary axi-symmetric systems in general relativity (Ph.D. Dissertation, Cambridge University).

Chandrasekhar, S. 1935 *M.N.* **95**, 207.

Chandrasekhar, S. 1964 *Phys. Rev. Lett.* **12**, 114, 437.

Dicke, R. H. 1962 *Phys. Rev.* **125**, 2163.

Geroch, R. P. 1966 *Phys. Rev. Lett.* **17**, 446

Geroch, R. P. 1968a *Ann. Phys.* **48**, 526.

Geroch, R. P. 1968b *J. Math. Phys.* **9**, 1739.

Hawking, S. W. 1966a *Proc. Roy. Soc. Lond.* A **294**, 511

Hawking, S. W. 1966b *Proc. Roy. Soc. Lond.* A **295**, 490.

Hawking, S. W. 1966c Singularities and the Geometry of space-time (Adams Prize Essay, Cambridge University.)

Hawking, S. W. 1967 *Proc. Roy. Soc. Lond.* A **300**, 187.

Hawking, S. W. & Ellis, G. F. R. 1968 *Astrophys J.* **152**, 25.

Hicks, N. J. 1965 *Notes on differential geometry*. Princeton: D. van Nostrand Inc.

Hoyle, F. & Narlikar, J. V. 1963 *Proc. Roy. Soc. Lond.* A **273**, 1.

Kronheimer, E. H. & Penrose, R. 1967 *Proc. Camb. Phil. Soc. Lond.* **63**, 481.

Lifshitz, E. M. & Khalatnikov, I. M. 1963 *Adv. Phys.* **12**, 185.

Lindquist, R. W. & Wheeler, J. A. 1957 *Rev. Mod. Phys.* **29**, 432.

Milnor, J. 1963 *Morse theory*. Princeton University Press, Princeton.

Myers, S. B. 1941 *Duke Math. J.* **8**, 401.

Penrose, R. 1965a *Phys. Rev. Lett.* **14**, 57.

Penrose, R. 1965b *Revs. Mod. Phys.* **37**, 215.

Penrose, R. 1966 An analysis of the structure of space-time (Adams Prize Essay, Cambridge University).

Penrose, R. 1968 in *Battelle Rencontres, 1967 Lectures in Mathematics and Physics* (Ed. De Witt, C. M. & Wheeler, J. A.) New York: W. A. Benjamin Inc.)

Penrose, R. 1969 in *Contemporary physics: Trieste Symposium* 1968 (paper SMR/63).

Raychaudhuri, A. 1955 *Phys. Rev.* **98**, 1123.

Seifert, H. J. 1967 *Z. Naturforsch.* **22a**, 1356.

Sexl, R. U. & Urbantke, H. 1967 *Acta Phys. Austriaca* **26**, 339.

Steenrod, N. 1951 *The topology of fibre bundles* (Princeton University Press).

Appendix

We wish to show that there is enough matter on the past light-cone of our present location p to imply that the divergence of this cone changes sign somewhere to the past of p. A sufficient condition for this to be so is that there should be (affine) distances R_1 and R_2 such that along every past-directed null geodesic from p,

$$\tfrac{1}{2}\mathrm{K}R_1 \int_{R_1}^{R_2} T_{ab} l^a l^b \, \mathrm{d}r > 1. \tag{A1}$$

(This formula can be obtained by using a variational approach similar to that used in Hawking (1966a).) As in (3.5), $\mathrm{K} = 8\pi G$, where G (= 7.41×10^{-29} cm g^{-1}) is the gravitational constant. (Length and time units are related via $c = 1$, i.e. 3×10^{10} cm = 1 s.)

In this integral, the vector l^a is a future-directed tangent to the null geodesic and r is a corresponding affine parameter ($l^a \nabla_a r = -1$). Here l^a is parallelly propagated along the null geodesic and is such that $r = 0$ at p and $l^a U_a = 1$, where U^a is the future-directed unit timelike vector representing the local standard of rest at p.

In a recent paper (Hawking & Ellis 1968) it was shown that, with certain assumptions, observations of the microwave background radiation indicate that not only do the past directed null geodesics from us start 'converging again' but so also do the timelike ones. As we are concerned only with the null geodesics, the assumptions we shall need will be weaker.

The observations show that between the wavelengths of 20 cm and 2 mm the background radiation is isotropic to within 1 % and has a spectrum close to that of a black body at 2.7 K. We shall assume that this spectrum and its isotropy indicate not that the radiation was necessarily created with this form, but that it has undergone repeated scattering. (We do not assume that the radiation is necessarily primeval.) Thus there must be sufficient matter on each past directed null geodesic from p to make the optical depth large in that direction. We shall show that this matter will be sufficient to cause the inequality (A1) to be satisfied.

The smallest ratio of density to opacity at these wavelengths will be obtained if the matter consists of ionised hydrogen in which case there would be scattering by free electrons. The optical depth to distance R would be

$$\int_0^R \frac{\sigma}{m} \rho \, l^a V_a \, \mathrm{d}r,$$

where σ is the Thomson scattering cross-section, m the mass of a hydrogen atom, ρ the density, measured in g cm^{-3}, of the ionised gas and V^a the local velocity of the gas. The red-shift Z of the gas is given by ($l^a V_a - 1$). We assume that this increases down our past-light cone. As galaxies are observed with red-shifts of 0.46 most of the scattering must occur at red-shifts greater than this (in fact if the quasars really are at cosmological distances, the scattering must occur at red-shifts

of greater than 2). With a Hubble constant of 100 km s^{-1} Mpc^{-1}, a red-shift of
0.4 corresponds to a distance of about 3×10^{27} cm. Taking R_1 to be this distance,
the contribution of the gas density to the integral in (A 1) is

$$2.6 \int_{R_1}^{R_2} \rho \, (l^a V_a)^2 \, dr$$

while the optical depth of gas at red-shifts greater than 0.4 is

$$0.4 \int_{R_1}^{R_2} \rho \, l^a V_a \, dr.$$

As $l^a V_a$ will be greater than 1.4 for $r > R_1$ it can be seen that the inequality
(A 1) will be satisfied at an optical depth of about 0.1. If the optical depth of the
Universe were less than this, one would not expect either a black body spectrum
or a high degree of isotropy, as the photons would not suffer sufficient collisions.
Even if the radiation arose from an isotropic distribution of black-body emitters
at a higher temperature but covering less than $\frac{1}{10}$ of the sky, what one would see
would then be a dilute 'grey' body spectrum which could agree with the observations
between 20 and 2 cm but which would not fit those at 9 and 2 mm. Thus we can be
fairly certain that the required condition is satisfied in the observed Universe.

THE EVENT HORIZON

STEPHEN W. HAWKING
Institute of Astronomy
Cambridge, Great Britain

Contents

Introduction

We know from observations during eclipses and radio measurements of quasars passing behind the sun that light is deflected by gravitational fields. One would therefore imagine that if there were a sufficient amount of matter in a certain region of space, it would produce such a strong gravitational field that light from the region would not be able to escape to infinity but would be "dragged back". However one cannot really talk about things being dragged back in general relativity since there are not in general any well defined frames of reference against which to measure their progress. To overcome this difficulty one can use the following idea of Roger Penrose. Imagine that the matter is transparent and consider a flash of light emitted at some point near the centre of the region. As time passes, a wavefront will spread out from the point (Fig. 1). At first this wavefront will be nearly spherical and its area will be proportional to the square of the time since the flash was emitted. However the gravitational attraction of the matter through which the light is passing will deflect neighbouring rays towards each other and so reduce the rate at which they are diverging from each other. In other words, the light is being focused by the gravitational effect of the matter. If there is a sufficient amount of matter, the divergence of neighbouring rays will be reduced to zero and then turned into a convergence. The area of the wavefront will reach a maximum and start to decrease. The effect of passing through any more matter is further to step up the rate of decrease of the area of the wavefront. The wavefront therefore will not expand and reach infinity

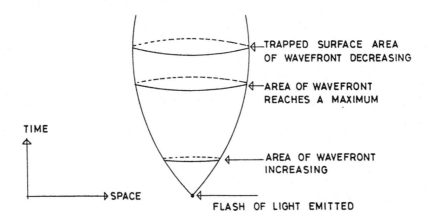

Fig. 1. The wavefront from a flash of light being focused and dragged back by a strong gravitational field.

since, if it were to do so, its area would have to become arbitrarily large. Instead, it is "trapped" by the gravitational field of the matter in the region.

We shall take this existence of a wavefront which is moving outward yet decreasing in area as our criterion that light is being "dragged back". In fact it does not matter whether or not the wavefront originated at a single point. All that is important is that it should be a *closed* (i.e. compact) surface, that it should be *outgoing* and that at each point of the wavefront neighbouring rays should be *converging* on each other. In more technical language, such a wavefront is a compact space like 2-surface [without edges] such that the family of outgoing future-directed null geodesics orthogonal to it is converging at each point of the surface. I shall call this an *outer trapped surface* (or simply, a trapped surface). This differs from Penrose's definition (Penrose, 1965a) in that he required the *ingoing* future-directed null geodesics orthogonal to the surface to be converging as well. The behaviour of the ingoing null geodesics is of importance in proving the occurrence of a spacetime singularity in the trapped region. However, in this course we are primarily interested in what can be seen by observers at a safe distance. Modulo certain reservations which will be discussed in Sec. 2, the existence of a closed outgoing wavefront (or null hypersurface) which is decreasing in area implies that information about what happens behind the wavefront cannot reach such observers. In other words, there is a region of spacetime from which it is not possible to escape to infinity. This is a *black hole*. The boundary of this region is formed by a wavefront or null hypersurface which just does not escape to infinity; its rays are asymptotically parallel and its area is asymptotically constant. This is the *event horizon*.

To show how event horizon and black holes can occur I shall now discuss the one situation that we can treat exactly, spherical symmetry.

1. Spherically Symmetric Collapse

Consider a non-rotating star. After its formation from an interstellar gas cloud, there will be a long period (10^9–10^{11} years) in which it will be in an almost stationary state burning hydrogen into helium. During this period the star will be supported against its own gravity by thermal pressure and will be spherically symmetric. The metric outside the star will be the Schwarzschild solution — the only empty spherically symmetric solution

$$ds^2 = \left(1 - \frac{2M}{r}\right) dt^2 - \left(1 - \frac{2M}{r}\right)^{-1} dr^2 - r^2(d\theta^2 + \sin^2\theta d\phi^2) \qquad (1.1)$$

This is the form of the metric for r greater than some value r_0 corresponding to the surface of the star. For $r < r_0$ the metric has some different forms depending on the distribution of density in the star. The details do not concern us here.

When the star has exhausted its nuclear fuel, it begins to lose its thermal energy and to contract. If the mass M is less than about 1.5–2$M_\odot$, this contraction can be halted by degeneracy pressure of electrons or neutrons resulting in a white dwarf or neutron star respectively. If, on the other hand, M is greater than this limit, contraction cannot be halted. During this spherical contraction the metric outside the star remains of the form (1.1) since this is the only spherically symmetric empty solution. There is an apparent difficulty when the surface of the star gets down to the Schwarzschild radius $r = 2M$ since the metric (1.1) is singular there. This however is simply because the coordinate system goes wrong here. If one introduces an advanced time coordinate v defined by

$$v = t + r + 2M \log(r - 2M) \qquad (1.2)$$

the metric takes the Eddington–Finkelstein form

$$ds^2 = \left(1 - \frac{2M}{r}\right) dv^2 - 2dvdr - r^2(d\theta^2 + \sin^2\theta d\phi^2) \qquad (1.3)$$

This metric is perfectly regular at $r = 2M$ but still has a singularity of infinite curvature at $r = 0$ which cannot be removed by coordinate transformation. The orientation of the light-cones in this metric is shown in Fig. 2. At large values of r they are like the light-cones in Minkowski space and they allow a particle or photon following a nonspacelike (i.e., timelike or null) curve to move outwards or inwards. As r decreases the light-cones tilt over until for $r < 2M$ all nonspacelike curves necessarily move inwards and hit the singularity at $r = 0$. At $r = 2M$ all nonspacelike curves except one move inwards. The exception is the null geodesic r, θ, ϕ constant which neither moves inwards nor outwards. From the behaviour it follows that light emitted from points with $r > 2M$ can escape to infinity whereas that from $r \leq 2M$ cannot. In particular the singularity at $r = 0$ cannot be seen by observers who remain outside $r = 2M$. This is an important feature about which I shall have more to say later.

The metric (1.3) holds only outside the surface of the star which will be represented by a timelike surface which crosses $r = 2M$ and hits the singularity at $r = 0$. Inside the star the metric will be different but the details again do not matter. One

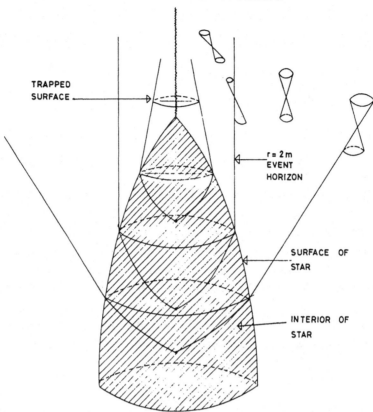

Fig. 2. The collapse of a spherical star leading to the formation of trapped surfaces, event horizon and spacetime singularity.

can analyse the important qualitative features by considering the behaviour of a series of flashes of light emitted from the centre of the star which again is taken to be transparent. In the early stage of the collapse when the density is still low, the divergence of the outgoing light rays or null geodesics will not be reduced much by the focusing effect of the matter. The wavefront will therefore continue to increase in area and will reach infinity. As the collapse continues and the density increases, the focusing effect will get bigger until there will be a critical wavefront whose rays emerge from the surface of the star with zero divergence. Outside the star the area of this wavefront will remain constant and it will be the surface $r = 2M$ in the metric (1.3). Wavefronts corresponding to flashes of light emitted after this critical time will be focused so much by the matter that their rays will begin to *converge* and their area to decrease. They will then form *trapped surfaces*. Their area will continue to decrease, reaching zero when they hit the singularity at $r = 0$.

The critical wavefront which just avoids being converged is the *event horizon*, the boundary of the region of spacetime from which it is not possible to escape to infinity along a future directed nonspacelike curve. It is worth noting certain properties of the event horizon for future reference.

30

(1) The event horizon is a null hypersurface which is generated by null geodesic segments which have no future end-points but which do have past end-points (at the point of emission of the flash).

(2) The divergence of these null geodesic generators is positive during the collapse phase and is zero in the final time-independent state. It is never negative.

(3) The area of a 2-dimensional cross-section of the horizon increases monotonically from zero to a final value of $16\pi M^2$.

We shall see that the event horizon in the general case without spherical symmetry will also have these properties with a couple of small modifications. The first modification is that in general the null geodesic generators will not all have their past end-points at the same point but will have them on some caustic or crossing surface. The second modification is that if the collapsing star is rotating, the final areas of the event horizon will be

$$8\pi[M^2 + (M^4 - L^2)^{\frac{1}{2}}] \qquad (1.4)$$

where L is the final angular momentum of the black hole, i.e., that part of the original angular momentum of the star that is not carried away by gravitational radiation during the collapse. This formula (1.4) will play an important role later on.

In the example we have been considering the event horizon has another property in the time-independent region outside the star. It is the boundary of the part of spacetime containing trapped surfaces. This is not true however in the time-dependent region inside the star. There has in the past been some confusion between the event horizon and the boundary of the region containing trapped surfaces, so it is worth spending a little time to clarify the distinction. Let us introduce a family of spacelike surfaces $S(\tau)$ labelled by a parameter τ which we shall interpret as some sort of time coordinate. In the example we are considering τ could be chosen to be $v - r$ but the react form is not important. Given a particular surface $S(\tau)$, one can find whether there are any trapped surfaces which lie in $S(\tau)$. The boundary of the region of $S(\tau)$ containing trapped surfaces lying in $S(\tau)$ will be called the *apparent horizon* in $S(\tau)$. This is not necessarily the same as the intersection of the event horizon with $S(\tau)$ which is the boundary of the region of $S(\tau)$ from which it is not possible to escape to infinity. To see the differences consider a situation which is similar to the previous example of a collapsing spherical star of mass M but where there is also a thin spherical shell of matter of mass δM which collapses from infinity at some later time and hits the singularity at $r = 0$ (Fig. 3). Between the surface of the star and the shell the metric is of the form (1.3) while outside the shell it is of the form (1.3) with M replaced by $M + \delta M$. The apparent horizon in $S(\tau_1)$, the boundary of the trapped surfaces in $S(\tau_1)$, will be at $r = 2M$. It will remain at $r = 2M$ until the surface $S(\tau_2)$ when it will suddenly jump out to $r = 2(M + \delta M)$. On the other hand, the event horizon, the boundary of the points from which it is not possible to escape to infinity, will intersect $S(\tau_1)$ just outside $r = 2M$. It will

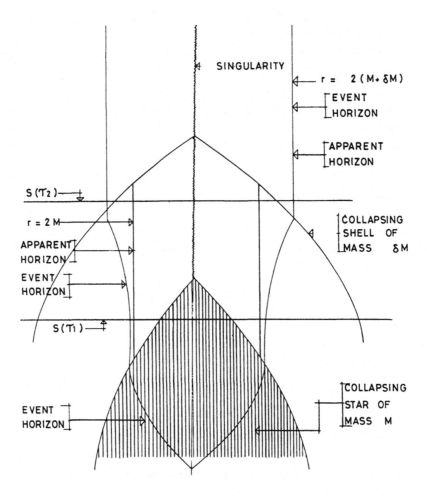

Fig. 3. The collapse of a star followed by the collapse of a thin shell of matter. The apparent horizon moves outwards discontinuously but the event horizon moves in a continuous manner.

move out continuously reach $r = 2(M + \delta M)$ at the surface $S(\tau_2)$. Thereafter it will remain at this radius provided no more shells of matter fall in from infinity.

The apparent horizon has the practical advantage that one can locate it on a given surface $S(\tau)$ knowing the solution only on that surface. On the other hand one has to know the solution at all times to locate the event horizon. However, the event horizon has the mathematical advantage of being a null hypersurface with nice properties like the area always increasing whereas the apparent horizon is not in general null and can move discontinuously. In this course I shall therefore concentrate on the event horizon. I shall show that it will always coincide with or be outside the apparent horizon. During periods when the solution is nearly time independent and nothing is just about to fall into the black hole, the two horizons will nearly coincide and their areas will be almost equal. If the black hole now undergoes some interaction and settles down to another almost stationary state, the area of the event horizon will have increased. Thus the area of the apparent horizon will also have increased. I shall show how the area increase can be used to measure the amounts of energy and angular momentum which fell into the black hole.

2. Nonspherical Collapse

No real star is exactly spherical; they all are rotating a bit and have magnetic fields. One must therefore ask whether their collapse will show the same features as the spherical case we discussed before. One would not expect this necessarily to be the case if the departure from spherical symmetry were too large. For example a rapidly rotating star would not collapse to within $r = 2M$ but would form a thin rotating disc, maintaining itself by centrifugal force against the gravitational attraction. However one might hope that the picture would be qualitatively similar to the spherical case for departures from spherical symmetry that are small initially. One can divide this question of stability under small perturbations of the initial conditions into three parts.

(1) Is the occurrence of a singularity a stable feature?
(2) Is the form of the singularity stable?
(3) Is the fact that the singularity cannot be seen from infinity stable?

The Einstein equations being a well behaved system of differential equations have the property of local stability. The solution at nonsingular points depends continuously on the initial data (see Hawking and Ellis, 1973. I shall refer to this as HE). In other words, given a compact nonsingular region V in the Cauchy development of an initial surface S, one can find a perturbation of the initial data on S which is sufficiently small that the solution on V changes by less than a given amount. One can apply this result to show that small initial departure from spherical symmetry will not affect the fact that the wavefronts corresponding to flashes of light emitted from the centre of the star will be focused and made to start to reconverge. It follows from a theorem of Penrose and myself (Hawking and Penrose, 1970) that the existence of such a reconverging wavefront implies the occurrence of a spacetime singularity provided that certain other reasonable conditions like positive energy density and causality are satisfied. Thus the answer to question (1) is "yes"; the occurrence of a singularity is a stable feature of gravitational collapse.

As the local stability result holds only at non-singular points it cannot be used to answer question (2): is the form of the singularity stable? In fact the answer is "no". For example adding a small amount of electric charge to the star changes the singularity from that in the Schwarzschild solution to that in the Reissner–Nordström solution which is completely different. It is reasonable to expect that a small departure from spherical symmetry would also completely change the singularity. This makes it very difficult to study singularities since one does not know what a "generic" singularity would look like. The work of Liftshitz, Belinsky and Khalatnikov suggests that it is probably very complicated. Fortunately we do not have to worry about this in this course provided we have an affirmative answer to question (3): is the fact that the singularity cannot be seen from infinity stable?

One cannot use the local stability result to answer this since it applies only to the behaviour of perturbations over a finite interval of time. The question of whether the singularities can be seen from infinity depends on the behaviour of the solution at arbitrarily large times and at such times the perturbations might have grown large. In fact this question which is absolutely fundamental to the whole study of black holes has not yet been properly answered. However there are grounds for optimism. The first of these is that linearized perturbation studies of spherical collapse by Regge and Wheeler (1957), Doroshkevich, Zeldovich and Novikov (1965), Price (1972) and others have shown that all perturbations except one die away with time. The one exception corresponds to a rotational perturbation which changes the Schwarzschild solution into a linearized Kerr solution. In this the singularities are also hidden from infinity. These perturbation calculations do not completely answer the stability question since they are only first order: one would need to show that the perturbations of the second and higher orders also die away and that the perturbation series converged.

The second ground for believing that the singularities are hidden is that Penrose and Gibbons have tried and failed to devise situations in which they are not. The idea was to try and obtain a contradiction with the result that the area of the event horizon increases which is a consequence of the assumption that the singularities are hidden. However they failed. Of course their failure does not prove anything but it does strengthen my personal conviction that the singularities in gravitational collapse will not be visible from infinity. One has to be slightly careful how one states this because one can always devise situations where there are naked singularities of a sort. For example, if one has pressure-free matter (dust), one can arrange the flow-lines to intersect on caustics which will be three dimensional surfaces of infinite density. However such singularities are really trivial in the sense that the addition of a small amount of pressure or a slight variation in the initial conditions would remove them. I believe that if one starts from a non-singular, asymptotically flat initial surface there will not be any non-trivial singularities which can be seen from infinity.

If there are non-trivial singularities which are naked, i.e., which can be seen from infinity, we may as well all give up. One cannot predict the future in the presence of a spacetime singularity since the Einstein equations and all the known laws of physics break down there. This does not matter so much if the singularities are all safely hidden inside black holes but if they are not we could be in for a shock every time a star in the galaxy collapsed. People working in General Relativity have a strong vested interest in believing that singularities are hidden.

In order to investigate this in more detail one needs precise notions of infinity and of causality relations. These will be introduced in the next two sections.

3. Conformal Infinity

What can be seen from infinity is determined by the light-cone structure of spacetime. This is unchanged by a conformal transformation of the metric, i.e., $g_{ab} \rightarrow$

$\Omega^2 g_{ab}$ where Ω is some suitably smooth positive function of position. It is therefore helpful to make a conformal transformation which squashes everything up near infinity and brings infinity up to a finite distance. To see how this can be done consider Minkowski space:

$$ds^2 = dt^2 - dr^2 - r^2(d\theta^2 + \sin^2\theta\phi^2) \tag{3.1}$$

Introduce retarded and advanced time coordinates, $w = t - r$, $v = t + r$. The metric then takes the form

$$ds^2 = dvdw - r^2(d\theta^2 + \sin^2\theta d\phi^2) \tag{3.2}$$

Now introduce new coordinates p and q defined by $\tan p = v$, $\tan q = w$, $p - q \geq 0$. The metric then becomes

$$ds^2 = \sec^2 p \sec^2 q \left[dpdq - \frac{1}{4}\sin^2(p - q)(d\theta^2 + \sin^2\theta d\phi^2) \right] \tag{3.3}$$

This is of the form $ds^2 = \Omega^{-2} d\tilde{s}^2$ where $d\tilde{s}^2$ is the metric within the square brackets. In new coordinates $t' = \frac{1}{2}(p + q)$, $r' = \frac{1}{2}(p - q)$ the conformal metric $d\tilde{s}^2$ becomes

$$d\tilde{s}^2 = dt'^2 - dr'^2 - \frac{1}{4}\sin^2 2r'(d\theta^2 + \sin^2\theta d\phi^2) \tag{3.4}$$

This is the metric of the Einstein universe, the static spacetime where space sections are 3-spheres. Minkowski space is conformal to the region bounded by the null surface $t' - r' = -\pi/2$ [this can be regarded as the future light-cone of the point $r' = 0$, $t' = -(\pi/2)$] and the null surface $t' + r' = \pi/2$ (the past light-cone of $r' = 0$, $t' = \pi/2$) (Fig. 4). Following Penrose (1963, 1965b) these null surfaces will be denoted by $\mathcal{I}^-$ and $\mathcal{I}^+$ respectively. The point $r' = 0$, $t' = \pm\pi/2$ will be denoted by $t^\pm$ and the points $r' = \pi/2$, $t' = 0$ will be denoted by i^0. (It is a point because $\sin^2 2r'$ is zero there.) Penrose originally used capital I's for these points but this would cause confusion with the symbol for the timelike future which will be introduced in the next section.

All timelike geodesics in Minkowski space start at i^- which represents past timelike infinity and end at i^+ which represents future timelike infinity. Spacelike geodesics start and end at i^0 which represents spacelike infinity. Null geodesics, on the other hand, start at some point on the null surface $\mathcal{I}^-$ and end at some point on $\mathcal{I}^+$. These surfaces represent past and future null infinity respectively (Fig. 5).

When one says that spacetime is asymptotically flat one means that near infinity it is like Minkowski space in some sense. One would therefore expect the conformal structure of its infinity to be similar to that of Minkowski space. In fact it turns out that the conformal metric is singular in general at the points corresponding to $i^- i^+ i^0$. However it is regular on the null surfaces $\mathcal{I}^- \mathcal{I}^+$. This led Penrose (1963, 1965b) to adopt this feature as a *definition* of asymptotic flatness. A manifold M with a metric g_{ab} is said to be *asymptotically simple* if there exists a manifold $\tilde{M}$ with a metric $\tilde{g_{ab}}$ such that

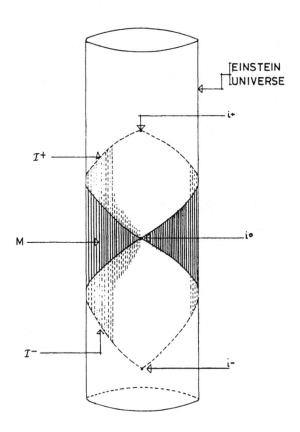

Fig. 4. Minkowski space M conformally imbedded in the Einstein Static Universe. The conformal boundary is formed by the two null surfaces $\mathcal{I}^+$, $\mathcal{I}^-$ and the points $i^+ i^0$ and i^-.

(1) M can be imbedded in $\tilde{M}$ as a manifold with boundary ∂M

(2) On M, $\tilde{g}_{ab} = \Omega^2 g_{ab}$

(3) On ∂M, $\Omega = 0$, $\Omega_{;a} \neq 0$

(4) Every null geodesic in M has past and future end-points on ∂M

(5) The Einstein equations hold in M which is empty or contains only an electromagnetic field near ∂M (Penrose did not actually include this last condition in the definition but it is useful really only if this condition holds)

Condition (3) implies that the conformal boundary ∂M is at infinity from the point of view of someone in the manifold M. Penrose showed that conditions (4) and (5) implied that ∂M consisted of two disjoint null hypersurfaces, labelled $\mathcal{I}^-$ and $\mathcal{I}^+$, which each had topology $R^1 \times S^2$. An example of an asymptotically simple space would be a solution containing a bounded object such as a star which did not undergo gravitational collapse. However the definition is too strong to apply to solutions containing black holes because condition (4) requires that every null geodesic should escape to infinity in both directions. To overcome this difficulty Penrose (1968) introduced the notion of a *weakly asymptotically simple* space. A manifold M with a metric g_{ab} is said to be weakly asymptotically simple if there

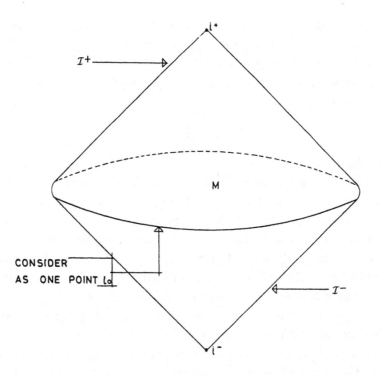

Fig. 5. Another picture of Conformal Infinity as two light-cones $\mathcal{I}^-$ and $\mathcal{I}^+$ joined by a rim which represents the point i^0.

exists an asymptotically simple spacetime M', g'_{ab} such that a neighbourhood of $\mathcal{I}^+$ and $\mathcal{I}^-$ in M' is isometric with a similar neighbourhood in M. This will be the definition of asymptotic flatness I shall use to discuss black holes. Since condition (4) no longer holds for the whole of M there can be points from which it is not possible to reach future null infinity $\mathcal{I}^+$ along a future directed timelike or null curve. In other words these points are not in the past of $\mathcal{I}^+$. The boundary of these points, the event horizon, is the boundary of the past of $\mathcal{I}^+$. I shall discuss properties of such boundaries in the next section.

Exercise

Show that the Schwarzschild solution is weakly asymptotically simple.

4. Causality Relations

I shall assume that one can define a consistent distinction between past and future at each point of spacetime. This is a physically reasonable assumption. Even if it did not hold in the actual spacetime manifold M, there would be a covering manifold in which it did hold (Markus 1955).

Given a point p, I shall denote by $I^+(p)$ the *timelike or chronological future* of p, i.e., the set of all points which can be reached from p by future directed timelike curves. Similarly $I^-(p)$ will denote the past of p. Many of the definitions I shall give will have duals in which future is replaced by past and plus by minus. I shall regard such duals as self-evident. Note that p itself is not contained in $I^+(p)$ unless there is a timelike curve from p which returns to p. Let q be a point in $I^+(p)$ and let $\lambda(v)$ be a future directed timelike curve from p to q. The condition that $\lambda(v)$ is timelike is an inequality:

$$g_{ab} \frac{dx^a}{dv} \frac{dx^b}{dv} > 0$$

where $\frac{dx^a}{dv}$ is the tangent vector to $\lambda(v)$. One can deform the curve $\lambda(v)$ slightly without violating the inequality to obtain a future directed timelike curve from p to any point in a small neighbourhood of q. Thus $I^+(p)$ is an open set.

The *causal future* of p, $J^+(p)$, is defined as the union of p with the set of points that can be reached from p by future directed nonspacelike, i.e., timelike or null curves. If one considers only a small neighbourhood of p, then $I^+(p)$ is the interior of the future light-cone of p and $J^+(p)$ is $I^+(p)$ with the addition of the future light-cone itself including the vertex. Note that the boundary of $I^+(p)$, which I shall denote by $\dot{I}^+(p)$, is the same as $\dot{J}^+(p)$, the boundary of $J^+(p)$, and is generated by null geodesic segments with past end-points at p.

When one is dealing with regions larger than a small neighbourhood, there is the possibility that some of the null geodesics through p may reintersect each other and the forms of $I^+(p)$ and $J^+(p)$ may be more complicated. To see the general relationship between them consider a future directed curve from a point p to some point $q \in J^+(p)$. If this curve is not a null geodesic from p, one can deform it slightly to obtain a timelike curve from p to q. From this one can deduce the following:

(a) If q is contained in $J^+(p)$ and r is contained in $I^+(q)$, then r is contained in $I^+(p)$. The same is true if q is in $I^+(p)$ and r is in $J^+(q)$.

(b) The set $E^+(p)$, defined as $J^+(p) - I^+(p)$, is contained in (not necessarily equal to) the set of points lying on future directed null geodesics from p.

(c) $\dot{I}^+(p)$ equals $\dot{J}^+(p)$. It is not necessarily the same as $E^+(p)$.

A simple example of a space in which $E^+(p)$ does not contain the whole of the future directed null geodesics from p is provided by a 2-dimensional cylinder with the time direction along the axis of the cylinder and the space direction round the circumference (Fig. 6). The null geodesics from the point p meet up again at the point q. After this they enter $I^+(p)$. An example in which $E^+(p)$ does not form all of $\dot{I}^+(p)$ is 2-dimensional Minkowski space with a point r removed (Fig. 7). The null geodesic in $\dot{I}^+(p)$ beyond r does not pass through p and is not in $J^+(p)$.

The definitions of timelike and causal futures can be extended from points to sets: for a set S, $I^+(S)$ is defined to be the union of $I^+(p)$ for all $p \in S$. Similarly for $J^+(S)$. They will have the same properties (a), (b) and (c) as the futures of points. Suppose there were two points q, r on the boundary $\dot{I}^+(S)$ of the future of a set S with a future directed timelike curve λ from q to r. One could deform λ slightly to give a timelike curve from a point x in $I^+(S)$ near q to a point y in $M - I^+(S)$ near r. This would be a contradiction since $I^+(x)$ is contained in $I^+(S)$. Thus one has

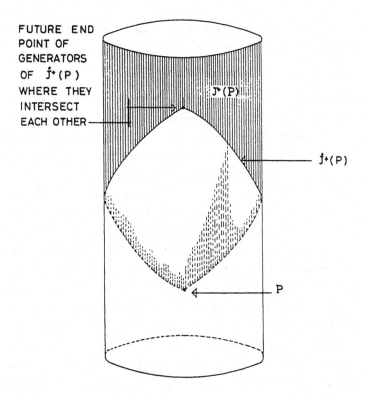

Fig. 6. A space in which the future directed null geodesics from a point P have future end-points as generators of $\dot{J}^+(P)$.

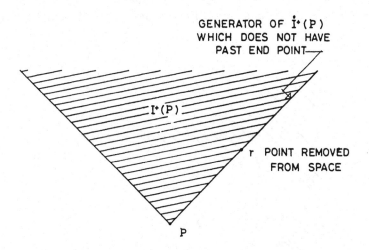

Fig. 7. The point r has been removed from two-dimensional Minkowski space.

(d) $\dot{I}^+(S)$ does not contain any pair of points with timelike separation. In other words, the boundary $\dot{I}^+(S)$ is null or spacelike at each point.

Consider a point $q \in \dot{I}^+(S)$. One can introduce normal coordinates x^1, x^2, x^3, x^4 (x^4 timelike) in a small neighbourhood of q. Each timelike curve $x^i = $ constant ($i =$

1, 2, 3) will intersect $\dot{I}^+(S)$ once and once only. These curves will give a continuous map of a small region of $\dot{I}^+(S)$ to the 3-plane $x^4 = 0$. Thus

(e) $I^+(S)$ is a manifold (not necessarily a differentiable one).

Now consider a point q in $\dot{I}^+(S)$ but not in S itself, or its topological closure $\bar{S}$. One can thus find a small convex neighbourhood U of q which does not intersect $\bar{S}$. In U one can find a sequence $\{y_n\}$ of points in $I^+(S)$ which converge to the point q (Fig. 8). From each y_n there will be a past directed timelike curve λ_n to S. The intersections of the $\{\lambda_n\}$ with the boundary $\dot{U}$ of U must have some limit point z since $i\dot{U}$ is compact. Any neighbourhood of z will intersect an infinite number of the $\{\lambda_n\}$. Thus z will be in $\bar{I}^+(S)$. The point z cannot be spacelike separated from q since, if it were, it would not be near timelike curves from points y_n near q. It cannot be timelike separated from q since if it were one could deform one of the λ_n passing near z to give a timelike curve from S to q which would then have to be in the interior of $I^+(S)$ and not on boundary. Thus z must lie on a past directed null geodesic segment γ from q. Each point of γ between q and z will be in $\dot{I}^+(S)$. One can now repeat the construction at z and obtain a past directed null geodesic segment μ from z which lies in $\dot{I}^+(S)$. If the direction of μ were differed from that of γ one could join points of μ to points of γ by timelike curves. This would contradict property (d) which says that no two points of $\dot{I}^+(S)$ have timelike separation. Thus μ will be a continuation of γ. One can continue extending γ to the past in $\dot{I}^+(S)$ unless and until it intersects S.

If there are two past directed null geodesic segments γ_1 and γ_2 lying in $\dot{I}^+(S)$ from a point $q \in \dot{I}^+(S)$, there can be no future directed such segment from q since

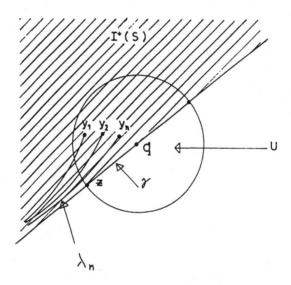

Fig. 8. The points y_n converge to the point q in the boundary of $I^+(S)$. From each y_n there is past directed timelike curve λ_n to S. These curves converge to the past directed dull geodesic segment γ through q.

if there were, it would be in a different direction to and be timelike separated from, either γ_1 or γ_2. One therefore has

(f) $\dot{I}^+(S)$ (and also $\dot{J}^+(S)$) is generated by null geodesic segments which have future end-points where they intersect each other but which can have past end-points only if and when they intersect S.

The example of 2-dimensional Minkowski space with a point removed shows that there can be null geodesic generators which do not intersect S and which do not have past end-points in the space.

The region of spacetime from which one can escape to infinity along a future directed nonspacelike curve is $J^-(\mathcal{I}^+)$ the causal past of future null infinity. Thus $\dot{J}^-(\mathcal{I}^+)$ is the *event horizon*, the boundary of the region from which one cannot escape to infinity (Fig. 9). Interchanging future and past in the results above, one sees that the event horizon is a manifold which is generated by null geodesic segments which may have past end-points but which could have future end-points only if they intersected $\mathcal{I}^+$. Suppose there were some generator γ of $\dot{J}^-(\mathcal{I}^+)$ which intersected $\mathcal{I}^+$ at some point q. Let λ be the generator of the null surface $\mathcal{I}^+$ which passes through q. Since the direction of λ would be different from that of γ, one could join points on λ to the future of q by timelike curves to points on γ the past of q. This would contradict the assumption that γ was in $\dot{J}^-(\mathcal{I}^+)$. Thus the null geodesic generators of the event horizon *have no future end-points*. This is one of the fundamental properties of the event horizon. The other fundamental property, that neighbouring generators are never converging, will be described in Sec. 6.

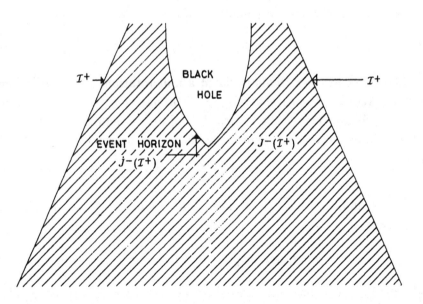

Fig. 9. The event horizon $\dot{J}^-(\mathcal{I}^+)$ is the boundary of the region from which one cannot escape to $\mathcal{I}^+$.

41

5. The Focusing Effect

The most obvious feature of gravity is that it is attractive rather than repulsive. A theoretical statement of this is that gravitational mass is always positive. By the principle of equivalence the positive character of gravitational mass is related to the positive definiteness of energy density which in turn is normally considered to be a consequence of local quantum mechanics. There are possible modifications to this positive definiteness in the very strong fields near singularities. However these will not worry us if, as we shall assume, the singularities are safely hidden behind an event horizon. We shall be concerned, in this course, only with the region outside and including the event horizon.

The fact that gravity is always attractive means that a gravitational field always has a net focusing (i.e., converging) effect on light rays. To describe this effect in more detail, consider a family of null geodesics. Let $l^a = dx^a/dv$ denote the null tangent vectors to these geodesics where v is some parameter along the geodesic. At each point one can introduce a pair of unit spacelike vectors a^a and b^a which are orthogonal to each other and to l^a. It turns out to be more convenient to work with the complex conjugate vectors

$$\sqrt{2}m^a = a^a + ib^a, \qquad \sqrt{2}\bar{m}^a = a^a - ib^a$$

These are actually null vectors in the sense that $m^a m_a = \bar{m}^a \bar{m}_a = 0$, they are orthogonal to l^a, $l^a m_a = l^a \bar{m}_a = 0$ and they satisfy $m^a \bar{m}_a = -1$. These conditions determine m^a up to a spatial rotation

$$m^a \to m^a e^{i\phi} \tag{5.1}$$

and up to the addition of a complex multiple of l^a

$$m^a \to m^a + cl^a \tag{5.2}$$

where c is a complex number. This is called a null rotation. Given m^a there is a unique real null vector n^a such that $l^a n_a = 1$, $n^a m_a = n^a \bar{m}_a = 0$. The vectors $(1^a, n^a, m^a, \bar{m}^a)$ form what is called a null tetrad or vierbein (Fig. 10).

Using this null tetrad one can express the fact that the curves of the family are geodesics as

$$l_{a;b} m^a l^b = 0 \tag{5.3}$$

where semi-colon indicates covariant derivative. One can also define complex quantities ρ and σ as

$$\rho = l_{a;b} m^a \bar{m}^b, \qquad \sigma = l_{a;b} m^a m^b \tag{5.4}$$

The imaginary part of ρ measures the twist or rate of rotation of neighbouring null geodesics. It is zero if and only if the null geodesics lie in 3-dimensional null hypersurfaces. This will always be the case in what follows so I shall henceforth take

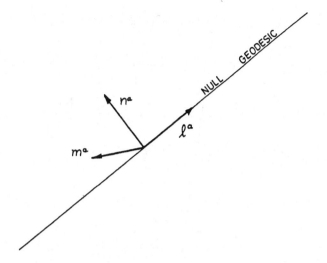

Fig. 10. The null vector l^a lies along the null geodesic. The null vector n^a is such that $l^a n_a = 1$. The null vector m^a is complex combination of two spacelike vectors orthogonal to l^a, n^a and to each other.

ρ to be real. The real part of ρ measures the average rate of convergence of nearby null geodesics. To see what this means consider a null hypersurface N generated by null geodesics with tangent vectors l^a. Let ΔT be a small element of a spacelike 2-surface in N (Fig. 11). One can move each point of ΔT a parameter distance δv up the null geodesics. As one does so the area of ΔT changes by an amount

$$\delta A = -2A\rho\delta v \tag{5.5}$$

The quantity σ measures the rate of distortion or shear of the null geodesics, that is, the difference between the rates of convergence of neighbouring geodesics in the two spacelike directions orthogonal to l^a. The effect of shear is to make a small 2-surface which was spatially circular, become elliptical as it is moved up the null geodesic.

The rate of change of the quantities ρ and σ along the null geodesics is given by two of the Newman–Penrose (1962) equations

$$\frac{dp}{dv} = \rho^2 + \sigma\bar{\sigma} + (\epsilon + \bar{\epsilon})\,\rho + \phi_{00} \tag{5.6}$$

$$\frac{d\sigma}{dv} = 2\rho\sigma + (3\epsilon - \bar{\epsilon})\,\sigma + \psi_0 \tag{5.7}$$

where

$$\epsilon = \frac{1}{2}(l_{a;b}n^a l^b + \bar{m}_{a;b}m^a l^b)$$

$$\phi_{00} = \frac{1}{2}R_{ab}l^a l^b$$

$$\psi_0 = C_{abcd}l^a m^b l^c m^d$$

43

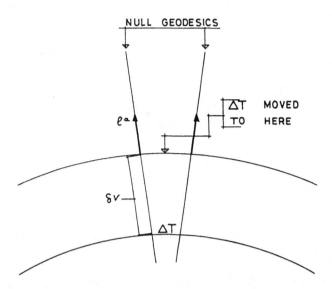

Fig. 11. The area A of two surface element ΔT increases by an amount $-2\rho A\delta v$ when ΔT is moved a parameter distance δv along the null geodesics.

(Note that my definitions of the Ricci and Weyl tensors have the opposite sign to those of Newman and Penrose.)

The imaginary part of ϵ is the rate of spatial rotation of the vectors m^a and $\bar{m}^a$ relative to a parallelly transported frame as one moves along the null geodesics. In what follows m^a will always be chosen so that $\epsilon - \bar{\epsilon} = 0$. The real part of ϵ measures the rate at which the tangent vector l^a changes in magnitude compared to a parallelly transported vector as one moves along the null geodesics. It is zero if $l^a = dx^a/dv$ where v is an affine parameter. It is convenient however in some situations to choose v not to be an affine parameter.

The Ricci tensor term ϕ_{00} in equation (5.6) represents the focusing effect of the matter. By the Einstein equations

$$R_{ab} - \frac{1}{2}g_{ab}R = 8\pi T_{ab} \tag{5.8}$$

it is equal to $4\pi T_{ab}l^a l^b$. The local energy density of matter (i.e., non-gravitational) fields measured by an observer with velocity vector v^a is $T_{ab}v^a v^b$. It seems reasonable from local quantum mechanics to assume that this is always non-negative. It then follows from continuity that $T_{ab}w^a w^b \geq 0$ for any null vector w^a. I shall call this the *weak energy condition* (Penrose 1965a, Hawking and Penrose 1970, HE) and shall assume it in what follows. With this assumption one can see from equation (5.6) that the effect of matter is always to increase the average convergence ρ, i.e., to focus the null geodesic.

The Weyl tensor term ψ_0 can be thought of as representing, in a sense, the gravitational radiation crossing the null hypersurface N. One can see from equation (5.7) that it has the effect of inducing shear in the null geodesic. This shear then induces convergence by equation (5.6). Thus both matter and pure gravitational fields have a focusing effect on null geodesics.

To see the significance of this, consider the boundary $\dot{I}^+(S)$ of the future of a set S. As I showed in the last section this will be generated by null geodesic segments. Suppose that the convergence of neighbouring segments has some positive value ρ_0 at a point $q \in \dot{I}^+(S)$ on a generator γ. Then choosing v to be an affine parameter, one can see from equation (5.6) that ρ will increase and become infinite at a point r on the null geodesic γ within an affine distance of $1/\rho_0$ to the future of q. The point r will be a *focal point* where neighbouring null geodesics intersect. We saw in the last section that the generators of $\dot{I}^+(S)$ have future end-points where they intersect other generators. Strictly speaking, this was shown only for generators which intersect each other at a finite angle but it is true also for neighbouring generators which intersect at infinitesimal angles (see HE for proof). Thus the generator γ through q will have an end-point at or before the point r. (It may be before r because γ may intersect some other generator at a finite angle.) In other words, once the generators of $\dot{I}^+(S)$ start converging, they are destined to have future end-points within a finite affine distance. They may not, however, attain this distance because they may run into a singularity first.

The importance of this result will be seen in the next section.

6. Predictability

A 3-dimensional spacelike surface S without edges will be said to be a *partial Cauchy surface* if it does not intersect any nonspacelike curve more than once. Given suitable data on such a surface one can solve the Cauchy problem and predict the solution on a region denoted by the $D^+(S)$ and called the *future Cauchy development of S*. This can be defined as the set of all points q such that every past directed nonspacelike curve from q intersects S if continuted far enough. Note that this definition is not the same as the one used in Penrose (1968) and Hawking and Penrose (1970) where nonspacelike is replaced by timelike. However the difference affects only whether points on the boundary of $D^+(S)$ are considered to be in $D^+(S)$ or not.

When one is dealing with the gravitational collapse of a local object such as a star or even a galaxy, it is reasonable to neglect the curvature of the universe and the "big-bang" singularity 10^{10} years ago and to consider spacetime to be asymptotically flat and initially nonsingular. As I said earlier in Sec. 4, I shall take asymptotically flat to mean that the spacetime manifold M and metric g_{ab} are weakly asymptotically simple. This means that there are well-defined past and future null infinities $\mathcal{I}^-$ and $\mathcal{I}^+$. The assumption that we are implicitly making in this Summer School that one can predict the future, at least in the region far away from the collapsing object, can now be expressed as the assumption that there is a partial Cauchy surface S such that points near $\mathcal{I}^+$ lie in $D^+(S)$ (Fig. 12). ($\mathcal{I}^+$ cannot lie actually in $D^+(S)$ since its null geodesic generators do not intersect S. However the solution on $D^+(S)$ determines the conformal structure of $\mathcal{I}^+$ by continuity.) I shall say that a weakly asymptotically simple spacetime M, g_{ab} which admits such a partial Cauchy surface

S is (*future*) *asymptotically predictable*. This definition, and a slightly stronger version which I shall introduce shortly, will form the basis of my course. Asymptotic predictability implies that every past directed nonspacelike curve from points near $\mathcal{I}^+$ continues back to *S* and does not run into a singularity on the way. One can think of this as a precise statement to the effect that there are no singularities to the future of *S* which are naked, i.e., visible from $\mathcal{I}^+$.

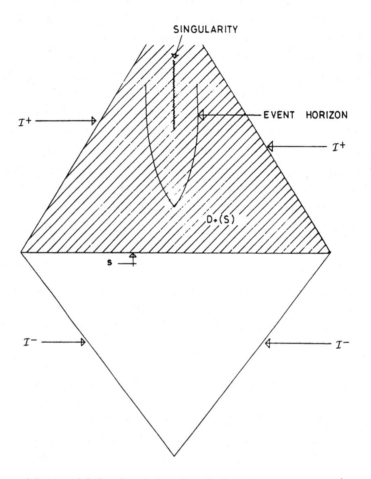

Fig. 12. A space with a partial Cauchy surface *S* such that the points near $\mathcal{I}^+$ are contained in the future Cauchy development $D^+(S)$.

Asymptotic predictability implies that the future Cauchy development $D^+(S)$ contains $J^+(S) \cap J^-(\mathcal{I}^+)$, i.e., it contains all points to the future of *S* which are outside the event horizon. Suppose there were a point *p* on the event horizon to the future of *S* which was not contained in $D^+(S)$. Then there would be a past directed nonspacelike curve λ (in fact a null geodesic) from *p* which did not intersect *S* but ran into some sort of singularity instead. This singularity would be "nearly naked" in that the slightest variation of the metric could result in it being visible from $\mathcal{I}^+$. Since we are assuming that the non-existence of naked singularities is a stable property, we would wish to rule out such an unstable situation. One can also argue that the metric of spacetime is some classical limit of an underlying quantum reality. This would mean that the metric could not be defined so exactly as to distinguish between nearly naked singularities and those which are actually naked. These considerations motivate a slightly stronger version of asymptotic predictability. I shall say that a weakly asymptotically simple spacetime M, g_{ab} is *strongly (future) asymptotically predictable* if there is a partial Cauchy surface S such that

(a) $\mathcal{I}^+$ lies in the boundary of $D^+(S)$,
(b) $J^+(S) \cap \dot{J}^-(\mathcal{I}^+)$ is contained in $D^+(S)$.

Suppose that at some time after the initial surface S, a star starts collapsing and gives rise to a trapped surface T in $D^+(S)$. Recall that a trapped surface is defined to be a compact spacelike 2-surface such that the future directed outgoing null geodesics orthogonal to it have positive convergence ρ. This definition assumes that one can define which direction is outgoing. I shall assume that the 2-surface is orientable and shall require that the initial surface S has the property:

(α) S is simply connected.

Physically, one is interested only in black holes which develop from non-singular situations. In such cases the partial Cauchy surface S can be chosen to be R^3 and so will be simply connected. It is however convenient to frame the definitions so that they can be applied also to spaces like the Schwarzschild and Kerr solutions which are not initially non-singular but which may approximate the form of initially non-singular solutions at late times. In these solutions also one can find partial Cauchy surfaces S which are simply connected.

Given a compact orientable spacelike 2-surface T in the future Cauchy development $D^+(S)$ one can define which direction is outwards. To do this one uses the fact that on any manifold M with a metric g_{ab} of Lorentz signature one can find a vector field X^a which is everywhere nonzero and timelike. Using the integral curves of this vector field, one can map the 2-surface T onto a 2-surface $\hat{T}$ in S. Since S is simply connected, this 2-surface $\hat{T}$ separates S into two regions. One can label the region which contains the part of S near infinity in the asymptotically flat space as the outer region and the other as the inner region. The side of $\hat{T}$ facing the outer region is then the outer side and carrying this up the integral curves of the vector field X^a one can define which is the outgoing direction on T.

Now suppose that one could escape from a point on T to infinity, i.e., suppose that T intersected $J^-(\mathcal{I}^+)$ (Fig. 13). Then there would be some point $q \in \mathcal{I}^+$ which was in $J^+(T)$. Proceeding to the past along the null geodesic generator λ of $\mathcal{I}^+$ through q one would eventually leave $J^+(T)$. Thus λ must countain a point r of $\dot{J}^+(T)$. The null geodesic generator γ of $\dot{J}^+(T)$ through r would enter the physical manifold M. If it did not have a past end-point it would intersect the partial Cauchy surface S. This is impossible since it lies in the boundary of the future of T and T is to the future of S. Thus it would have to have a past end-point which, from Sec. 4, would have to be on T. It would have to intersect T orthogonally as otherwise one

47

could join points of T to points of γ by timelike curves. However the outgoing null geodesics orthogonal to T are converging because T is a trapped surface. As we saw in the last section, this implies that neighbouring null geodesics would intersect γ within a finite affine distance. This means that the generator γ of $\dot{J}^+(T)$ would have a future end-point and would not remain in $\dot{J}^+(T)$ all the way out to $\mathcal{I}^+$. This establishes a contradiction which shows that the supposition that T intersects $J^-(\mathcal{I}^+)$ must be false. In other words, every point on or inside a trapped surface really is trapped: one cannot escape to $\mathcal{I}^+$ along a future directed nonspacelike curve.

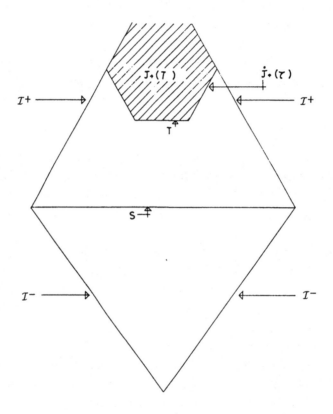

Fig. 13. If a trapped surface T intersected $J^-(\mathcal{I}^+)$, there would be a null geodesic generator of $\dot{J}^+(T)$ from T to $\mathcal{I}^+$. This would be impossible as all null geodesics orthogonal to T contain a conjugate point within a finite affine distance of T.

The same applies to a compact orientable 2-surface T which is *marginally trapped*, i.e., which is such that the outgoing future directed null geodesics orthogonal to T have zero convergence ρ at T. For suppose T intersected $J^-(\mathcal{I}^+)$, then $\dot{J}^+(T)$ would intersect $\mathcal{I}^+$. The area of this intersection would be infinite since it is at infinity. However the generators of $\dot{J}^+(T)$ start off with zero convergence and therefore cannot ever be diverging. Thus the area of $\dot{J}^+(T) \cap \mathcal{I}^+$ could not be greater than that of T. This shows that the marginally trapped surfaces in $D^+(S)$ cannot intersect $J^-(\mathcal{I}^+)$.

What has been shown is that a trapped surface implies *either* a breakdown of asymptotic predictability (i.e., the occurrence of naked singularities) *or* the existence of an event horizon. I shall assume that the first alternative does not occur and shall concentrate on the second. As was shown in Sec. 4, the event horizon will be generated by null geodesic segments which have no future end-points. If one assumed that these generators were geodesically complete in future directions it would follow that the convergence of neighbouring generators could not be positive anywhere on the horizon since, if it were, neighbouring generators would intersect and have future end-points within a finite affine distance. In examples such as the Kerr solution, the generators *are* geodesically complete in the future direction but there does not seem to be any a priori reason why this should always be the case. I shall now show, however, that asymptotic predictability itself without any assumption of completeness of the horizon is sufficient to prove that ρ is non-positive.

Consider a spacelike 2-surface F lying in the event horizon to the future of S. The null geodesic generators of the horizon will intersect F orthogonally. Suppose their convergence ρ was positive at some point $p \in F$. In a small neighbourhood of p one could deform the 2-surface F slightly outwards into $J^-(\mathcal{I}^+)$ so that the convergence ρ of the outgoing null geodesics orthogonal to F was still positive (Fig. 14). This would lead to a contradiction similar to the one we have just considered. The null geodesics in $J^-(\mathcal{I}^+)$ which are orthogonal to F would intersect each other within a

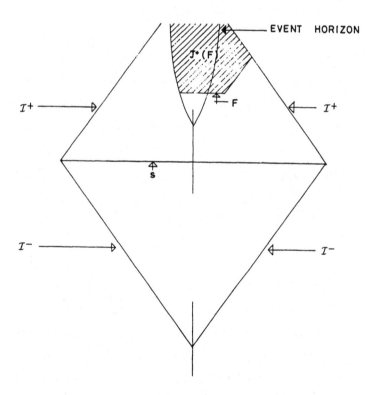

Fig. 14. If the null geodesics orthogonal to a two surface F in the event horizon were converging, one could deform F outwards slightly and obtain a contradiction similar to that in Fig. 13.

finite affine distance and hence could not be generators of $\dot{J}^+(F)$ all the way out to $\mathcal{I}^+$, which being at infinity is at an infinite affine distance.

This shows that the convergence ρ of neighbouring generators of the event horizon cannot be positive anywhere to the future of S. Together with the result that the generators of the event horizon do not have future end-points, this implies that the area of a two-dimensional cross section of the horizon must increase with time. This will be discussed further in the next section.

7. Black Holes

In order to describe the formation and evolution of black holes, one needs a suitable time coordinate. The usual coordinate t in the Schwarzschild and Kerr solutions is no good because all the surfaces of constant t intersect the horizon at the same place (see Carter's lectures). What one wants is a coordinate τ such that the surfaces of constant τ cover the future Cauchy development $D^+(S)$. By the assumption of strong future asymptotic predictability the event horizon to the future of S will be contained in $D^+(S)$ and so will be covered by the surfaces of constant τ. I shall denote the surface $\tau = \tau_0$ by $S(\tau_0)$ with $S(0) = S$. Near infinity the surfaces $S(\tau)$ for $\tau > 0$ could be chosen to be asymptotically flat spacelike surfaces like S which approached spacelike infinity i^0 and which were such that $\mathcal{I}^+$ lay in the boundary

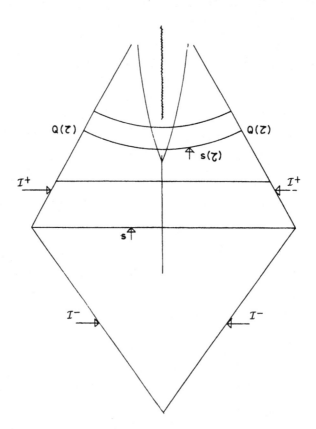

Fig. 15. The surface $S(\tau)$ of constant τ intersect $\mathcal{I}^+$ in the two-spheres $Q(\tau)$.

of $D^+(S(\tau))$ for each $\tau \geq 0$. However it is somewhat more convenient to choose the surface $S(\tau)$ for $\tau > 0$ so that they intersect $\mathcal{I}^+$ (Fig. 15). This means that asymptotically they tend to null surfaces of constant retarded time. The advantage of such a choice of surfaces $S(\tau)$ is that the gravitational radiation emitted during the formation and interaction of black holes will escape to $\mathcal{I}^+$ and will not intersect the surfaces $S(\tau)$ for τ sufficiently large. When the solution settles down to a nearly stationary state, one can relate the properties of the event horizon at the time τ to the values of the mass and angular momentum measured on the intersection of $\mathcal{I}^+$ and $S(\tau)$. There is no unique choice of the surface $S(\tau)$ and of the correspondence between points on the horizon and points on $\mathcal{I}^+$ at the same values of τ. This arbitrariness does not matter provided one relates the properties of the event horizon to the mass and angular momentum measured on $\mathcal{I}^+$ only during periods when the system is nearly stationary. I shall be concerned with relations between initial and final quasi-stationary states.

It turns out that one can always find such a time coordinate τ if the solution is strongly asymptotically predictable, i.e., if there exists a partial Cauchy surface S such that

(a) $\mathcal{I}^+$ lies in the boundary of $D^+(S)$,

(b) $J^+(S) \cap \dot{J}^-(\mathcal{I}^+)$ lies in $D^+(S)$.

More precisely, one can find a function $\tau \geq 0$ on $D^+(S)$ such that the surfaces $S(\tau)$ of constant τ are spacelike surfaces without edges in M and satisfy

(i) $S(0) = S$,

(ii) $S(\tau_2)$ lies to the future of $S(\tau_1)$ for $\tau_2 > \tau_1$,

(iii) Each $S(\tau)$ for $\tau > 0$ intersects $\mathcal{I}^+$ in a 2-sphere $Q(\tau)$. The $\{Q(\tau)\}$ for $\tau > 0$ cover $\mathcal{I}^+$,

(iv) Every future directed nonspacelike curve from any point in the region of $D^+(S)$ between S and $S(\tau)$ intersects either $\mathcal{I}^+$ or $S(\tau)$ if continued far enough,

(v) $S(\tau)$ minus the boundary 2-sphere $Q(\tau)$ is topologically equivalent to S.

The point that one can find such a time function τ is somewhat technical so I shall just give an outline here. Full details are in HE. It is based on an idea of Geroch (1968). One first chooses a volume measure $d\mu$ on M so that the total volume of M in this measure is finite. In the case of a weakly asymptotically simple space such as I am considering, this volume measure could be that defined by the conformal metric $\tilde{g}_{ab}$ which is regular on $\mathcal{I}^-$ and $\mathcal{I}^+$. For a point $p \in D^+(S)$ one can then define a quantity $f(\rho)$ which is the volume of $J^+(p) \cap D^+(S)$ evaluated in the measure $d\mu$. Now choose a family $\{Q(\tau)\}$, $\tau > 0$ of 2-spheres which cover $\mathcal{I}^+$ and which are such that $Q(\tau_2)$ lies to the future of $Q(\tau_1)$ for $\tau_2 > \tau_1$. Then, given $p \in D^+(S)$ one can define a quantity $h(p, \tau)$ as the volume in the measure of $d\mu$ of $D^+(S) \cap [J^-(p) - J^-(Q(\tau))]$. The functions $f(p)$ and $h(p, \tau)$ are continuous in p and τ. The surface $S(\tau)$ can now be defined as the set of points p for which

$h(p, \tau) = \tau f(p)$. Properties (i)–(v) can easily be verified.

With the time function τ one can describe the evolution of black holes. Suppose that a star collapses and gives rise to a trapped surface T. As was shown in the last section, the assumption of strong asymptotic predictability implies that one cannot escape from T to $\mathcal{I}^+$. There must thus be an event horizon $\dot{J}^-(\mathcal{I}^+)$ to the future of S. Also by the assumption of strong asymptotic predictability, $J^+(S) \cap \dot{J}^-(\mathcal{I}^+)$ will be contained in $D^+(S)$. For sufficiently large τ, the surface $S(\tau)$ will intersect the horizon and the set $B(\tau)$ defined as $S(\tau) - J^-(\mathcal{I}^+)$ will be nonempty. I shall define a *black hole* on the surface $S(\tau)$ to be a connected component of $B(\tau)$. In other words, it is a connected region of the surface $S(\tau)$ from which one cannot escape to $\mathcal{I}^+$. As τ increases, black holes may grow or merge together and new black holes may be formed by further stars collapsing but a black hole, once formed, *cannot disappear, nor can it bifurcate*. To see that it cannot disappear is easy. Consider a black hole $B_1(\tau_1)$ on a surface $S(\tau_1)$. Let p be a point of $B_1(\tau_1)$. By property (iv), every future directed nonspacelike curve λ from p will intersect either $\mathcal{I}^+$ or $S(\tau_2)$ for any $\tau_2 > \tau_1$. The former is impossible since p is not in $J^-(\mathcal{I}^+)$. This also implies that λ must intersect $S(\tau_2)$ at some point q which is not in $J^-(\mathcal{I}^+)$. Thus q must be contained in some black hole $B_2(\tau_2)$ on the surface $S(\tau_2)$ which will be said to be *descended from* the black hole $B_1(\tau_1)$. Since black holes can merge together, $B_2(\tau_2)$ may be descended from more than one black hole on the surface $S(\tau_1)$. Alternatively, a black hole on $S(\tau_2)$ may not be descended from any on $S(\tau_1)$ but have formed between τ_1 and τ_2 (Fig. 16). The result that a black hole cannot bifurcate can be expressed by saying that $B_1(\tau_1)$ cannot have more than one descendant on a later surface $S(\tau_2)$. This follows from the fact that any future directed nonspacelike curve from a point $p \in B_1(\tau_1)$ can be continuously deformed through a sequence of such curves into any other future directed nonspacelike curve from p. Since all these curves will intersect $S(\tau_2)$, their intersection with $S(\tau_2)$ will form a continuous curve in $S(\tau_2)$. Thus $J^+(p) \cap S(\tau_2)$ will be connected. Similarly $J^+(B_1(\tau_1)) \cap S(\tau_2)$ will be connected. It must be contained in $B(\tau_2)$ and so will be contained in only one connected component of $B(\tau_2)$. There will thus be only one black hole on $S(\tau_2)$ which is descended from $B_1(\tau_1)$.

The boundary $\partial B_1(\tau_1)$ in $S(\tau_1)$ of a black hole $B_1(\tau_1)$ is formed by part of the intersection of the event horizon with the surface $S(\tau_1)$. Since we are assuming that the initial surface S is simply connected, it follows from property (v) that each of the surfaces $S(\tau)$ is also simply connected. This implies that the boundary $\partial B_1(\tau_1)$ is connected. For suppose that $\partial B_1(\tau_1)$ consisted of two components $\partial_1 B_1(\tau_1)$ and $\partial_2 B_1(\tau_1)$. One could join a point $q_1 \in \partial_1 B_1(\tau_1)$ to a point $q_2 \in \partial_2 B_1(\tau_1)$ by a curve μ lying in $B_1(\tau_1)$ and a curve λ lying in $S(\tau_1) - B_1(\tau_1)$. Joining μ and λ, one would obtain a closed curve in $S(\tau_1)$ which could not be deformed to zero in $S(\tau_1)$ since it crossed the closed surface $\partial_1 B_1(\tau_1)$ only once. This would contradict the fact that $S(\tau_1)$ is simply connected.

If the black holes are formed by collapses in a space which is nonsingular initially, the surface S can be chosen to have a topology of Euclidean 3-space R^3. By

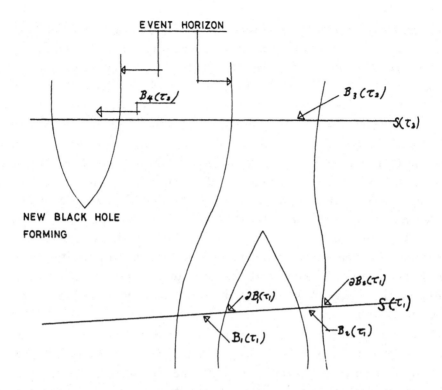

Fig. 16. The two black holes $B_1(\tau_1)$ and $B_2(\tau_1)$ on the surface $S(\tau_1)$ merge to form the black hole $B_3(\tau_2)$ on the surface $S(\tau_2)$. A new black hole $B_4(\tau_2)$ is formed between $S(\tau_1)$ and $S(\tau_2)$.

property (v) each surface $S(\tau)$ minus the bounding 2-sphere $Q(\tau)$ on $\mathcal{I}^+$ will also have this topology. It then follows that the boundary $\partial B_1(\tau)$ of a black hole $B_1(\tau)$ will be compact and that the topology of $S(\tau) \cap \bar{J}^-(\mathcal{I}^+)$, the space outside and including the horizon, will have the topology of R^3 minus a number of open sets with compact closure. As I said earlier, it is sometimes convenient to consider black hole solutions which are not initially nonsingular but which may outside the event horizon approximate the behaviour of initially nonsingular solutions at large times. If they are to do this it is not necessary that the surfaces $S(\tau) - Q(\tau)$ have the topology R^3 (indeed they do not in the Schwarzschild and Kerr solutions), but they should have the same topology outside the event horizon. One can ensure this by requiring that the initial surface S has the property:

(β) $S \cap J^-(\mathcal{I}^+)$ has the topology of R^3 minus a finite number of open sets with a compact closure.

It is easy to show that if S has the property (β) then each surface $S(\tau) - Q(\tau)$ has the property (β) also.

I showed earlier that the null geodesic generator of the event horizon did not have any future end-points and had negative or zero convergence ρ. It follows from this that the area of the boundary $\partial B_1(\tau)$ of a black hole $B_1(\tau)$ cannot decrease with increasing τ. If two black holes $B_1(\tau_1)$ and $B_2(\tau_2)$ on a surface $S(\tau_1)$ later collide and merge to form a black hole $B_3(\tau_2)$ on the surface $S(\tau_2)$, the area of $\partial B_3(\tau_2)$ must be at least as great as the sum of the areas of the boundaries $\partial B_1(\tau_1)$ and $\partial B_2(\tau_1)$ of the original black holes. In fact it must be strictly greatly because $\partial B_3(\tau_2)$ will contain two disjoint closed sets corresponding to generators which intersected $\partial B_1(\tau_1)$ and $\partial B_2(\tau_1)$ respectively. Since $\partial B_3(\tau_2)$ is connected, it must also contain an open set corresponding to generators which had past end-points between $S(\tau_1)$ and $S(\tau_2)$.

The area of the boundaries of black holes has strong analogies to the concept of entropy in thermodynamics: it never decreases and it is additive. We shall see later that the area will remain constant only if the black hole is in a stationary state. When the black hole interacts with anything else the area will always increase. Under favourable circumstances one can arrange that the increase is arbitrarily small. This corresponds to using nearly reversible transformations in thermodynamics. I shall show later how the area of a black hole in a stationary state is related to its mass and angular momentum. The fact that the area cannot decrease will impose certain inequalities on the change of the mass and angular momentum of the black hole as a result of interaction.

I shall denote by $T(\tau)$ the region of the surface $S(\tau)$ that contains trapped or marginally trapped surfaces lying in $S(\tau)$. I shall call the boundary $\partial T(\tau)$ of $T(\tau)$, the *apparent horizon* in the surface $S(\tau)$. In the last section it was shown that trapped or marginally trapped surfaces cannot intersect $J^-(\mathcal{I}^+)$. Thus $T(\tau)$ must be contained in $B(\tau)$ and the apparent horizon must lie behind or coincide with the event horizon. The apparent horizon $\partial T(\tau)$ will be a *marginally trapped surface*. That is, it is a spacelike 2-surface such that the convergence ρ of the outgoing full geodesics orthogonal to it is zero. As τ increases, these null geodesics may be focused by matter or gravitational radiation and the position of the apparent horizon will move outwards on the surface $S(\tau)$ at or faster than the speed of light. As the example of the spherical collapsing shell shows, it can move outwards discontinuously. When the solution is in a quasi-stationary state, the apparent horizon will lie just inside the event horizon and the area of $\partial T(\tau)$ will be nearly equal to that of $\partial B(\tau)$. In the transition from one quasi-stationary state to another the area of $\partial B(\tau)$ will increase and so the area of $\partial T(\tau)$ must be greater in the final state than in the initial one. I have not been able to show, however, that the area of $\partial T(\tau)$ increases monotonically though I believe it probably does.

It is interesting to see the behaviour of the event and apparent horizon in the case of two black holes which collide and merge together. Suppose two stars a long way apart collapse to form black holes $B_1(\tau)$ and $B_2(\tau)$ which have settled down to a quasi-stationary state by the surface $S(\tau_1)$ (Fig. 17). Just inside the two components $\partial B_1(\tau_1)$ and $\partial B_2(\tau_1)$ of the event horizon there will be two components $\partial T_1(\tau_1)$ and $\partial T_2(\tau_1)$ of the apparent horizon. The 2-surfaces $\partial T_1(\tau_1)$ and $\partial T_2(\tau_1)$ will be smooth but the 2-surfaces $\partial B_1(\tau_1)$ and $\partial B_2(\tau_1)$ will each have a slight cusp on the side facing the other. As the black holes approach each other, these cusps will become more pronounced and will join up to give a single component $\partial B_3(\tau)$ of the event horizon. The apparent horizon $\partial T_1(\tau)$ and $\partial T_2(\tau)$ on the other hand,

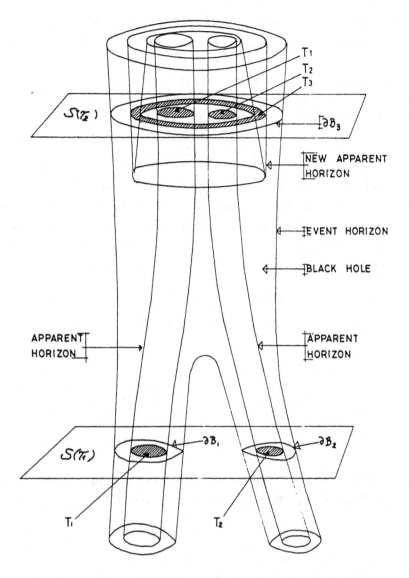

Fig. 17. The collison of two black holes. The event horizons ∂B_1 and ∂B_2 merge to form the event horizon ∂B_3. The apparent horizons ∂T_2 do not merge but are enveloped by a new apparent horizon ∂T_3.

will not join up. As they approach each other there will be some surfaces $S(\tau_2)$ on which there will be a third component $\partial T_3(\tau_2)$ which surrounds both $\partial T_1(\tau_2)$ and $\partial T_2(\tau_2)$.

I shall now show that each component of the apparent horizon $\partial T(\tau)$ must have the topology of a 2-sphere. I originally developed this proof for the event horizon in the stationary situations considered in the next section but I am grateful to G. W. Gibbons for pointing out that it can be applied to apparent horizon at any time. The idea is to show that if a connected component $\partial T_1(\tau)$ of the apparent horizon had any topology other than that of a 2-sphere, one could deform it to give a trapped or marginally trapped surface just outside $\partial T(\tau)$ This would be a contradiction of the fact that the apparent horizon is the outer boundary of such surfaces.

Let u^a be the unit timelike vector field orthogonal to the surface $S(\tau)$. Let l^a and n^a be respectively the future directed outgoing and ingoing null vector fields, orthogonal to $\partial T_1(\tau)$ and normalized so that

$$l^a u_a = 2^{-\frac{1}{2}}, \qquad n^a u_a = 2^{-\frac{1}{2}}, \qquad l^a n_a = 1$$

The complex null vectors m^a and $\bar{m}^a$ will then lie in the 2-surface $\partial T_1(\tau)$. The vector $\omega^a = 2^{-\frac{1}{2}}(l^a - n^a)$ will be the unit outward spacelike vector in $S(\tau)$ orthogonal to $\partial T_1(\tau)$. Suppose one now moves each point of $\partial T_1(\tau)$ a parameter distance h outwards along the vector field $y^a = \omega^a e^f$ where f is some function on $\partial T_1(\tau)$. To maintain the orthogonality of l^a and n^a to the 2-surface requires

$$\kappa - \tau - \delta f + \bar{\alpha} + \beta = 0 \tag{7.1}$$
$$\nu - \pi + \bar{\delta} f + \alpha + \bar{\beta} = 0 \tag{7.2}$$

where

$$\kappa = l_{a;b}m^a l^b, \qquad \tau = l_{a;b}m^a n^b$$
$$\nu = -n_{a;b}\bar{m}^a n^b, \qquad \pi = -n_{a;b}\bar{m}^a l^b$$
$$\delta f = m^a f_{;a} \qquad \text{and} \quad \bar{\alpha} + \beta = l_{a;b}n^a m^b$$

Under this movement of the 2-surface $\partial T_1(\tau)$, the change in the convergence ρ at the outgoing orthogonal null geodesics can be evaluated from the Newman–Penrose equations:

$$\frac{d\rho}{du} = 2^{-\frac{1}{2}}e^f[\sigma\bar{\sigma} + \phi_{00} + (\kappa - \tau)(\bar{\kappa} - \bar{\tau}) + \rho(\rho + \epsilon + \bar{\epsilon} - \bar{\mu} - \gamma - \bar{\gamma})$$
$$+ \bar{\eth}\delta f - \bar{\eth}(\bar{\alpha} + \beta) + \sigma\lambda + \psi_2 + 2\Lambda] \tag{7.3}$$

where

$$\lambda = -n_{a;b}\bar{m}^a \bar{m}^b, \qquad \mu = -n_{a;b}\bar{m}^a m^b, \qquad \gamma = -\frac{1}{2}(n_{a;b}l^a n^b - \bar{m}_{a;b}m^a n^b),$$

$$\psi_2 = -\frac{1}{2}C_{abcd}(l^a n^b l^c n^d - l^a u^b m^c \bar{m}^d),$$

$$\Lambda = \frac{-R}{24} \qquad \text{and} \qquad \bar{\eth} = \bar{\delta} - (\alpha - \bar{\beta})$$

where $\alpha - \bar{\beta} = \bar{m}_{a;b}m^a \bar{m}^b$. The first three terms on the right of equation (7.3) are non-negative. The term $\bar{\eth}\delta f$ is the Laplacian of f in the 2-surface. One can choose f so that the sum of the last five terms on the right of equation (7.3) is constant over the 2-surface. The sign of this constant value will be determined by that of the integral of $(\sigma\lambda + \psi_2 + 2\Lambda)$ over the 2-surface $(\eth(\bar{\alpha} + \beta)$ being a divergence, has zero integral). This integral can be evaluated from another Newman–Penrose equation which can be written as

$$\eth(\alpha + \bar{\beta}) - \bar{\eth}(\bar{\alpha} + \beta) + \eth(\alpha - \beta) + \bar{\eth}(\bar{\alpha} - \bar{\beta})$$
$$= -2\sigma\lambda - 2\psi_2 + 2\Lambda + 2\phi_{11} \qquad (7.4)$$

where

$$\phi_{11} = \frac{1}{4} R_{ab}(l^a n^b + m^a \bar{m}^b)$$

When integrated over the 2-surface the terms in $\bar{\alpha} + \beta$ disappear but there is in general a contribution from the $\bar{\alpha} - \beta$ terms because the vector field m^a will have singularities on the 2-surface. The contribution from these singularities is determined by the Euler number χ of the 2-surface. Thus

$$\int (-\sigma\lambda - \psi_2 + \phi_{11} + \Lambda)dA = 2\pi\chi \qquad (7.5)$$

(The real part of the equation is in fact the Gauss–Bonnet theorem.) Therefore

$$-\int (\sigma\lambda + \psi_2 + 2\Lambda)dA = 2\pi\chi - \int (\phi_{11} + 3\Lambda)dA \qquad (7.6)$$

Any reasonable form of matter will obey the Dominant Energy condition (Hawking 1971): $-T^{00} \geq |T^{ab}|$ in any orthonormal tetrad. This and the Einstein equations imply that $\phi_{11} + 3\Lambda \geq 0$. The Euler number χ is +2 for a sphere, 0 for a torus and negative for any other compact orientable 2-surface. $(\partial T_1(\tau)$ has to be orientable as it is a boundary.) Suppose $\partial T_1(\tau)$ was not a sphere. Then one could choose f so that the right hand side of equation (7.3) was everywhere positive or zero. This would mean that there would be a trapped or marginally trapped surface just outside $\partial T(\tau)$, which is supposed to be the outer boundary of such surfaces. Thus each component of the apparent horizon has the topology of a 2-sphere.

In the next section I shall show that the event horizon will coincide with the apparent horizon in the final stationary state of the solution. Thus each connected component $\partial B_1(\tau)$ of the event horizon will have spherical topology at late times. It might, however, have some other topology during the earlier, time-dependent phase of the solution.

8. The Final State of Black Holes

During the formation of a black hole in a stellar collapse, the solution will change rapidly with time. Gravitational radiation will propagate out to $\mathcal{I}^+$ and across the event horizon into the black hole. By the conservation law for asymptotically flat space (Bondi *et al.* 1962, Penrose 1963), the energy of the gravitational radiation reaching $\mathcal{I}^+$ will reduce the mass of the system as measured from $\mathcal{I}^+$. The radiation crossing the event horizon will cause the area of the horizon to increase. The amount of energy that can be radiated to $\mathcal{I}^+$ or down the black hole is presumably bounded by the original rest mass of the star. Thus one might expect that the area of the horizon and the mass measured on $\mathcal{I}^+$ might eventually tend to constant values and the solution outside the horizon settle down to a stationary state. Although we cannot at the moment describe in detail the time-dependent formation phase, it seems that we probably can find all these final stationary states. In this section therefore I shall consider stationary black hole solutions in the expectation that outside the horizon they will approximate to time-dependent solutions at late times.

More precisely, I shall consider spacetimes M, g_{ab} which satisfy

(1) M, g_{ab} is strongly asymptotically predictable.
(2) M, g_{ab} is stationary, i.e., there exists a one parameter isometry group ϕ_t : $M \to M$ whose Killing vector K^a is timelike near $\mathcal{I}^-$ and $\mathcal{I}^+$. (Note that it may be spacelike near the black hole.)

Since these stationary spaces are not necessarily nonsingular initially, the partial Cauchy surface S may not have the topology R^3. In fact, in most cases it will be $R^1 \times S^2$. However, one wants these spaces to approximate physical initially nonsingular solutions in the region outside and including the horizon at late times, i.e., on $S(\tau) \cap \bar{J}^-(\mathcal{I}^+)$ for large τ. Thus $S(\tau) \cap \bar{J}^-(\mathcal{I}^+)$ must have the same topology as it would have in an initially nonsingular solution. One can ensure this by requiring the property

(β) $S \cap \bar{J}^-(\mathcal{I}^+)$ has the topology of R^3 minus a finite number of open sets with compact closure.

It is also convenient (but not essential) to require

(α) S is simply connected.

Finally, one is interested only in black holes that one could fall into from infinity. Thus it is reasonable to require

(γ) There is some τ_0 such that for $\tau \geq \tau_0$, $S(\tau) \cap J^-(\mathcal{I}^+)$ is contained in $J^+(\mathcal{I}^-)$.

I shall call a space satisfying (1), (α), (β), (γ) a *regular predictable* space. If, in addition, (2) is satisfied, I shall call it a *stationary regular predictable space*. I shall show that in such a space the convergence ρ and shear σ of the generators of the horizon are zero. It then follows that the Ricci tensor term $\phi_{00} = 4\pi T_{ab}l^a l^b$ and Weyl tensor term $\psi_0 = C_{abcd}l^a m^b l^c m^d$ must be zero on the horizon. One can interpret this as saying that no matter or gravitational radiation is crossing the horizon.

The fact that ρ is zero implies that each connected component $\partial B_i(\tau)$ of the event horizon is a marginally trapped surface. Since there are no trapped or marginally trapped surfaces outside the event horizon $\partial B_i(\tau)$ must coincide with a component $\partial T_i(\tau)$ of the apparent horizon. Thus all stationary black holes are topologically spherical; there are no toroidal ones. There could be several components $\partial B_i(\tau)$ of the event horizon corresponding to black holes which maintain themselves at constant distances from each other. This is possible in the limiting case of non-rotating black holes carrying electric charge equal to their mass (Hartle and Hawking 1972): the electric repulsion just cancels the gravitational repulsion. It seems probable but has not yet been proved that these solutions are the only stationary regular predictable spaces containing more than one black hole.

Assuming there is only one black hole, the question of the final state has two branches according as to whether or not the solution is static. A stationary solution is said to be *static* if the Killing vector K^a is hypersurface orthogonal, i.e., if the twist $\omega^a = \frac{1}{2}\eta^{abcd}K_b K_{a;d}$ is zero. In a static regular predictable space which is empty or contains only an electromagnetic field one can apply Israel's theorem (Israel, 1968) to show that the space must be the Schwarzschild or Reissner–Nordström solution.

If the solution is not static but only stationary, I shall show (modulo one point) that the black hole must be rotating. I shall prove that a stationary regular predictable space containing a rotating black hole must be axisymmetric. One can then appeal to Carter's theorem (see his lectures) to show that such spaces, if empty, can depend only on two parameters; the mass and angular momentum. One two parameter family is known, the Kerr solutions for $a^2 \leq m^2$ (the Kerr solutions for $a^2 > m^2$ contain naked singularities). It seems unlikely that there are any others. Thus it appears that the final state of a black hole is a Kerr solution. In the case where the collapsing star carries a net electric charge one would expect it to be a Newman–Kerr solution.

I shall only give outlines of the results mentioned above. The full gory details will be found in HE.

To show that the convergence and shear of the generators of the event horizon are zero, consider a compact spacelike 2-surface F lying in the horizon. Under the time translation ϕ_t the surface F will be moved into another 2-surface $\phi_t(F)$ in the event horizon. Assuming that $\phi_t(F)$ lies to the future of F on the event horizon for $t > 0$, one can compare their areas by moving each element of F up the generators of the horizon to $\phi_t(F)$. I showed earlier that the generators had no future endpoints and did not have positive convergence ρ. If any of them had past end-points or negative convergence between F and $\phi_t(F)$, the area of $\phi_t(F)$ would be greater than that of F. But the area of $\phi_t(F)$ must be the same as that of F since ϕ_t is an isometry. Thus the generators of the event horizon cannot have any past end-points and must have zero convergence ρ. From the Newman–Penrose equations

$$\frac{d\rho}{dv} = \rho^2 + \sigma\bar{\sigma} + (\epsilon + \bar{\epsilon})\rho + \phi_{00}$$

$$\frac{d\sigma}{dv} = 2\rho\sigma + (3\epsilon - \bar{\epsilon})\sigma + \psi_0$$

it follows that the shear σ, the Ricci tensor term ϕ_{00} and the Weyl tensor term ψ_0 are zero on the horizon.

The only complication in this proof comes from the fact that the Killing vector K^a which represents infinitesimal time translations, may be spacelike on and near the horizon. (I shall have more to say about this later.) This means that for an arbitrary 2-surface F in the horizon these may be some points of $\phi_t(F)$ for $t > 0$ which lie to the past of F. However one can construct a 2-surface F for which $\phi_t(F)$ lies wholly to the future of F in the following way. Choose a compact spacelike 2-sphere C on $\mathcal{I}^-$. The Killing vector K^a will be directed along the null geodesic generators on $\mathcal{I}^-$. Thus $\phi_t(C)$ will lie to the future of C for $t > 0$. The intersection of $\dot{J}^+(C)$, the boundary of the future of C, with the event horizon will define a 2-surface F with the required properties.

If the solution is static, one can apply Israel's theorem. If the solution is only stationary but not static one can apply a generalization of the Lichnerowicz theorem (cf. Carter) to show that the Killing vector K^a is spacelike in a non-zero region (called the *ergosphere*) part of which lies outside the horizon. The non-trivial part of this generalization consists of showing that a certain surface integral over the horizon would be zero if K^a were not spacelike there. Details are given in HE.

There are now two possibilities: either the ergosphere intersects the horizon or it does not. The horizon is mapped into itself by the time translation ϕ_t. In the former case the Killing vector K^a will be spacelike on part of the horizon and so some null geodesic generators will be mapped into other ones. The generators form a 2-dimensional space Q which is topologically a 2-sphere, and which has a metric corresponding to the constant separation of the generators. The time translation ϕ_t which moves generators into generators can be regarded as an isometry group on Q. Thus its action corresponds to rotating Q about an axis. One can interpret this as follows. A point of Q represents a generator of the horizon. As one moves along a generator one is moving relative to the stationary frame defined by the integral curves of K^a, i.e., relative to infinity. Thus the horizon would be *rotating* with respect to infinity. I shall show that such a rotating black hole must be axisymmetric.

The other possibility is that the ergosphere might be disjoint from the horizon. Hajicek (1972) has shown that in general the ergosphere must intersect the horizon if the region outside the horizon is null geodesically complete in both the future and the past directions. However, these stationary spaces approximate to physical solutions only at late times. There is thus no physically compelling reason why they should not contain geodesics in the exterior region which are incomplete in the past direction. I shall therefore give an alternative intuitive argument to show that the ergosphere must intersect the horizon.

When there is an ergosphere one can extract energy from the solution by the Penrose process (Penrose 1969). This consists of sending a particle with energy $E_1 = P_1^a K_a$ from infinity into the ergosphere. It then splits into two particles with energies E_2 and E_3. By local conservation $E_1 = E_2 + E_3$. Since the Killing vector K^a is spacelike in the ergosphere, one can choose the momentum p_2^a of the second particle such that E_2 is negative. Thus E_3 is greater than E_1. The particle 3 can escape to infinity where its total energy (the rest mass + kinetic energy) will be greater than that of the original particle 1. Thus one has extracted energy. Particle 2, having negative energy, must remain in the region where K^a is spacelike.

Suppose that the ergosphere did not intersect the horizon. Then particle 2 would have to remain outside the horizon. One could repeat the process and extract more energy. As one did so the solution would presumably change gradually. However the ergosphere could not disappear because there has to be somewhere for the negative energy particles to exist. If the ergosphere remained disjoint from the horizon one could extract an arbitrarily large amount of energy. This does not seem reasonable physically. On the other hand, if the ergosphere moved so that it intersected the horizon, the solution would have to become axisymmetric. At the moment the ergosphere touched the horizon one would have a stationary, non-static, axisymmetric black hole solution. This could not be a Kerr solution because in a non-static Kerr solution the ergosphere actually intersects and does not merely touch the horizon. However it appears from the results of Carter that the Kerr solutions are the only stationary axisymmetric black hole solutions. Thus it seems that one ends up with a contradiction if one supposes that the ergosphere is disjoint from the horizon. I shall therefore assume that any stationary, non-static black hole is rotating.

My original proof (Hawking 1972) that a stationary rotating black hole must be axisymmetric had the great advantage of simplicity. However it involved the assumption that as well as the *future* event horizon $\dot{J}^-(\mathcal{I}^+)$ there was a *past* event horizon $\dot{J}^+(\mathcal{I}^-)$ and that the two horizons intersected in a compact spacelike 2-surface. Penrose pointed out that there is no necessity for this assumption to hold. These stationary spaces represent physical solutions only at large times. There would be a past horizon if the solution were time-symmetric. By the Papapetrou theorem (see Carter) time-symmetry is a *consequence* of stationary and axial-symmetry. It should not be assumed to prove axial symmetry. I therefore developed another proof of axial symmetry which depends only on the future horizon. Unfortunately, this proof is rather long and messy. I shall try to give an intuitive picture of it here and shall give the full details in HE.

Consider a rotating black hole. Let t_1 be the period of rotation of the horizon. This means that for a point p on a generator λ of the horizon $\phi_{t_1}(p)$ is also on λ (Fig. 18). One can choose a parameter v on λ so that $l^a = dx^a/dv$ satisfies

$$l_{a;b}l^b = 2\epsilon l_a$$

where ϵ is constant on λ and so that difference between the values of v at p and at $\phi_{t_1}(p)$ is t_1. This fixes the scaling of l^a. One can now form the vector field

$$\tilde{K}^a = \frac{t_1}{2\pi}(l^a - K^a)$$

on the horizon. The orbits of $\tilde{K}^a$ will be closed spacelike curves in the horizon. The aim will be to show that they correspond to rotations of the solution about an axis of symmetry. Choose a spacelike 2-surface F in the horizon tangent to $\tilde{K}^a$. Let N be the null surface generated by the ingoing null geodesics orthogonal to F (Fig. 19). The idea of the proof is to consider the Cauchy problem for the region to

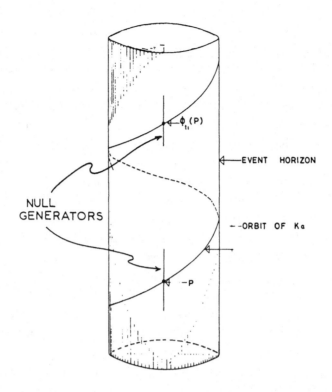

Fig. 18. The time translation ϕ_{t_1} moves a point P on the horizon along the orbit of K^a to the point $\phi_{t_1}(P)$ on the same generator of the horizon.

the past of both the horizon and N. The Cauchy data for the empty space Einstein equations in this situation consists of ψ_0 on the horizon $\psi_4 = C_{abcd}n^a\bar{m}^b n^c \bar{m}^d$ on N where n^a is the null vector tangent to N and ρ, $\mu = -n_{a;b}\bar{m}^a m^b$ and $\psi_2 = \frac{1}{2}C_{abcd}(l^a n^b l^c n^d - l^a n^b m^c m^d)$ on the 2-surface F. If there are other fields present (e.g., an electromagnetic field) one has to give additional data for them. I shall consider only the empty case but similar arguments hold in the presence of any fields obeying well-behaved hyperbolic equations.

By the stationarity of the horizon, p and ψ_0 are zero and one can show from the Newman–Penrose equations that ψ_2 is constant along the generators of the horizon. Thus the only non-trivial Cauchy data are that on the null surface N. The idea now is to show that these Cauchy data are unchanged if one moves N by moving each point of the 2-surface F an equal parameter distance down the generators of the horizon. If this is the case, it follows from the uniqueness of the Cauchy problem that the solution admits a Killing vector $\hat{K}^a$ which coincides with l^a on the horizon. Then $\tilde{K}^a$ defined as $t_1/2\pi(\hat{K}^a - K^a)$ will also be a Killing vector. Since the orbits of $\tilde{K}^a$ are closed curves on the horizon, they will be closed everywhere and so will correspond to rotations about an axis of symmetry.

To show that the data on N are unchanged on moving each point of F down the generators of the horizon, I assume that the solution is analytic though this is almost certainly not necessary. The data on N can then be represented by their partial

62

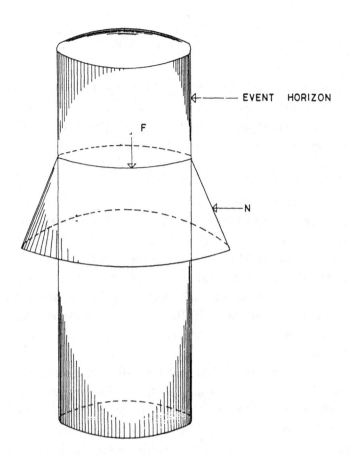

EVENT HORIZON

F

N

Fig. 19. The event horizon and the null surface N intersected in the spacelike surface F.

derivatives at F in the direction along N. From the Newman–Penrose equations one can evaluate the derivatives along a generator λ of the horizon of these and certain other quantities. If one takes them in a certain order one obtains equations of the form

$$\frac{dx}{dv} = ax + b$$

where x is the quantity in question and a and b are constant along λ.

Now moving F a parameter distance t_1 (the period of rotation of the black hole) to the past along the generators of the horizon is the same as moving F by the time translation ϕ_{-t_1}. Since ϕ_{-t_1} is an isometry, the quantity x will be unchanged under it. Thus x must be periodic along the generator λ with period t_1. This is possible only if x is constant along λ and equal to $-(b/a)$. One then uses this to calculate the derivative along λ of another quantity and shows that it is constant by a similar argument. Proceeding by induction one shows that all the derivatives at the horizon of the Cauchy data on N are constant along the generators of the horizon.

The first quantity x that one considers is $\bar{\alpha} + \beta$. The Newman–Penrose equation for this is

$$\frac{d}{dt}(\bar{\alpha} + \beta) = \delta(\epsilon + \bar{\epsilon}) + \psi_1 \,.$$

By construction $\delta(\epsilon + \bar{\epsilon})$ is constant along the generators of the horizon and by another Newman–Penrose equation, $\psi_1 = 0$ on the horizon. Therefore in order for $\bar{\alpha} + \beta$ to be periodic it has to be constant along the generators and $\delta(\epsilon + \bar{\epsilon})$ has to be zero. This means that $\epsilon + \bar{\epsilon}$ must be constant over the whole horizon. In the next section we shall see $\epsilon + \bar{\epsilon}$ can be interpreted as the restoring force or effective surface gravity of the black hole.

One now applies similar arguments to show that $(\bar{\alpha} - \beta)$, μ, λ, ψ_3 and ψ_4 are constant along the generators of the horizon. One then repeats the arguments to show that the first and higher derivatives of all quantities along the vector n^a are constant along the generators. This completes the proof.

It turns out that if ϵ is nonzero (as it is in general) the solution is completely determined by a knowledge of ψ_2 on each generator. I shall use this fact in one of the applications in the next section. It holds true even if the space outside the black hole is not empty but contains, say, a ring of matter (in which case the space would not be a Kerr solution).

The proof of axial symmetry implies that a rotating black hole cannot be exactly stationary unless all distance matter and all fields are arranged axisymmetrically. In real life this will never be the case. Thus a rotating black hole can never be exactly stationary, it must be slowing down. However, calculations by Press (1972), Hawking and Hartle (1972), and Hartle (1972) have shown that the rate of slowing down is very small in most cases. I shall discuss this further in the next section.

9. Applications

In this final section I shall outline some of the ways in which the theory described so far can be used to obtain quantitative results, which is what most people want. I shall discuss three applications:

- (A) The limits that can be placed from the area theorem on the amount of energy that can be extracted from black holes.
- (B) The change in the mass and angular momentum of a nearly stationary black hole produced by small perturbations.
- (C) Time-symmetric black holes. (These are not very realistic but they provide some concrete examples.)

A. Energy Limits

In view of the last section it seems reasonable to assume that a black hole settles down to a Kerr solution or, if carrying an electric charge, to a Newman–Kerr solution. The area of the event horizon of such a solution is

$$A = 4\pi[2M^2 - e^2 + 2(M^4 - M^2 e^2 - L^2)^{\frac{1}{2}}] \tag{9.1}$$

64

where M is the mass, e the electric charge and L the angular momentum of the black hole. (All in units such that $G = c = 1$.) Now suppose that the black hole, having settled down by the surface $S(\tau_1)$ to a nearly stationary state with parameters M_1, e_1, L_1, now undergoes some interaction with external particles or fields and then settles down again by the surface $S(\tau_2)$ to a nearly stationary state with parameters M_2, e_2, L_2. Since the area of the horizon cannot decrease

$$A_2 \geq A_1 \qquad (9.2)$$

where A_1 and A_2 are given by equation (9.1) with the appropriate values of M, e and L. In fact (9.2) is a strict inequality if there is any disturbance at the horizon. It puts an upper limit on $M_1 - M_2$, which represents the amount of energy extracted from the black hole by the interaction. To see what this limit is, it is convenient to express equation (9.1) in the form:

$$M^2 = \frac{A}{16\pi} + \frac{4\pi L^2}{A} + \frac{\pi e^4}{A} + \frac{e^2}{2} \qquad (9.3)$$

The first term on the right can be regarded as the "irreducible" part of M^2, the part that is irretrievably lost down the black hole. The second term can be regarded as the contribution of the rotational energy of the black hole and the third and fourth terms as the contribution of the electrostatic energy. Christodoulou (1970) has shown that one can extract an arbitrarily large fraction of the rotational energy by the Penrose process of sending a particle from infinity into the ergosphere where it splits into two particles one of which returns to infinity with more than the original energy while the other falls through the horizon and reduces the mass and angular momentum of the black hole. Similarly, using charged particles, one can extract an arbitrarily large fraction of the electrostatic energy.

Note that it is M^2 and not M which has an irreducible part. This distinction does not matter when there is only one black hole but it means that one can extract energy, other than rotational or electrostatic energy, by allowing black holes to collide and merge. Consider two black holes $B_1(\tau)$ and $B_2(\tau)$ a long way apart which have settled down to nearly stationary states. One can neglect the interaction between them and regard the solution near each as a Kerr solution with the parameters M_1, e_1, L_1 and M_2, e_2 and L_2 respectively. The areas A_1 and A_2 of $\partial B_1(\tau)$ and $\partial B_2(\tau)$ will be given by equation (9.1). Suppose that at some later time the two black holes come together and merge to form a single black hole $B_3(\tau)$ which settles down to a nearly stationary state with parameters M_3, e_3 and L_3. During the collision process a certain amount of gravitational and possibly electromagnetic radiation will be emitted to infinity. The energy of this radiation will be $M_1 + M_2 - M_3$. This is limited by the requirement that the area A_3 of $\partial B_3(\tau)$ must be greater than the sum of A_1 and A_2. The fraction $\epsilon = (M_1 + M_2)^{-1}(M_1 + M_2 - M_3)$ of the total mass that can be radiated is always less than $1 - 2^{-\frac{1}{2}}$, i.e., about 65%. If the black holes are uncharged or carry the same sign of charge, the fraction is less than a half, i.e., 50%. If the black holes are also non-rotating the fraction is less than $1 - 2^{-\frac{1}{2}}$, i.e., about 29%.

By the conservation of charge $e_3 = e_1 + e_2$. Angular momentum, on the other hand, can be carried away by the radiation. This cannot happen, however, if the situation is axisymmetric, i.e., if the rotation axes of the black holes are aligned along their direction of approach to each other. Then $L_3 = L_1 + L_2$. One can see from equation (9.3) that M_3 can be smaller, i.e., there can be more energy radiated, if the rotations of the black holes are in opposite directions than if they are in the same direction. This suggests that there may be an orientation dependent force between black holes analogous to that between magnetic dipoles. Unlike the electromagnetic case, the force is repulsive if the orientations are the same and attractive if they are opposite. Even in the limiting case when $L_1 = M_1^2$ and $L_2 = M_2^2$, there is still energy available to be radiated. Thus it seems that the force can never be sufficiently repulsive to prevent the black holes colliding.

B. Perturbations of Black Holes

To perform dynamic calculations about black holes seems to require the use of a computer in general. However there are a number of situations that can be treated as small perturbations of stationary black holes, i.e., Kerr solutions. The general idea in these calculations is to solve the linearized equations for a perturbation field (scalar, electromagnetic or gravitational) in a Kerr background and to try to find the radiation emitted to infinity and the rate of change of the mass and angular momentum of the black hole. In the case of the scalar and electromagnetic field these latter can be evaluated by integrating the appropriate components of the energy-momentum tensor of the field over the horizon. For gravitational perturbations, however, there is no well defined local energy-momentum tensor. Instead I shall show how one can determine the change in the mass and angular momentum of the black hole by calculating the change in the area of the horizon and the quantity ψ_2 on the horizon. It turns out that these depend only on the Ricci tensor terms $\phi_{00} = 4\pi T_{ab} l^a l^b$ and $\phi_{01} = 4\pi T_{ab} l^a m^b$ and the Weyl tensor term ψ_0 on the horizon. This is fortunate because it seems that the full equation for gravitational perturbations in a Kerr background are not solvable by separation of functions but Teukolsky (1972) has obtained decoupled separable equations for the quantities ψ_0 and ψ_4.

The mass, the magnitude of the angular momentum and its orientation make up four parameters in all. However, in many uses there are constraints which make it sufficient to caluclate the change in only one function of these four parameters. The simplest such function is the area of the horizon which is given by equation (9.1). The rate of charge of this area can be calculated from the Newman–Penrose equations

$$\frac{d\rho}{dv} = \rho^2 + \sigma\bar{\sigma} + 2\epsilon\rho + \phi_{00} \tag{9.4}$$

$$\frac{d\sigma}{dv} = 2\rho\sigma + 2\epsilon\sigma + \psi_0 \tag{9.5}$$

Choose a spacelike surface S which intersects the event horizon of the background Kerr solutions in $J^+(\mathcal{I}^-)$ and is tangent to the rotation Killing vector $\tilde{K}^a$. Then one can define a family $S(t)$ of such surfaces by moving S under the time translation ϕ_t, i.e., by moving each point of S a parameter distance t along orbits of the Killing vector K^a of the unperturbed metric. This defines a time coordinate t on the horizon. It is convenient to choose the parameter v along the generators of the horizon to be equal to t. Then in the unperturbed Kerr metric

$$\epsilon = \frac{y}{4M(M^2 + y)}, \tag{9.6}$$

where

$$y = (M^4 - L^2)^{\frac{1}{2}}$$

There are two kinds of perturbations one can consider, those in which there is some matter fields like the scalar or electromagnetic field on the horizon with energy–momentum tensor T_{ab} and those in which the perturbations at the horizon are purely gravitational and are produced by matter at a distance from the black hole. Consider first a matter field perturbation where the field is proportional to a small parameter λ. The energy–momentum tensor and so the perturbation in the metric and in ψ_0 will be proportional to λ^2. Thus ρ and σ will be proportional to λ^2 and to order λ^2 equation (9.4) becomes

$$\frac{d\rho}{dt} = 2\epsilon\rho + 4\pi T_{ab}l^a l^b \tag{9.7}$$

where ϵ is given by (9.6). Suppose that the perturbation field is turned off after some time t_1. The black hole will then settle down to a stationary state with $\rho = 0$.

Thus the solution of (9.7) for ρ is

$$\rho = -4\pi \int_t^\infty \exp\{2\epsilon(t - t')\} T_{ab}l^a l^b dt'. \tag{9.8}$$

The rate of increase of area of the horizon is

$$\frac{dA}{dt} = -2 \int \rho dA$$

where the integral is taken over the two surface $\partial B(t)$ which is the intersection of the event horizon with the surface $S(t)$. Substituting from equation (9.8) and performing a partial integration with respect to time one finds that total area increase of the horizon is

$$\delta A = \frac{4\pi}{\epsilon} \int T_{ab}l^a d\Sigma^b \tag{9.9}$$

where $d\Sigma^b = l^b dA dt$ is the 3-surface element of the event horizon. The null vector l^a tangent to the horizon can be expressed in terms of K^a and $\tilde{K}^a$ the Killing vectors

67

of the background Kerr metric which correspond to time translations and spatial rotations respectively.

$$l^a = K^a + \omega \tilde{K}^a + 0(\lambda^2), \tag{9.10}$$

where

$$\omega = \frac{L}{2M(M^2 + y)} \tag{9.11}$$

is the angular velocity of the black hole. The vectors $T_{ab}K^a$ and $-T_{ab}\tilde{K}^a$ represent the flow of energy and angular momentum respectively in the matter fields. They are conserved in the background Kerr metric and their fluxes across the horizon give change of mass and angular momentum of the black hole.

Thus

$$\delta A = \frac{4\pi}{\epsilon}[\delta M - \omega \delta L]. \tag{9.12}$$

This is just the change needed to preserve the formula (9.1) for the area of the horizon of Kerr solution. It is therefore consistent with the idea that the perturbation changes the black hole from one Kerr solution to one with slightly different parameters.

The case of purely gravitational perturbations is rather more interesting because one does not have an energy–momentum tensor from which to compute the fluxes of energy and angular momentum into the black hole. Instead one can use the area increase as a measure of a certain combination of them. One takes the gravitational perturbation field to be proportional to a small parameter λ. Then from equations (9.4), (9.5) σ will be proportional to λ and ρ to λ^2

$$\frac{d\rho}{dt} = \sigma\bar{\sigma} + 2\epsilon\rho, \tag{9.13}$$

$$\frac{d\sigma}{dt} = 2\epsilon\sigma + \psi_0. \tag{9.14}$$

From (9.14)

$$\sigma = -\int_t^\infty \exp\{2\epsilon(t - t')\}\psi_0 dt' \tag{9.15}$$

and

$$\delta A = \frac{1}{\epsilon}\int \sigma\bar{\sigma}dAdt. \tag{9.16}$$

One can apply this formula in at least two situations. First there are stationary gravitational perturbations induced by distant matter which is stationary or nearly stationary. In such perturbations there will be no radiation at infinity and the energy of the sources of the perturbation will be nearly constant. Thus there can be no energy flow into or out of the black hole and its mass must remain constant. From equation (9.1) it then follows that the increase in the area A of the horizon must be accompanied by a decrease in the angular momentum of the black hole. In other words, the effect of stationary perturbation is to slow down the rotation of the black hole. What is happening is that the rotational energy part of M^2 in equation (9.3) is being dissipated into the irreducible part of M^2 represented by A.

There is a strong analogy between this process and ordinary tidal friction in a shallow sea covering a rotating planet. A nearly stationary external body such as a moon will raise tides in the sea. As the planet rotates, the shape of a fluid element will change and so the fluid will be shearing. There will be dissipation of energy at a rate proportional to the coefficient of viscosity times the square of the shear. This energy must come from the rotational energy of the planet. Thus the planet will slow down.

Similarly one can regard the perturbation field of a stationary external object as tidally distorting the horizon of the black hole (Fig. 20) with consequent shearing as the black hole rotates and dissipation of rotational energy at a rate proportional to the square of the shear. The dimensionless analogue of the viscosity in this case is of order unity. Hartle (1972) has calculated the rate of slowing down of a slowly rotating black hole caused by a stationary object of mass M' at coordinates r and θ. For r/M large he finds

$$\frac{dL}{dt} = -2/5 \frac{L}{M} \left(\frac{M'}{M}\right)^2 \sin^2 \theta \left(\frac{M}{r}\right)^6$$

 OBJECT

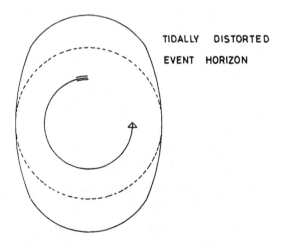

TIDALLY   DISTORTED
EVENT   HORIZON

Fig. 20. The gravitational of an external object tidally distorts the event horizon.

Because of the last factor, this seems too small ever to be of astrophysical significance. This situation might be different, however, for a rapidly rotating black hole with L nearly equal to M^2. In this case the quantity ϵ which acts as a restoring force in equations (9.13) and (9.14) is very small. In a sense the black hole is rotating with nearly break up velocity so centrifugal force almost balances gravity and a small object can raise a large tide on the horizon. For maximum effect, the object should be orbiting the black hole near the horizon with nearly the same angular velocity as that of the black hole. Under these circumstances the black hole would lose energy and angular momentum at a significant rate to the object. The object would also be losing energy and angular momentum in radiation to infinity. It is possible that the rates would balance to give what is called a *floating orbit*. To find out whether this could happen, it would be sufficient to calculate the rate of increase of the area of the horizon and the rate of radiation of energy and angular momentum to infinity since an object in a circular orbit can gain or lose energy and angular momentum only in a certain ratio.

For other problems it would be helpful to be able to calculate separately the rate of change of the mass and the three components of angular momentum. In the last section we saw that a stationary black hole solution is in general determined by a knowledge of the quantity ψ_2 on a 2-dimensional section of the horizon. In the case of a Kerr black hole, the angular momentum is represented by the imaginary $l = 1$ part of $\psi_2^{-\frac{1}{3}}$. From the Newman–Penrose equations one can calculate the change in ψ_2 produced by the perturbation.

$$
\frac{d}{dt}[(\psi_2 + \sigma\lambda - \mu\rho - \Phi_{11} - \Lambda)dA]
$$
$$
= \bar{\eth}\bar{\eth}\sigma - \bar{\eth}\eth\rho + \bar{\eth}[(\bar{\alpha} + \beta)\rho + \Phi_{01}] - \eth[(\alpha + \bar{\beta})\rho + \Phi_{10}]
$$

where $\bar{\eth}$ acting on a spin weight s quantity is $\bar{\delta} + s(\alpha - \bar{\beta})$. Further details will be given elsewhere.

C. Time-Symmetric Black Holes

The last application I shall describe is largely based on the work of G. W. Gibbons. Some of it is about to be published (Gibbons 1972) and more will be in his Ph.D. thesis.

To calculate the evolution of a section of the Einstein equation one requires initial data on a partial Cauchy surface S. The Cauchy data on a spacelike surface can be represented by two symmetric 3-dimensional tensor fields h_{ij} and χ_{ij}. The negative definite tensor h_{ij} is the first fundamental form or induced metric of the 3-surface S imbedded in the 4-dimensional spacetime manifold M. It is equal to $g_{ij} - u_j u_j$ where u_i is the unit timelike vector orthogonal to S. The tensor χ_{ij} is the second fundamental form or extrinsic curvature of S imbedded in M. It is equal to $u_{k;l}h_i^k h_i^l$. The fields h_{ij} and χ_{ij} have to obey the constraint equations:

$$\chi_{ab\|c}h^{bc} - \chi_{bc\|a}h^{bc} = 8\pi T^{bd}h_{ab}u_d$$

$$\frac{1}{2}[^{(3)}R + (\chi_{ab}h^{ab})^2 - \chi_{ab}\chi_{cd}h^{ac}h^{bd}] = 8\pi T^{ab}u_a u_b$$

where $\|$ indicates covariant differentiation with respect to the 3-dimensional metric h_{ij} in the surfaces. The constraint equations are non-linear and difficult to solve in general. However the problem is much simpler if the solution is time-symmetric. The solution is said to be *time-symmetric* about the surface S if there is an isometry which leaves the surface S pointwise fixed but reverses the direction of time, i.e., it moves a point to the future of S on a timelike geodesic orthogonal to S to the point on the some geodesic an equal distance to the past of S. The time symmetry isometry maps χ_{ij} to $-\chi_{ij}$ since it reverses the direction of the normal u_i to S. Thus $\chi_{ij} = 0$. The first constraint is trivially satisfied and the second one becomes in the empty case

$$^{(3)}R = 0$$

The convergence of the outgoing null geodesics orthogonal to a 2-surface F in S is

$$\rho = 2^{-\frac{1}{2}}m^i \bar{m}^j (u_{i;j} + w_{i;j})$$

where w_i is the unit spacelike vector in S orthogonal to F. The first term is zero because $\chi_{ij} = 0$. Thus if F is a marginally trapped surface, the convergence of its normals in S must be zero. This means that it is an extremal surface, i.e., its area is unchanged to first order under a small deformation. In fact F must be a minimal surface if it is an apparent horizon, i.e., if it is the outer boundary of a region containing closed trapped surfaces. Conversely any minimal 2-surface in S is an apparent horizon.

One can write down an explicit family of solutions of the remaining constraint equation by taking the metric h_{ij} on S to be $V^4 \eta_{ij}$ where η_{ij} is the three-dimensional flat metric and V satisfies the Laplace's equation in this metric

$$\nabla^2 V = 0.$$

I shall consider solutions of the form $V = 1 + \Sigma M_i / 2r_i$ representing the field of a number of point masses M_i where the distance from the ith mass is r_i.

The solution with only one mass is the Schwarzschild solution expressed in isotropic coordinance. The minimal surface, which in this case is both the apparent and event horizon, is at $r = \frac{1}{2}M$ and has area $16\pi M^2$. Now consider the case of two equal mass points M_1 and M_2. If they are far apart the minimal surfaces around each will be almost at $r_1 = \frac{1}{2}M$ and $r_2 = \frac{1}{2}M$ and their areas will be nearly $16\pi M^2$. Each surface will however be slightly distorted by the field of the other points and their areas will be slightly greater than $16\pi M^2$. As the solution evolves the two black holes containing these two apparent horizons will fall towards each other and will merge to form a single black hole which will settle down to a Schwarzschild solution with mass M'. The energy of the gravitational radiation emitted in this process will be the initial mass $2M$ of the system minus the final mass M'. This is limited by the fact that the area $16\pi M'^2$ of the event horizon of the final black hole must be greater than the sum of the areas of the event horizons around the two original black holes. The area of these event horizons must be greater than those of the corresponding apparent horizons since these are minimal surfaces. Thus the upper limit on the fraction ϵ of the initial mass that can be radiated is somewhat less than $1 - 2^{-\frac{1}{2}}$. If the two mass points are moved nearer to each other in the initial surface S the minimal surfaces around them become more distorted and their area increases. Thus the upper limit on the fraction of energy that can be radiated becomes less. This is what one would expect since the available energy of each black hole is reduced by the negative gravitational potential of the other. In fact to first order, the reduction in the upper limit on ϵ just corresponds to the Newtonian gravitational interaction energy of the two point masses. When the two mass points are moved close to each other the area of the minimal surface around each becomes greater than $32\pi M^2$. This seems to indicate that the amount of energy that could be radiated would be negative which would be a contradiction. However before the two mass points are close enough for this to happen, it seems that a third minimal surface will be formed which surrounds them both and has area less than $64\pi M^2$ (Fig. 21).

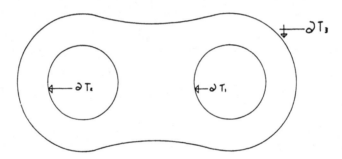

Fig. 21. The two apparent horizons ∂T_1 and ∂T_2 are surrounded by another apparent horizon ∂T_3.

Gibbons (1972) has shown that any minimal surface in a conformally flat initial surface must have an area greater than

$$2\pi \left(\frac{1}{2\pi} \int V_{\|a} dA^a \right)^2$$

where the integral is taken over the minimal surface. The expression in the brackets represents the contribution to the total mass on the initial surface arising from points within the minimal surface. The solution that evolves from the initial surface will eventually settle down to a Schwarzschild solution with an event horizon of area $16\pi M'^2$. Since this area must be greater than the area of the event horizon on the initial surface which in turn must be greater than the area of the minimal surface, the difference between the initial mass M and the final mass M' must be less than $(1 - 2^{-3/2})M$. This means that a single distorted black hole on a surface of time symmetry cannot radiate more than 65% of its initial mass M in relaxing to a spherical black hole.

The black holes that have been considered so far in this subsection are non-rotating. This is because the condition that the solution be invariant under $t \to -t$ rules out any rotation. However, one can include rotation in a simple way if the solution is invariant under the simultaneous transformation $t \to -t$, $\varphi \to -\varphi$. I shall call such a solution (t, φ) symmetric. To obtain such a solution the initial data must be of the form

$$\chi_{ab} = \frac{2J_{(a}\tilde{K}_{b)}}{\tilde{K}_c\tilde{K}^c}$$

where J_a is an axisymmetric vector field orthogonal to the Killing vector $\tilde{K}^b$ which corresponds to rotations about the axis of symmetry. The first constraint equation then becomes

$$J^a_{\|a} = 8\pi T_{ab}u^a\tilde{K}^b .$$

One can integrate this equation to obtain the total angular momentum within a given 2-surface

$$L = -\frac{1}{8\pi}\int J_a dA^a .$$

In the empty case, to which I shall now restrict myself, the angular momentum will arise from singularities of the field J^a. The solution will be asymptotically predictable and will represent black holes if those singularities are contained within apparent horizons. From the form of χ_{ab} it follows that the apparent horizons in the initial surface of (t, φ) symmetry are minimal 2-surfaces. Note that this is the case only in a surface of time symmetry or (t, φ) symmetry. It is not true in later space-like surfaces.

In the empty case the second constraint equation becomes

$$^{(3)}R = \frac{2J_aJ^a}{\tilde{K}_a\tilde{K}^a} .$$

This equation can be solved by a technique of Lichnerowicz. Choose a spatial metric h_{ab}. Then choose a spatial vector field J_a which is axisymmetric, orthognal to $\tilde{K}^a$ and which satisfies $J^a_{\|a} = 0$ in the metric h_{ab}. One then makes a conformal transformation $\tilde{h}_{ab} = V^4 h_{ab}$. The first constraint equation will remain satisfied if J_a transforms as $\tilde{J}_a = V^{-2}J_a$. The second constraint equation will be satisfied if V is chosen so that

$$V^{(3)}R - 8V_{\|ab}h^{ab} = 2V^{-7}\frac{J_aJ^a}{\tilde{K}_b\tilde{K}^b}$$

where the covariant derivatives are with respect to the metric h_{ab}. This equation is non-linear so one cannot write down explicit solutions even in the case where the metric h_{ab} is chosen to be flat. However one can note certain qualitative features. One of these is that the addition of angular momentum tends to increase the total mass of the solution. Thus it seems that the rotational energy of black holes is positive as one would expect. Calculations by Gibbons in the case of two black holes indicate that the ratio of the area of the apparent horizons to the square of the total mass is bigger when the angular momenta of the black holes are in opposite directions than when they are in the same direction. This indicates that there is less energy available to the radiated in the former case than in the latter which is consistent with the idea that there is a spin-dependent force between black holes which is attractive in the case of opposite angular momenta and repulsive in the other case.

The calculations of Gibbons indicate that when the black holes are far apart the force is proportional to the inverse fourth power of the separation which is what one would expect from the analogy with magnetic dipoles.

References

H. Bondi, M. G. J. Van der Burg and A. W. K. Metzner, *Proc. Roy. Soc.* (London) **A269**, 21 (1962).

D. Christodoulou, *Phys. Rev. Letters* **25**, 1596 (1970).

A. G. Doroshkevich, Ya, B. Zel'dovich and I. D. Novikov, *Sov. Phys. JETP* **22**, 122 (1966).

R. P. Geroch, *J. Math. Phys.* **9**, 1739 (1968).

G. W. Gibbons, *Commun. Math. Phys.* **27**, 87 (1972).

J. B. Hartle, 1972 (preprint).

J. B. Hartle and S. W. Hawking, *Commun. Math. Phys.* **26**, 87 (1972).

S. W. Hawking and R. Penrose, *Proc. Roy. Soc.* **A314**, 529 (1970).

S. W. Hawking, *Commun. Math. Phys.* **18**, 301 (1970).

S. W. Hawking and J. Hartle, *Commun. Math. Phys.* **27**, 283 (1972).

S. W. Hawking, *Commun. Math. Phys.* **25**, 152 (1972).

S. W. Hawking and G. F. R. Ellis, *The Large Scale Structure of Space-Time* (Cambridge University Press, Cambridge, to be published).

W. Israel, *Phys. Rev.* **164**, 1776 (1967).

L. Markus, *Ann. Math.* **62**, 411 (1955).

E. T. Newman and R. Penrose, *J. Math. Phys.* **3**, 566 (1962).

R. Penrose, *Phys. Rev. Letters* **10**, 66 (1963).

R. Penrose, *Phys. Rev. Letters* **14**, 57 (1965a).

R. Penrose, *Proc. Roy. Soc.* **A284**, 159 (1965b).

R. Penrose, *Battelle Rencontres*, 1967 *Lectures in Mathematics and Physics*, edited by C. DeWitt and J. A. Wheeler (W. A. Benjamin, Inc., New York 1968).

R. Penrose, *Riv. Nuovo Cimento* **1** (Num. spec.) 252 (1969).

W. H. Press, *AP.J.* **175**, 243 (1972).

R. Price, *Phys. Rev.* **D5**, 2419 (1972).

T. Regge and J. A. Wheeler, *Phys. Rev.* **108**, 1063 (1957).

S. Teukolsky, 1972 (Caltech preprint).

Commun. math. Phys. 31, 161–170 (1973)
© by Springer-Verlag 1973

The Four Laws of Black Hole Mechanics

J. M. Bardeen*

Department of Physics, Yale University, New Haven, Connecticut, USA

B. Carter and S. W. Hawking

Institute of Astronomy, University of Cambridge, England

Received January 24, 1973

Abstract. Expressions are derived for the mass of a stationary axisymmetric solution of the Einstein equations containing a black hole surrounded by matter and for the difference in mass between two neighboring such solutions. Two of the quantities which appear in these expressions, namely the area A of the event horizon and the "surface gravity" κ of the black hole, have a close analogy with entropy and temperature respectively. This analogy suggests the formulation of four laws of black hole mechanics which correspond to and in some ways transcend the four laws of thermodynamics.

1. Introduction

It is generally believed that a gravitationally collapsing body will give rise to a black hole and that this black hole will settle down to a stationary state. If the black hole is rotating, the stationary state must be axisymmetric [1] (An improved version of this theorem involving weaker assumptions is outlined in [2] and is given in detail in [3]). It has been shown that stationary axisymmetric black hole solutions which are empty outside the event horizon fall into discrete families each of which depends on only two parameters, the mass M and the angular momentum J [4–6]. The Kerr solutions for $M^4 > J^2$ are one such family. It seems unlikely that there are any others. It also seems reasonable to suppose that the Newman-Kerr solutions for $M^4 > J^2 + M^2Q^2$, where Q is the electric charge, are the only stationary axisymmetric black hole solutions which are empty outside the event horizon apart from an electromagnetic field. On the other hand there will be an infinite dimensional family of stationary axisymmetric solutions in which there are rings of matter orbiting the black hole. In Sections 2 and 3 of this paper we shall derive formulae for the mass of such a solution and for the difference in mass of two nearby solutions. These formulae

---

* Research supported in part by the National Science Foundation.

generalise the expressions found by Smarr [7] and Beckenstein [8] for the Kerr and Newman-Kerr solutions. We show that the quantities appearing in the formulae have well-defined physical interpretations. Of particular interest are the area A of the event horizon and the "surface gravity" κ, which appear together. These have strong analogies to entropy and temperature respectively. Pursuing this analogy we are led in Section 4 to formulate four laws of black hole mechanics which are similar to, but distinct from, the four laws of thermodynamics.

2. The Integral Formula

In a stationary axisymmetric asymptotically flat space, there is a unique time translational Killing vector K^a which is timelike near infinity with $K^a K_a = -1$ and a unique rotational Killing vector $\tilde{K}^a$ whose orbits are closed curves with parameter length 2π. These Killing vectors obey equations

$$K_{a;b} = K_{[a;b]}, \qquad \tilde{K}_{a;b} = \tilde{K}_{[a;b]}, \tag{1}$$

$$K_{a;b}\tilde{K}^b = \tilde{K}_{a;b}K^b, \tag{2}$$

$$K^{a;b}{}_b = -R^a{}_b K^b, \tag{3}$$

$$\tilde{K}^{a;b}{}_b = -R^a{}_b \tilde{K}^b, \tag{4}$$

where a semicolon denotes the covariant derivatives, square brackets around indices imply antisymmetrization and $R_{ab} = R_{acb}{}^c$ with

$$v_{d;[bc]} = \tfrac{1}{2} R_{adbc} v^a$$

for any vector v^a. Since $K_{a;b}$ is antisymmetric, one can integrate Eq. (3) over a hypersurface S and transfer the volume on the left to an integral over a 2-surface ∂S bounding S:

$$\int\limits_{\partial S} K^{a;b} d\Sigma_{ab} = -\int\limits_{S} R^a_b K^b d\Sigma_a, \tag{5}$$

where $d\Sigma_{ab}$ and $d\Sigma_a$ are the surface elements of ∂S and S respectively. We shall choose the surface to be spacelike, asymptotically flat, tangent to the rotation Killing vector $\tilde{K}^a$, and to intersect the event horizon [1] in a 2-surface ∂B. The boundary ∂S of S consists of ∂B and a 2-surface ∂S_∞ at infinity. For an asymptotically flat space, the integral over ∂S_∞ in equation (5) is equal to $-4\pi M$, where M is the mass as measured from infinity. Thus

$$M = \int\limits_{S} (2 T_a^b - T\delta_a^b) K^a d\Sigma_b + \frac{1}{4\pi} \int\limits_{\partial B} K^{a;b} d\Sigma_{ab}, \tag{6}$$

where

$$R_{ab} - \tfrac{1}{2} R g_{ab} = 8\pi T_{ab}.$$

The first integral on the right can be regarded as the contribution to the total mass of the matter outside the event horizon, and the second integral may be regarded as the mass of the black hole. One can integrate Eq. (4) similarly to obtain an expression for the total angular momentum J as measured asymptotically from infinity,

$$J = -\int_S T^a{}_b \tilde{K}^b d\Sigma_a - \frac{1}{8\pi} \int_{\partial B} \tilde{K}^{a;b} d\Sigma_{ab} \,. \tag{7}$$

The first integral on the right is the angular momentum of the matter, and the second integral can be regarded as the angular momentum of the black hole.

One can introduce a time coordinate t which measures the parameter distance from S along the integral curves of K^a (i.e. $t_{;a}K^a = 1$). The null vector $l^a = dx^a/dt$, tangent to the generators of the horizon, can be expressed as

$$l^a = K^a + \Omega_H \tilde{K}^a \,. \tag{8}$$

The coefficient Ω_H is the angular velocity of the black hole and is the same at all points of the horizon [9]. Thus one can rewrite Eq. (6) as

$$M = \int_S (2T_a{}^b - T\delta_a^b) K^a d\Sigma_b + 2\Omega_H J_H + \frac{1}{4\pi} \int_{\partial B} l^{a;b} d\Sigma_{ab} \,, \tag{9}$$

where

$$J_H = -\frac{1}{8\pi} \int_{\partial B} \tilde{K}^{a;b} d\Sigma_{ab}$$

is the angular momentum of the black hole. One can express $d\Sigma_{ab}$ as $l_{[a}n_{b]}dA$, where n_a is the other null vector orthogonal to ∂B, normalized so that $n_a l^a = -1$, and dA is the surface area element of ∂B. Thus the last term on the right of Eq. (9) is

$$\frac{1}{4\pi} \int_{\partial B} \kappa dA \,,$$

where $\kappa = -l_{a;b}n^a l^b$ represents the extent to which the time coordinate t is not an affine parameter along the generators of the horizon. One can think of κ as the "surface gravity" of the black hole in the following sense: a particle outside the horizon which rigidly corotates with the black hole has an angular velocity Ω_H, a four-velocity $v^a = v^t(K^a + \Omega_H \tilde{K}^a)$, and an acceleration four-vector $v^a{}_{;b}v^b$. The magnitude of the acceleration, multiplied by a factor $1/v^t$ to convert from change in velocity per unit proper time to change in velocity per unit coordinate time t, tends to κ when the particle is infinitesimally close to the event horizon.

We shall now show that κ is constant over the horizon. Let $m^a, \bar{m}^a$ be complex conjugate null vectors lying in ∂B and normalised so that

$m^a \bar{m}_a = 1$. Then

$$
\begin{aligned}
\kappa_{;a} m^a &= -(l_{a;b} n^a l^b)_{;c} m^c \\
&= -l_{a;bc} n^a l^b m^c - l_{a;b} n^a{}_{;c} l^b m^c - l_{a;b} n^a l^b{}_{;c} m^c .
\end{aligned}
\tag{10}
$$

Since l^a is a Killing vector, $l_{a;bc} = R_{dcba} l^d$. The normalization of the null tetrad on the horizon, from which

$$
g_{ab} = -n_a l_b - l_a n_b + m_a \bar{m}_b + \bar{m}_a m_b ,
$$

is used to put the second term in the form $\kappa l_{a;c} n^a m^c$. The third term is $-\kappa l_{a;c} n^a m^c$ as a result of the vanishing of the shear and convergence of the generators of the horizon, $l_{a;b} m^a \bar{m}^b = 0 = l_{a;b} m^a m^b$. Thus

$$
\kappa_{;a} m^a = -R_{abcd} l^a m^b l^c n^d .
\tag{11}
$$

But on the horizon

$$
\begin{aligned}
0 &= (l_{a;b} m^a \bar{m}^b)_{;c} m^c \\
&= R_{dabc} l^d m^a \bar{m}^b m^c \\
&= -R_{db} l^d m^b + R_{abcd} l^a m^b l^c n^d .
\end{aligned}
\tag{12}
$$

By the Einstein equations $R_{bd} l^b m^d = 8\pi T_{bd} l^b m^d$.

If energy-momentum tensor obeys the Dominant Energy Condition [10], $T_{bd} l^b$ will be a non-spacelike vector. However $T_{bd} l^b l^d = 0$ on the horizon since the shear and convergence of the horizon are zero. This shows that $T_{bd} l^b$ must be zero or parallel to l^d and that $T_{bd} l^b m^d = 0$. Thus $\kappa_{;a} m^a$ is zero and κ is constant on the horizon.

The integral mass formula becomes

$$
M = \int\limits_S (2 T_a^b - T \delta_a^b) K^a d\Sigma_b + 2\Omega_H J_H + \frac{\kappa}{4\pi} A ,
\tag{13}
$$

where A is the area of a 2-dimensional cross section of the horizon. When T_{ab} is zero, i.e. when the space outside the horizon is empty, this formula reduces to that found by Smarr [7] for the Kerr solution. In the Kerr solution,

$$
\Omega_H = \frac{J_H}{2M(M^2 + (M^4 - J_H^2)^{1/2})} ,
\tag{14}
$$

$$
\kappa = \frac{(M^4 - J_H^2)^{1/2}}{2M(M^2 + (M^4 - J_H^2)^{1/2})} ,
\tag{15}
$$

$$
A = 8\pi(M^2 + (M^4 - J_H^2)^{1/2}) .
\tag{16}
$$

For a Kerr solution with a zero angular momentum, the total mass is represented by the last term in equation (13). As the angular momentum increases, the surface gravity decreases until it is zero in the limiting case, $J_H^2 = M^4$. The mass is then all represented by the rotational term

$2\Omega_H J_H$. The reduction of the surface gravity with angular momentum can be thought of as a centrifugal effect. When the angular momentum is near the limiting value, the horizon is, in a sense, very loosely bound and a small perturbation can raise a large tide [11].

3. The Differential Formula

In this section we shall use the integral mass formula to derive an expression for the difference δM between the masses of two slightly different stationary axisymmetric black hole solutions. For simplicity we shall consider only the case in which the matter outside the horizon is a perfect fluid in circular orbit around the black hole. The differential mass formula for rotating stars without the blackhole terms is discussed in [12]. A treatment including electromagnetic fields, which allows the matter to be an elastic solid, is given in [6].

A perfect fluid may be described by an energy density ε which is a function of the particle number density n and entropy density s. The temperature θ, chemical potential μ and pressure p are defined by

$$\theta = \frac{\partial \varepsilon}{\partial s}, \tag{17}$$

$$\mu = \frac{\partial \varepsilon}{\partial n}, \tag{18}$$

$$p = \mu n + \theta s - \varepsilon. \tag{19}$$

The energy momentum tensor is

$$T_{ab} = (\varepsilon + p)\, v_a v_b + p g_{ab}, \tag{20}$$

where $v^a = (-u_b u^b)^{-1/2} u^a$ is the unit vector tangent to the flow lines and $u^a = K^a + \Omega \tilde{K}^a$, where Ω is the angular velocity of the fluid. The angular momentum, entropy and number of particles of the fluid can be expressed as

$$-\int T^a{}_b \tilde{K}^b d\Sigma_a,$$
$$\int s v^a d\Sigma_a,$$

and

$$\int n v^a d\Sigma_a \quad \text{respectively.}$$

When comparing two slightly different solutions there is a certain freedom in which points are chosen to correspond. We shall use this freedom to make the surfaces S, the event horizons, and the Killing vectors K^a and $\tilde{K}^a$ the same in the two solutions. Thus

$$\delta K^a = \delta \tilde{K}^a = 0 \tag{21}$$

and

$$\delta K_a = h_{ab} K^b, \qquad \delta \tilde{K}_a = h_{ab} \tilde{K}^b, \tag{22}$$

where $h_{ab} = \delta g_{ab} = -g_{ac}g_{bd}\delta g^{cd}$. Then

$$\delta l^a = \delta\Omega_H \tilde{K}^a, \tag{23}$$

$$\delta l_a = h_{ab}l^b + g_{ab}\delta\Omega_H \tilde{K}^b. \tag{24}$$

Since the event horizons are in the same position in the two solutions, the covariant vectors normal to them must be parallel,

$$\delta l_{[a}l_{b]} = 0, \qquad \delta n^{[a}n^{b]} = 0. \tag{25}$$

Also, the Lie derivative of δl_a by l^b is zero, $(\delta l_a)_{;b}l^b + \delta l_a l^a_{;b} = 0$. Therefore

$$\begin{aligned}
\delta\kappa &= \tfrac{1}{2}(\delta l_a l^a + l_a \delta l^a)_{;c} n^c + \tfrac{1}{2}(l_a l^a)_{;c}\delta n^c \\
&= \tfrac{1}{2}(\delta l_a)_{;b}(l^a n^b + n^a l^b) + \delta l_a l^a_{;b} n^b \\
&\quad + \delta\Omega_H \tilde{K}^a_{;b}l_a n^b + \delta n^b l^a_{;b} l_a \\
&= \tfrac{1}{2}(\delta l_a)_{;b}(l^a n^b + n^a l^b) + \delta\Omega_H \tilde{K}^a_{;b}l_a n^b.
\end{aligned} \tag{26}$$

As δl_a is proportional to l_a on the horizon, $(\delta l_a)_{;b}m^a \bar{m}^b$ is zero. Thus

$$\begin{aligned}
\delta\kappa &= -\tfrac{1}{2}(\delta l_a)^{;a} + \delta\Omega_H \tilde{K}^a_{;b}l_a n^b \\
&= -\tfrac{1}{2}h_{ab}{}^{;a}l^b + \delta\Omega_H \tilde{K}^a_{;b}l_a n^b.
\end{aligned} \tag{27}$$

To evaluate δM, we express the mass formula derived in the previous section in the form

$$M = \int_S \left(2T_a^b + \frac{1}{8\pi}R\delta_a^b\right)K^a d\Sigma_b + 2\Omega_H J_H + \frac{\kappa}{4\pi}A. \tag{28}$$

The variation of the term involving the scalar curvature, R, gives

$$-\frac{1}{8\pi}\int_S \left\{\left(R_{cd} - \frac{1}{2}g_{cd}R\right)h^{cd} + 2h_{[c;d]}^{c}{}^{;d}\right\}K^a d\Sigma_a. \tag{29}$$

But

$$2h_{[c;d]}^{c}{}^{;d}K^a = 2(K^a h_c^{[c;d]} - K^d h_c^{[c;a]})_{;d}, \tag{30}$$

using $h_{cd;a}K^a + h_{ad}K^a_{;c} + h_{ac}K^a_{;d} = 0$. One can therefore transform the last term in (29) into the 2-surface integral

$$-\frac{1}{4\pi}\int_{\partial S}(K^a h_c^{[c;d]} - K^d h_c^{[c;a]})\,d\Sigma_{ad}. \tag{31}$$

The integral over ∂S_∞ gives $-\delta M$ and, by Eq. (27), the integral over ∂B gives $-\dfrac{\delta\kappa}{4\pi}A - 2\delta\Omega_H J_H$.

The variation of the energy-momentum tensor term in (28) is

$$\begin{aligned}
2\delta\int T_a^b K^a d\Sigma_b &= -2\int\Omega\delta\{T_a^b \tilde{K}^a d\Sigma_b\} + 2\delta\int p K^a d\Sigma_a \\
&\quad + 2\int u^a \delta\{(\varepsilon + p)(-u^c u^d g_{cd})^{-1}u_a K^b d\Sigma_b\}.
\end{aligned} \tag{32}$$

But $\varepsilon + p = \mu n + \theta s$, $\delta p = \delta \mu n + \delta \theta s$, and $u^a \delta\{(-u^c u^d g_{cd})^{-1/2} u_a\} = \frac{1}{2} v^c v^d h_{cd}$. Therefore

$$
\begin{aligned}
2\delta \int T_a{}^b K^a d\Sigma_b = & \int T^{cd} h_{cd} K^a d\Sigma_a + 2\int \Omega \delta dJ \\
& + 2\int \bar{\mu} \delta dN + 2\int \bar{\theta} \delta dS,
\end{aligned}
\tag{33}
$$

where $\delta dJ = -\delta\{T_a{}^b \tilde{K}^a d\Sigma_b\}$ is the change in the angular momentum of the fluid crossing the surface element $d\Sigma_b$,

$$
\delta dN = \delta\{n(-u_a u^a)^{-1/2} K^b d\Sigma_b\}
$$

is the change in the number of particles crossing $d\Sigma_b$,

$$
\delta dS = \delta\{s(-u_a u^a)^{-1/2} K^b d\Sigma_b\}
$$

is the change in the entropy crossing $d\Sigma_b$,

$$
\bar{\mu} = (-u_a u^a)^{1/2} \mu
$$

is the "red-shifted" chemical potential, and

$$
\bar{\theta} = (-u_a u^a)^{1/2} \theta
$$

is the "red-shifted" temperature. Thus

$$
\delta M = \int \Omega \delta dJ + \int \bar{\mu} \delta dN + \int \bar{\theta} \delta dS + \Omega_H \delta J_H + \frac{\kappa}{8\pi} \delta A.
\tag{34}
$$

This is the differential mass formula.

If an infinitesimal ring is added to a black hole slowly, without allowing any matter or radiation to cross the event horizon, the area and the angular momentum of the black hole are constant and the matter terms in the Eq. (34) give the net energy required to add the ring. Since Ω_H and κ *do* change to first order in the mass of the ring, the change in $M_H = 2\Omega_H J_H + \kappa A/4\pi$ must be taken into account in the integral mass formula of Eq. (13).

4. The Four Laws

In this section we shall pursue the analogy between black holes and thermodynamics and shall formulate four laws which correspond to and in some ways transcend the four laws of thermodynamics. We start with the most obvious analogy:

The Second Law [1]

The area A of the event horizon of each black hole does not decrease with time, i.e.

$$
\delta A \geqq 0.
$$

If two black holes coalesce, the area of the final event horizon is greater than the sum of the areas of the initial horizons, i.e.

$$A_3 > A_1 + A_2 \,.$$

This establishes the analogy between the area of the event horizon and entropy. The second law of black hole mechanics is slightly stronger than the corresponding thermodynamic law. In thermodynamics one can transfer entropy from one system to another, and it is required only that the *total* entropy does not decrease. However one cannot transfer area from one black hole to another since black holes cannot bifurcate ([1, 2, 3]). Thus the second law of black hole mechanics requires that the area of each individual black hole should not decrease.

The First Law

Any two neighboring stationary axisymmetric solutions containing a perfect fluid with circular flow and a central black hole are related by

$$\delta M = \frac{\kappa}{8\pi} \delta A + \Omega_H \delta J_H + \int \Omega \delta J + \int \bar{\mu} \delta N + \int \bar{\theta} \delta S \,.$$

It can be seen that $\frac{\kappa}{8\pi}$ is analogous to temperature in the same way that A is analogous to entropy. It should however be emphasized that $\frac{\kappa}{8\pi}$ and A are distinct from the temperature and entropy of the black hole.

In fact the effective temperature of a black hole is absolute zero. One way of seeing this is to note that a black hole cannot be in equilibrium with black body radiation at any non-zero temperature, because no radiation could be emitted from the hole whereas some radiation would always cross the horizon into the black hole. If the wavelength of the radiation were very long, corresponding to a low black body temperature, the rate of absorption of radiation would be very slow, but true equilibrium would be possible only if there were no radiation present at all, i.e. if the external black body radiation temperature were zero. Another way of seeing that the effective temperature of a black hole is zero is to note that the "red shifted" effective temperature $\bar{\theta}$ of any matter orbiting the black hole must tend to zero as the horizon is approached, because the time dilatation factor $(-u^a u_a)^{1/2}$ tends to zero on the horizon. The fact that the effective temperature of a black hole is zero means that one can in principle add entropy to a black hole without changing it in any way. In this sense a black hole can be said to transcend the second law of thermodynamics. In practise of course any addition of entropy to a black hole would cause some increase in the area of the event horizon. One might therefore suppose that by adding some multiple of the area to the total entropy of all matter outside the event horizon one could obtain a quantity which never decreased. However this is not possible since by careful management one can arrange that the area increase accompanying a given addition of entropy is arbitrarily small. One way of doing this would be to put the entropy into two containers and lower them on ropes down the axis towards the north and south poles. As the containers approach the black hole they would distort the horizon. The shear or rate of distortion of the horizon would be proportional to the rate at which the containers were being lowered. The rate of increase of area of the horizon would be proportional to the square of the shear, [2, 11], and so to the square of the rate at which the containers were being lowered. Thus by lowering the containers very slowly, one could ensure that the area increase was very small. When the containers reach the horizon, they would be moving parallel to the null vector l^a and so would not cause any area increase as they cross the horizon.

In a similar way the effective chemical potential $\bar{\mu}$ tends to zero on the horizon, which means that in principle one can also add particles to a black hole without changing it. In this sense a black hole transcends the law of conservation of baryons.

Continuing the analogy between $\dfrac{\kappa}{8\pi}$ and temperature, one has:

The Zeroth Law

The surface gravity, κ of a stationary black hole is constant over the event horizon.

This was proved in Section 2. Other proofs under slightly different assumptions are given in [6, 2].

Extending the analogy even further one would postulate:

The Third Law

It is impossible by any procedure, no matter how idealized, to reduce κ to zero by a finite sequence of operations.

This law has a rather different status from the others, in that it does not, so far at least, have a rigorous mathematical proof. However there are strong reasons for believing in it. For example if one tries to reduce the value of κ of a Kerr black hole by throwing in particles to increase the angular momentum, one finds that the decrease in κ per particle thrown in gets smaller and smaller as the mass and angular momentum tend to the critical ratio $J/M^2 = 1$ for which κ is zero. While idealized accretion processes do exist for which $J/M^2 \to 1$ with the addition of a finite amount of rest mass ([13, 14]), they require an infinite divisibility of the matter and an infinite time. Another reason for believing the third law is that if one could reduce κ to zero by a finite sequence of operations, then presumably one could carry the process further, thereby creating a naked singularity. If this were to happen there would be a breakdown of the assumption of asymptotic predictability which is the basis of many results in black hole theory, including the law that A cannot decrease.

This work was carried out while the authors were attending the 1972 Les Houches Summer School on Black Holes. The authors would like to thank Larry Smarr, Bryce de Witt and other participants of the school for valuable discussions.

References

1. Hawking, S. W.: Commun. math. Phys. **25**, 152—166 (1972).
2. Hawking, S. W.: The event horizon. In: Black Holes. New York, London, Paris: Gordon and Breach 1973 (to be published).
3. Hawking, S. W., Ellis, G. F. R.: The large scale structure of space-time. Cambridge: Cambridge University Press 1973 (to be published).
4. Carter, B.: Phys. Rev. Letters **26**, 331—333 (1971).
5. Carter, B.: (Preprint, Institute of Theoretical Astronomy, Cambridge, England).
6. Carter, B.: Properties of the Kerr metric. In: Black Holes. New York, London, Paris: Gordon and Breach 1973 (to be published).
7. Smarr, L.: Phys. Rev. Letters **30**, 71—73 (1973).
8. Beckenstein, J.: PhD Thesis. Princeton University, 1972.
9. Carter, B.: J. Math. Phys. **10**, 70—81 (1969).
10. Hawking, S. W.: Commun. math. Phys. **18**, 301—306 (1970).
11. Hawking, S. W., Hartle, J. B.: Commun. math. Phys. **27**, 283—290 (1972).
12. Bardeen, J. M.: Astrophys. J. **162**, 71—95 (1970).
13. Bardeen, J. M.: Nature **226**, 64—65 (1970).
14. Christodoulou, D.: Phys. Rev. Letters **25**, 1596—1597 (1970).

J. M. Bardeen
Department of Physics
Yale University
New Haven, Connecticut 06520
USA

B. Carter
S. W. Hawking
Institute of Astronomy
University of Cambridge
Cambridge, U.K.

Commun. math. Phys. 43, 199—220 (1975)

Particle Creation by Black Holes

S. W. Hawking

Department of Applied Mathematics and Theoretical Physics, University of Cambridge,
Cambridge, England

Received April 12, 1975

Abstract. In the classical theory black holes can only absorb and not emit particles. However it is shown that quantum mechanical effects cause black holes to create and emit particles as if they were hot bodies with temperature $\frac{\hbar\kappa}{2\pi k} \approx 10^{-6}\left(\frac{M_\odot}{M}\right)$ °K where κ is the surface gravity of the black hole. This thermal emission leads to a slow decrease in the mass of the black hole and to its eventual disappearance: any primordial black hole of mass less than about 10^{15} g would have evaporated by now. Although these quantum effects violate the classical law that the area of the event horizon of a black hole cannot decrease, there remains a Generalized Second Law: $S + \frac{1}{4}A$ never decreases where S is the entropy of matter outside black holes and A is the sum of the surface areas of the event horizons. This shows that gravitational collapse converts the baryons and leptons in the collapsing body into entropy. It is tempting to speculate that this might be the reason why the Universe contains so much entropy per baryon.

1.

Although there has been a lot of work in the last fifteen years (see [1, 2] for recent reviews), I think it would be fair to say that we do not yet have a fully satisfactory and consistent quantum theory of gravity. At the moment classical General Relativity still provides the most successful description of gravity. In classical General Relativity one has a classical metric which obeys the Einstein equations, the right hand side of which is supposed to be the energy momentum tensor of the classical matter fields. However, although it may be reasonable to ignore quantum gravitational effects on the grounds that these are likely to be small, we know that quantum mechanics plays a vital role in the behaviour of the matter fields. One therefore has the problem of defining a consistent scheme in which the space-time metric is treated classically but is coupled to the matter fields which are treated quantum mechanically. Presumably such a scheme would be only an approximation to a deeper theory (still to be found) in which space-time itself was quantized. However one would hope that it would be a very good approximation for most purposes except near space-time singularities.

The approximation I shall use in this paper is that the matter fields, such as scalar, electro-magnetic, or neutrino fields, obey the usual wave equations with the Minkowski metric replaced by a classical space-time metric g_{ab}. This metric satisfies the Einstein equations where the source on the right hand side is taken to be the expectation value of some suitably defined energy momentum operator for the matter fields. In this theory of quantum mechanics in curved space-time there is a problem in interpreting the field operators in terms of annihilation and creation operators. In flat space-time the standard procedure is to decompose the field into positive and negative frequency components. For example, if ϕ is a massless Hermitian scalar field obeying the equation $\phi_{:ab}\eta^{ab}=0$ one expresses ϕ as

$$\phi = \sum_i \{f_i a_i + \bar{f}_i a_i^\dagger\} \tag{1.1}$$

where the $\{f_i\}$ are a complete orthonormal family of complex valued solutions of the wave equation $f_{i:ab}\eta^{ab}=0$ which contain only positive frequencies with respect to the usual Minkowski time coordinate. The operators a_i and $a_i^\dagger$ are interpreted as the annihilation and creation operators respectively for particles in the ith state. The vacuum state $|0\rangle$ is defined to be the state from which one cannot annihilate any particles, i.e.

$$a_i|0\rangle = 0 \quad \text{for all } i.$$

In curved space-time one can also consider a Hermitian scalar field operator ϕ which obeys the covariant wave equation $\phi_{:ab}g^{ab}=0$. However one cannot decompose into its positive and negative frequency parts as positive and negative frequencies have no invariant meaning in curved space-time. One could still require that the $\{f_i\}$ and the $\{\bar{f}_i\}$ together formed a complete basis for solutions of the wave equations with

$$\tfrac{1}{2}i\int_S (f_i \bar{f}_{j:a} - \bar{f}_j f_{i:a}) d\Sigma^a = \delta_{ij} \tag{1.2}$$

where S is a suitable surface. However condition (1.2) does not uniquely fix the subspace of the space of all solutions which is spanned by the $\{f_i\}$ and therefore does not determine the splitting of the operator ϕ into annihilation and creation parts. In a region of space-time which was flat or asymptotically flat, the appropriate criterion for choosing the $\{f_i\}$ is that they should contain only positive frequencies with respect to the Minkowski time coordinate. However if one has a space-time which contains an initial flat region (1) followed by a region of curvature (2) then a final flat region (3), the basis $\{f_{1i}\}$ which contains only positive frequencies on region (1) will not be the same as the basis $\{f_{3i}\}$ which contains only positive frequencies on region (3). This means that the initial vacuum state $|0_1\rangle$, the state which satisfies $a_{1i}|0_1\rangle=0$ for each initial annihilation operator a_{1i}, will not be the same as the final vacuum state $|0_3\rangle$ i.e. $a_{3i}|0_1\rangle \neq 0$. One can interpret this as implying that the time dependent metric or gravitational field has caused the creation of a certain number of particles of the scalar field.

Although it is obvious what the subspace spanned by the $\{f_i\}$ is for an asymptotically flat region, it is not uniquely defined for a general point of a curved space-time. Consider an observer with velocity vector v^a at a point p. Let B be the least upper bound $|R_{abcd}|$ in any orthonormal tetrad whose timelike vector coincides with v^a. In a neighbourhood U of p the observer can set up a local inertial coordinate system (such as normal coordinates) with coordinate radius of the order of $B^{-\frac{1}{2}}$. He can then choose a family $\{f_i\}$ which satisfy equation (1.2) and which in the neighbourhood U are approximately positive frequency with respect to the time coordinate in U. For modes f_i whose characteristic frequency ω is high compared to $B^{\frac{1}{2}}$, this leaves an indeterminacy between f_i and its complex conjugate $\bar{f}_i$ of the order of the exponential of some multiple of $-\omega B^{-\frac{1}{2}}$. The indeterminacy between the annihilation operator a_i and the creation operator $a_i^\dagger$ for the mode is thus exponentially small. However, the ambiguity between the a_i and the $a_i^\dagger$ is virtually complete for modes for which $\omega < B^{\frac{1}{2}}$. This ambiguity introduces an uncertainty of $\pm\frac{1}{2}$ in the number operator $a_i^\dagger a_i$ for the mode. The density of modes per unit volume in the frequency interval ω to $\omega + d\omega$ is of the order of $\omega^2 d\omega$ for ω greater than the rest mass m of the field in question. Thus the uncertainty in the local energy density caused by the ambiguity in defining modes of wavelength longer than the local radius of curvature $B^{-\frac{1}{2}}$, is of order B^2 in units in which $G = c = \hbar = 1$. Because the ambiguity is exponentially small for wavelengths short compared to the radius of curvature $B^{-\frac{1}{2}}$, the total uncertainty in the local energy density is of order B^2. This uncertainty can be thought of as corresponding to the local energy density of particles created by the gravitational field. The uncertainty in the curvature produced via the Einstein equations by this uncertainty in the energy density is small compared to the total curvature of space-time provided that B is small compared to one, i.e. the radius of curvature $B^{-\frac{1}{2}}$ is large compared to the Planck length 10^{-33} cm. One would therefore expect that the scheme of treating the matter fields quantum mechanically on a classical curved space-time background would be a good approximation, except in regions where the curvature was comparable to the Planck value of 10^{66} cm^{-2}. From the classical singularity theorems [3–6], one would expect such high curvatures to occur in collapsing stars and, in the past, at the beginning of the present expansion phase of the universe. In the former case, one would expect the regions of high curvature to be hidden from us by an event horizon [7]. Thus, as far as we are concerned, the classical geometry–quantum matter treatment should be valid apart from the first 10^{-43} s of the universe. The view is sometimes expressed that this treatment will break down when the radius of curvature is comparable to the Compton wavelength $\sim 10^{-13}$ cm of an elementary particle such as a proton. However the Compton wavelength of a zero rest mass particle such as a photon or a neutrino is infinite, but we do not have any problem in dealing with electromagnetic or neutrino radiation in curved space-time. All that happens when the radius of curvature of space-time is smaller than the Compton wavelength of a given species of particle is that one gets an indeterminacy in the particle number or, in other words, particle creation. However, as was shown above, the energy density of the created particles is small locally compared to the curvature which created them.

Even though the effects of particle creation may be negligible locally, I shall show in this paper that they can add up to have a significant influence on black holes over the lifetime of the universe $\sim 10^{17}$ s or 10^{60} units of Planck time. It seems that the gravitational field of a black hole will create particles and emit them to infinity at just the rate that one would expect if the black hole were an ordinary body with a temperature in geometric units of $\kappa/2\pi$, where κ is the "surface gravity" of the black hole [8]. In ordinary units this temperature is of the order of $10^{26} M^{-1}$ °K, where M is the mass, in grams of the black hole. For a black hole of solar mass (10^{33} g) this temperature is much lower than the 3 °K temperature of the cosmic microwave background. Thus black holes of this size would be absorbing radiation faster than they emitted it and would be increasing in mass. However, in addition to black holes formed by stellar collapse, there might also be much smaller black holes which were formed by density fluctuations in the early universe [9, 10]. These small black holes, being at a higher temperature, would radiate more than they absorbed. They would therefore presumably decrease in mass. As they got smaller, they would get hotter and so would radiate faster. As the temperature rose, it would exceed the rest mass of particles such as the electron and the muon and the black hole would begin to emit them also. When the temperature got up to about 10^{12} °K or when the mass got down to about 10^{14} g the number of different species of particles being emitted might be so great [11] that the black hole radiated away all its remaining rest mass on a strong interaction time scale of the order of 10^{-23} s. This would produce an explosion with an energy of 10^{35} ergs. Even if the number of species of particle emitted did not increase very much, the black hole would radiate away all its mass in the order of $10^{-28} M^3$ s. In the last tenth of a second the energy released would be of the order of 10^{30} ergs.

As the mass of the black hole decreased, the area of the event horizon would have to go down, thus violating the law that, classically, the area cannot decrease [7, 12]. This violation must, presumably, be caused by a flux of negative energy across the event horizon which balances the positive energy flux emitted to infinity. One might picture this negative energy flux in the following way. Just outside the event horizon there will be virtual pairs of particles, one with negative energy and one with positive energy. The negative particle is in a region which is classically forbidden but it can tunnel through the event horizon to the region inside the black hole where the Killing vector which represents time translations is spacelike. In this region the particle can exist as a real particle with a timelike momentum vector even though its energy relative to infinity as measured by the time translation Killing vector is negative. The other particle of the pair, having a positive energy, can escape to infinity where it constitutes a part of the thermal emission described above. The probability of the negative energy particle tunnelling through the horizon is governed by the surface gravity κ since this quantity measures the gradient of the magnitude of the Killing vector or, in other words, how fast the Killing vector is becoming spacelike. Instead of thinking of negative energy particles tunnelling through the horizon in the positive sense of time one could regard them as positive energy particles crossing the horizon on past-directed world-lines and then being scattered on to future-directed world-lines by the gravitational field. It should be emphasized that these pictures of the mechanism responsible for the thermal emission and area decrease are heuristic only and should not be taken too literally. It should not be thought unreasonable that a black hole, which is an excited state of the gravitational field, should decay quantum mechanically and that, because of quantum fluctuation of the metric, energy should be able to tunnel out of the potential well of a black hole. This particle creation is directly analogous to that caused by a deep potential well in flat space-time [18]. The real justification of the thermal emission is the mathematical derivation given in Section (2) for the case of an uncharged non-rotating black hole. The effects of angular momentum and charge are considered in Section (3). In Section (4) it is shown that any renormalization of the energy-momentum tensor with suitable properties must give a negative energy flow down the black hole and consequent decrease in the area of the event horizon. This negative energy flow is non-observable locally.

The decrease in area of the event horizon is caused by a violation of the weak energy condition [5–7, 12] which arises from the indeterminacy of particle number and energy density in a curved space-time. However, as was shown above, this indeterminacy is small, being of the order of B^2 where B is the magnitude of the curvature tensor. Thus it can have a diverging effection a null surface like the event horizon which has very small convergence or divergence but it can not untrap a strongly converging trapped surface until B becomes of the order of one. Therefore one would not expect the negative energy density to cause a breakdown of the classical singularity theorems until the radius of curvature of space-time became 10^{-33} cm.

Perhaps the strongest reason for believing that black holes can create and emit particles at a steady rate is that the predicted rate is just that of the thermal emission of a body with the temperature $\kappa/2\pi$. There are independent, thermodynamic, grounds for regarding some multiple of the surface gravity as having a close relation to temperature. There is an obvious analogy with the second law of thermodynamics in the law that, classically, the area of the event horizon can never decrease and that when two black holes collide and merge together, the area of the final event horizon is greater than the sum of the areas of the two original horizons [7, 12]. There is also an analogy to the first law of thermodynamics in the result that two neighbouring black hole equilibrium states are related by [8]

$$dM = \frac{\kappa}{8\pi} dA + \Omega dJ$$

where M, Ω, and J are respectively the mass, angular velocity and angular momentum of the black hole and A is the area of the event horizon. Comparing this to

$$dU = TdS + pdV$$

one sees that if some multiple of A is regarded as being analogous to entropy, then some multiple of κ is analogous to temperature. The surface gravity is also analogous to temperature in that it is constant over the event horizon in equilibrium. Beckenstein [19] suggested that A and κ were not merely analogous to entropy and temperature respectively but that, in some sense, they actually were the entropy and temperature of the black hole. Although the ordinary second law of thermodynamics is transcended in that entropy can be lost down black holes, the flow of entropy across the event horizon would always cause some increase in the area of the horizon. Beckenstein therefore suggested [20] a Generalized Second Law: Entropy + some multiple (unspecified) of A never decreases. However he did not suggest that a black hole could emit particles as well as absorb them. Without such emission the Generalized Second Law would be violated by for example, a black hole immersed in black body radiation at a lower temperature than that of the black hole. On the other hand, if one accepts that black holes do emit particles at a steady rate, the identification of $\kappa/2\pi$ with temperature and $\frac{1}{4}A$ with entropy is established and a Generalized Second Law confirmed.

2. Gravitational Collapse

It is now generally believed that, according to classical theory, a gravitational collapse will produce a black hole which will settle down rapidly to a stationary axisymmetric equilibrium state characterized by its mass, angular momentum and electric charge [7, 13]. The Kerr-Newman solution represent one such family of black hole equilibrium states and it seems unlikely that there are any others. It has therefore become a common practice to ignore the collapse phase and to represent a black hole simply by one of these solutions. Because these solutions are stationary there will not be any mixing of positive and negative frequencies and so one would not expect to obtain any particle creation. However there is a classical phenomenon called superradiance [14–17] in which waves incident in certain modes on a rotating or charged black hole are scattered with increased amplitude [see Section (3)]. On a particle description this amplification must correspond to an increase in the number of particles and therefore to stimulated emission of particles. One would therefore expect on general grounds that there would also be a steady rate of spontaneous emission in these superradiant modes which would tend to carry away the angular momentum or charge of the black hole [16]. To understand how the particle creation can arise from mixing of positive and negative frequencies, it is essential to consider not only the quasi-stationary final state of the black hole but also the time-dependent formation phase. One would hope that, in the spirit of the "no hair" theorems, the rate of emission would not depend on details of the collapse process except through the mass, angular momentum and charge of the resulting black hole. I shall show that this is indeed the case but that, in addition to the emission in the super-radiant modes, there is a steady rate of emission in all modes at the rate one would expect if the black hole were an ordinary body with temperature $\kappa/2\pi$.

I shall consider first of all the simplest case of a non-rotating uncharged black hole. The final stationary state for such a black hole is represented by the Schwarzschild solution with metric

$$ds^2 = -\left(1 - \frac{2M}{r}\right)dt^2 - \left(1 - \frac{2M}{r}\right)^{-1} dr^2 + r^2(d\theta^2 + \sin^2\theta d\phi^2). \tag{2.1}$$

As is now well known, the apparent singularities at $r = 2M$ are fictitious, arising merely from a bad choice of coordinates. The global structure of the analytically extended Schwarzschild solution can be described in a simple manner by a Penrose diagram of the r-t plane (Fig. 1) [6, 13]. In this diagram null geodesics in the r-t plane are at $-45°$ to the vertical. Each point of the diagram represents a 2-sphere of area $4\pi r^2$. A conformal transformation has been applied to bring infinity to a finite distance: infinity is represented by the two diagonal lines (really null surfaces) labelled $\mathscr{I}^-$ and $\mathscr{I}^-$, and the points I^+, I^-, and I^0. The two horizontal lines $r = 0$ are curvature singularities and the two diagonal lines $r = 2M$ (really null surfaces) are the future and past event horizons which divide the solution up into regions from which one cannot escape to $\mathscr{I}^+$ and $\mathscr{I}^-$. On the left of the diagram there is another infinity and asymptotically flat region.

Most of the Penrose diagram is not in fact relevant to a black hole formed by gravitational collapse since the metric is that of the Schwarzschild solution

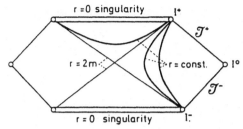

Fig. 1. The Penrose diagram for the analytically extended Schwarzschild solution

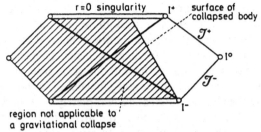

Fig. 2. Only the region of the Schwarzschild solution outside the collapsing body is relevant for a black hole formed by gravitational collapse. Inside the body the solution is completely different

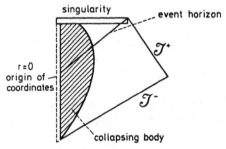

Fig. 3. The Penrose diagram of a spherically symmetric collapsing body producing a black hole. The vertical dotted line on the left represents the non-singular centre of the body

only in the region outside the collapsing matter and only in the asymptotic future. In the case of exactly spherical collapse, which I shall consider for simplicity, the metric is exactly the Schwarzschild metric everywhere outside the surface of the collapsing object which is represented by a timelike geodesic in the Penrose diagram (Fig. 2). Inside the object the metric is completely different, the past event horizon, the past $r=0$ singularity and the other asymptotically flat region do not exist and are replaced by a time-like curve representing the origin of polar coordinates. The appropriate Penrose diagram is shown in Fig. 3 where the conformal freedom has been used to make the origin of polar coordinates into a vertical line.

In this space-time consider (again for simplicity) a massless Hermitian scalar field operator ϕ obeying the wave equation

$$\phi_{:ab}g^{ab}=0 .\tag{2.2}$$

(The results obtained would be the same if one used the conformally invariant wave equation:

$$\phi_{;ab}g^{ab} + \tfrac{1}{6}R\phi = 0 .)$$

The operator ϕ can be expressed as

$$\phi = \sum_i \{ f_i a_i + \bar{f}_i a_i^\dagger \} . \tag{2.3}$$

The solutions $\{f_i\}$ of the wave equation $f_{i;ab}g^{ab} = 0$ can be chosen so that on past null infinity $\mathscr{I}^-$ they form a complete family satisfying the orthonormality conditions (1.2) where the surface S is $\mathscr{I}^-$ and so that they contain only positive frequencies with respect to the canonical affine parameter on $\mathscr{I}^-$. (This last condition of positive frequency can be uniquely defined despite the existence of "supertranslations" in the Bondi-Metzner-Sachs asymptotic symmetry group [21, 22].) The operators a_i and $a_i^\dagger$ have the natural interpretation as the annihilation and creation operators for ingoing particles i.e. for particles at past null infinity $\mathscr{I}^-$. Because massless fields are completely determined by their data on $\mathscr{I}^-$, the operator ϕ can be expressed in the form (2.3) everywhere. In the region outside the event horizon one can also determine massless fields by their data on the event horizon and on future null infinity $\mathscr{I}^+$. Thus one can also express ϕ in the form

$$\phi = \sum_i \{ p_i b_i + \bar{p}_i b_i^\dagger + q_i c_i + \bar{q}_i c_i^\dagger \} . \tag{2.4}$$

Here the $\{p_i\}$ are solutions of the wave equation which are purely outgoing, i.e. they have zero Cauchy data on the event horizon and the $\{q_i\}$ are solutions which contain no outgoing component, i.e. they have zero Cauchy data on $\mathscr{I}^+$. The $\{p_i\}$ and $\{q_i\}$ are required to be complete families satisfying the orthonormality conditions (1.2) where the surface S is taken to be $\mathscr{I}^+$ and the event horizon respectively. In addition the $\{p_i\}$ are required to contain only positive frequencies with respect to the canonical affine parameter along the null geodesic generators of $\mathscr{I}^+$. With the positive frequency condition on $\{p_i\}$, the operators $\{b_i\}$ and $\{b_i^\dagger\}$ can be interpreted as the annihilation and creation operators for outgoing particles, i.e. for particles on $\mathscr{I}^-$. It is not clear whether one should impose some positive frequency condition on the $\{q_i\}$ and if so with respect to what. The choice of the $\{q_i\}$ does not affect the calculation of the emission of particles to $\mathscr{I}^+$. I shall return to the question in Section (4).

Because massless fields are completely determined by their data on $\mathscr{I}^-$ one can express $\{p_i\}$ and $\{q_i\}$ as linear combinations of the $\{f_i\}$ and $\{\bar{f}_i\}$:

$$p_i = \sum_j (\alpha_{ij} f_j + \beta_{ij} \bar{f}_j) , \tag{2.5}$$

$$q_i = \sum_j (\gamma_{ij} f_j + \eta_{ij} \bar{f}_j) . \tag{2.6}$$

These relations lead to corresponding relations between the operators

$$b_i = \sum_j (\bar{\alpha}_{ij} a_j - \bar{\beta}_{ij} a_j^\dagger) , \tag{2.7}$$

$$c_i = \sum_j (\bar{\gamma}_{ij} a_j - \bar{\eta}_{ij} a_j^\dagger) . \tag{2.8}$$

The initial vacuum state $|0\rangle$, the state containing no incoming particles, i.e. no particles on $\mathscr{I}^-$, is defined by

$$a_i|0\rangle = 0 \quad \text{for all } i. \tag{2.9}$$

However, because the coefficients β_{ij} will not be zero in general, the initial vacuum state will not appear to be a vacuum state to an observer at $\mathscr{I}^+$. Instead he will find that the expectation value of the number operator for the ith outgoing mode is

$$\langle 0_-|b_i^\dagger b_i|0_-\rangle = \sum_j |\beta_{ij}|^2. \tag{2.10}$$

Thus in order to determine the number of particles created by the gravitational field and emitted to infinity one simply has to calculate the coefficients β_{ij}. One would expect this calculation to be very messy and to depend on the detailed nature of the gravitational collapse. However, as I shall show, one can derive an asymptotic form for the β_{ij} which depends only on the surface gravity of the resulting black hole. There will be a certain finite amount of particle creation which depends on the details of the collapse. These particles will disperse and at late retarded times on $\mathscr{I}^+$ there will be a steady flux of particles determined by the asymptotic form of β_{ij}.

In order to calculate this asymptotic form it is more convenient to decompose the ingoing and outgoing solutions of the wave equation into their Fourier components with respect to advanced or retarded time and use the continuum normalization. The finite normalization solutions can then be recovered by adding Fourier components to form wave packets. Because the space-time is spherically symmetric, one can also decompose the incoming and outgoing solutions into spherical harmonics. Thus, in the region outside the collapsing body, one can write the incoming and outgoing solutions as

$$f_{\omega'lm} = (2\pi)^{-\frac{1}{2}} r^{-1} (\omega')^{-\frac{1}{2}} F_{\omega'}(r) e^{i\omega' v} Y_{lm}(\theta, \phi), \tag{2.11}$$

$$p_{\omega lm} = (2\pi)^{-\frac{1}{2}} r^{-1} \omega^{-\frac{1}{2}} P_\omega(r) e^{i\omega u} Y_{lm}(\theta, \phi), \tag{2.12}$$

where v and u are the usual advanced and retarded coordinates defined by

$$v = t + r + 2M \log\left|\frac{r}{2M} - 1\right|, \tag{2.13}$$

$$u = t - r - 2M \log\left|\frac{r}{2M} - 1\right|. \tag{2.14}$$

Each solution $p_{\omega lm}$ can be expressed as an integral with respect to ω' over solutions $f_{\omega'lm}$ and $\bar{f}_{\omega'lm}$ with the same values of l and $|m|$ (from now on I shall drop the suffices l, m):

$$p_\omega = \int_0^\infty (\alpha_{\omega\omega'} f_{\omega'} + \beta_{\omega\omega'} \bar{f}_{\omega'}) d\omega'. \tag{2.15}$$

To calculate the coefficients $\alpha_{\omega\omega'}$ and $\beta_{\omega\omega'}$, consider a solution p_ω propagating backwards from $\mathscr{I}^+$ with zero Cauchy data on the event horizon. A part $p_\omega^{(1)}$ of the solution p_ω will be scattered by the static Schwarzchild field outside the collapsing body and will end up on $\mathscr{I}^-$ with the same frequency ω. This will give a $\delta(\omega' - \omega)$ term in $\alpha_{\omega\omega'}$. The remainder $p_\omega^{(2)}$ of p_ω will enter the collapsing body

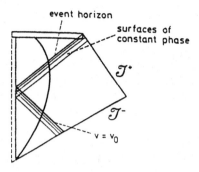

Fig. 4. The solution p_ω of the wave equation has an infinite number of cycles near the event horizon
and near the surface $v = v_0$

where it will be partly scattered and partly reflected through the centre, eventually
emerging to $\mathscr{I}^-$. It is this part $p_\omega^{(2)}$ which produces the interesting effects. Because
the retarded time coordinate u goes to infinity on the event horizon, the surfaces
of constant phase of the solution p_ω will pile up near the event horizon (Fig. 4).
To an observer on the collapsing body the wave would seem to have a very large
blue-shift. Because its effective frequency was very high, the wave would propagate by geometric optics through the centre of the body and out on $\mathscr{I}^-$. On
$\mathscr{I}^-$ $p_\omega^{(2)}$ would have an infinite number of cycles just before the advanced time
$v = v_0$ where v_0 is the latest time that a null geodesic could leave $\mathscr{I}^-$, pass through
the centre of the body and escape to $\mathscr{I}^+$ before being trapped by the event
horizon. One can estimate the form of $p_\omega^{(2)}$ on $\mathscr{I}^-$ near $v = v_0$ in the following
way. Let x be a point on the event horizon outside the matter and let l^a be a null
vector tangent to the horizon. Let n^a be the future-directed null vector at x which
is directed radially inwards and normalized so that $l^a n_a = -1$. The vector $-\varepsilon n^a$
(ε small and positive) will connect the point x on the event horizon with a nearby
null surface of constant retarded time u and therefore with a surface of constant
phase of the solution $p_\omega^{(2)}$. If the vectors l^a and n^a are parallelly transported along
the null geodesic γ through x which generates the horizon, the vector $-\varepsilon n^a$ will
always connect the event horizon with the same surface of constant phase of $p_\omega^{(2)}$.
To see what the relation between ε and the phase of $p_\omega^{(2)}$ is, imagine in Fig. 2 that
the collapsing body did not exist but one analytically continued the empty space
Schwarzchild solution back to cover the whole Penrose diagram. One could then
transport the pair (l^a, n^a) back along to the point where future and past event
horizons intersected. The vector $-\varepsilon n^a$ would then lie along the past event horizon.
Let λ be the affine parameter along the past event horizon which is such that at
the point of intersection of the two horizons, $\lambda = 0$ and $\dfrac{dx^a}{d\lambda} = n^a$. The affine parameter λ is related to the retarded time u on the past horizon by

$$\lambda = -C e^{-\kappa u} \tag{2.16}$$

where C is constant and κ is the surface gravity of the black hole defined by
$K^a_{;b} K^b = -\kappa K^a$ on the horizon where K^a is the time translation Killing vector.

$\left(\text{For a Schwarzchild black hole } \kappa = \dfrac{1}{4M}\right)$. It follows from this that the vector $-\varepsilon n^a$ connects the future event horizon with the surface of constant phase $-\dfrac{\omega}{\kappa}(\log \varepsilon - \log C)$ of the solution $p_\omega^{(2)}$. This result will also hold in the real space-time (including the collapsing body) in the region outside the body. Near the event horizon the solution $p_\omega^{(2)}$ will obey the geometric optics approximation as it passes through the body because its effective frequency will be very high. This means that if one extends the null geodesic γ back past the end-point of the event horizon and out onto $\mathscr{I}^-$ at $v = v_0$ and parallelly transports n^a along γ, the vector $-\varepsilon n^a$ will still connect γ to a surface of constant phase of the solution $p_\omega^{(2)}$. On $\mathscr{I}^-$ n^a will be parallel to the Killing vector K^a which is tangent to the null geodesic generators of $\mathscr{I}^-$:

$$n^a = DK^a.$$

Thus on $\mathscr{I}^-$ for $v_0 - v$ small and positive, the phase of the solution will be

$$-\frac{\omega}{\kappa}(\log(v_0 - v) - \log D - \log C). \tag{2.17}$$

Thus on $\mathscr{I}^-$ $p_\omega^{(2)}$ will be zero for $v > v_0$ and for $v < v_0$

$$p_\omega^{(2)} \sim (2\pi)^{-\frac{1}{2}} \omega^{-\frac{1}{2}} r^{-1} P_\omega^- \exp\left(-i\frac{\omega}{\kappa}\left(\log\left(\frac{v_0 - v}{CD}\right)\right)\right) \tag{2.18}$$

where $P_\omega^- \equiv P_\omega(2M)$ is the value of the radial function for P_ω on the past event horizon in the analytically continued Schwarzchild solution. The expression (2.18) for $p_\omega^{(2)}$ is valid only for $v_0 - v$ small and positive. At earlier advanced times the amplitude will be different and the frequency measured with respect to v, will approach the original frequency ω.

By Fourier transforming $p_\omega^{(2)}$ one can evaluate its contributions to $\alpha_{\omega\omega'}$ and $\beta_{\omega\omega'}$. For large values of ω' these will be determined by the asymptotic form (2.18). Thus for large ω'

$$\alpha_{\omega\omega'}^{(2)} \approx (2\pi)^{-1} P_\omega^-(CD)^{\frac{i\omega}{\kappa}} \exp(i(\omega - \omega')v_0)\left(\frac{\omega'}{\omega}\right)^{\frac{1}{2}} \Gamma\left(1 - \frac{i\omega}{\kappa}\right)(-i\omega')^{-1 + \frac{i\omega}{\kappa}}, \tag{2.19}$$

$$\beta_{\omega\omega'}^{(2)} \approx -i\alpha_{\omega(-\omega')}^{(2)}. \tag{2.20}$$

The solution $p_\omega^{(2)}$ is zero on $\mathscr{I}^-$ for large values of v. This means that its Fourier transform is analytic in the upper half ω' plane and that $p_\omega^{(2)}$ will be correctly represented by a Fourier integral in which the contour has been displaced into the upper half ω' plane. The Fourier transform of $p_\omega^{(2)}$ contains a factor $(-i\omega')^{-1 + \frac{i\omega}{\kappa}}$ which has a logarithmic singularity at $\omega' = 0$. To obtain $\beta_{\omega\omega'}^{(2)}$ from $\alpha_{\omega\omega'}^{(2)}$ by (2.20) one has to analytically continue $\alpha_{\omega - \omega'}^{(2)}$ anticlockwise round this singularity. This means that

$$|\alpha_{\omega\omega'}^{(2)}| = \exp\left(\frac{\pi\omega}{\kappa}\right)|\beta_{\omega\omega'}^{(2)}|. \tag{2.21}$$

Actually, the fact that $p_\omega^{(2)}$ is not given by (2.18) at early advanced times means that the singularity in $\alpha_{\omega\omega'}$ occurs at $\omega' = \omega$ and not at $\omega' = 0$. However the relation (2.21) is still valid for large ω'.

The expectation value of the total number of created particles at $\mathscr{I}^+$ in the frequency range ω to $\omega + d\omega$ is $d\omega \int_0^\infty |\beta_{\omega\omega'}|^2 d\omega'$. Because $|\beta_{\omega\omega'}|$ goes like $(\omega')^{-\frac{1}{2}}$ at large ω' this integral diverges. This infinite total number of created particles corresponds to a finite steady rate of emission continuing for an infinite time as can be seen by building up a complete orthonormal family of wave packets from the Fourier components p_ω. Let

$$p_{jn} = \varepsilon^{-\frac{1}{2}} \int_{j\varepsilon}^{(j+1)\varepsilon} e^{-2\pi i n \varepsilon^{-1} \omega} p_\omega d\omega \tag{2.22}$$

where j and n are integers, $j \geq 0$, $\varepsilon > 0$. For ε small these wave packets will have frequency $j\varepsilon$ and will be peaked around retarded time $u = 2\pi n \varepsilon^{-1}$ with width ε^{-1}. One can expand $\{p_{jn}\}$ in terms of the $\{f_\omega\}$

$$p_{jn} = \int_0^\infty (\alpha_{jn\omega'} f_{\omega'} + \beta_{jn\omega'} \bar{f}_{\omega'}) d\omega' \tag{2.23}$$

where

$$\alpha_{jn\omega'} = \varepsilon^{-\frac{1}{2}} \int_{j\varepsilon}^{(j+1)\varepsilon} e^{-2\pi i n \varepsilon^{-1} \omega} \alpha_{\omega\omega'} d\omega \quad \text{etc.} \tag{2.24}$$

For $j \gg \varepsilon$, $n \gg \varepsilon$

$$
\begin{aligned}
|\alpha_{jn\omega'}| &= \left| (2\pi)^{-1} P_\omega^- \omega^{-\frac{1}{2}} \Gamma\left(1 - \frac{i\omega}{\kappa}\right) \varepsilon^{-\frac{1}{2}} (\omega')^{-\frac{1}{2}} \right. \\
&\quad \left. \cdot \int_{j\varepsilon}^{(j+1)\varepsilon} \exp i\omega''(-2\pi n \varepsilon^{-1} + \kappa^{-1} \log \omega') d\omega'' \right| \\
&= \left| \pi^{-1} P_\omega^- \omega^{-\frac{1}{2}} \Gamma\left(1 - \frac{i\omega}{\kappa}\right) \varepsilon^{-\frac{1}{2}} (\omega')^{-\frac{1}{2}} z^{-1} \sin \tfrac{1}{2} \varepsilon z \right|
\end{aligned}
\tag{2.25}
$$

where $\omega = j\varepsilon$ and $z = \kappa^{-1} \log \omega' - 2\pi n \varepsilon^{-1}$. For wave-packets which reach $\mathscr{I}^+$ at late retarded times, i.e. those with large values of n, the main contribution to $\alpha_{jn\omega'}$ and $\beta_{jn\omega'}$ come from very high frequencies ω' of the order of $\exp(2\pi n \kappa \varepsilon^{-1})$. This means that these coefficients are governed only by the asymptotic forms (2.19, 2.20) for high ω' which are independent of the details of the collapse.

The expectation value of the number of particles created and emitted to infinity $\mathscr{I}^+$ in the wave-packet mode p_{jn} is

$$\int_0^\infty |\beta_{jn\omega'}|^2 d\omega'. \tag{2.26}$$

One can evaluate this as follows. Consider the wave-packet p_{jn} propagating backwards from $\mathscr{I}^+$. A fraction $1 - \Gamma_{jn}$ of the wave-packet will be scattered by the static Schwarzschild field and a fraction Γ_{jn} will enter the collapsing body.

$$\Gamma_{jn} = \int_0^\infty (|\alpha_{jn\omega'}^{(2)}|^2 - |\beta_{jn\omega'}^{(2)}|^2) d\omega' \tag{2.27}$$

where $\alpha_{jn\omega'}^{(2)}$ and $\beta_{jn\omega'}^{(2)}$, are calculated using (2.19, 2.20) from the part $p_{jn}^{(2)}$ of the wave-packet which enters the star. The minus sign in front of the second term on the right of (2.27) occurs because the negative frequency components of $p_{jn}^{(2)}$ make a negative contribution to the flux into the collapsing body. By (2.21)

$$|\alpha_{jn\omega'}^{(2)}| = \exp(\pi\omega\kappa^{-1})|\beta_{jn\omega'}^{(2)}|. \tag{2.28}$$

Thus the total number of particles created in the mode p_{jn} is

$$\Gamma_{jn}(\exp(2\pi\omega\kappa^{-1})-1)^{-1}.\qquad(2.29)$$

But for wave-packets at late retarded times, the fraction Γ_{jn} which enters the collapsing body is almost the same as the fraction of the wave-packet that would have crossed the past event horizon had the collapsing body not been there but the exterior Schwarzschild solution had been analytically continued. Thus this factor Γ_{jn} is also the same as the fraction of a similar wave-packet coming from $\mathscr{I}^-$ which would have crossed the future event horizon and have been absorbed by the black hole. The relation between emission and absorption cross-section is therefore exactly that for a body with a temperature, in geometric units, of $\kappa/2\pi$.

Similar results hold for the electromagnetic and linearised gravitational fields. The fields produced on $\mathscr{I}^-$ by positive frequency waves from $\mathscr{I}^+$ have the same asymptotic form as (2.18) but with an extra blue shift factor in the amplitude. This extra factor cancels out in the definition of the scalar product so that the asymptotic forms of the coefficients α and β are the same as in the Eqs. (2.19) and (2.20). Thus one would expect the black hole also to radiate photons and gravitons thermally. For massless fermions such as neutrinos one again gets similar results except that the negative frequency components given by the coefficients β now make a positive contribution to the probability flux into the collapsing body. This means that the term $|\beta|^2$ in (2.27) now has the opposite sign. From this it follows that the number of particles emitted in any outgoing wave packet mode is $(\exp(2\pi\omega\kappa^{-1})+1)^{-1}$ times the fraction of that wave packet that would have been absorbed by the black hole had it been incident from $\mathscr{I}^-$. This is again exactly what one would expect for thermal emission of particles obeying Fermi-Dirac statistics.

Fields of non-zero rest mass do not reach $\mathscr{I}^-$ and $\mathscr{I}^+$. One therefore has to describe ingoing and outgoing states for these fields in terms of some concept such as the projective infinity of Eardley and Sachs [23] and Schmidt [24]. However, if the initial and final states are asymptotically Schwarzschild or Kerr solutions, one can describe the ingoing and outgoing states in a simple manner by separation of variables and one can define positive frequencies with respect to the time translation Killing vectors of these initial and final asymptotic space-times. In the asymptotic future there will be no bound states: any particle will either fall through the event horizon or escape to infinity. Thus the unbound outgoing states and the event horizon states together form a complete basis for solutions of the wave equation in the region outside the event horizon. In the asymptotic past there could be bound states if the body that collapses had had a bounded radius for an infinite time. However one could equally well assume that the body had collapsed from an infinite radius in which case there would be no bound states. The possible existence of bound states in the past does not affect the rate of particle emission in the asymptotic future which will again be that of a body with temperature $\kappa/2\pi$. The only difference from the zero rest mass case is that the frequency ω in the thermal factor $(\exp(2\pi\omega\kappa^{-1})\mp1)^{-1}$ now includes the rest mass energy of the particle. Thus there will not be much emission of particles of rest mass m unless the temperature $\kappa/2\pi$ is greater than m.

One can show that these results on thermal emission do not depend on spherical symmetry. Consider an asymmetric collapse which produced a black hole which settled to a non-rotating uncharged Schwarzchild solution (angular momentum and charge will be considered in the next section). The fact that the final state is asymptotically quasi-stationary means that there is a preferred Bondi coordinate system [25] on $\mathcal{I}^+$ with respect to which one can decompose the Cauchy data for the outgoing states into positive frequencies and spherical harmonics. On $\mathcal{I}^-$ there may or may not be a preferred coordinate system but if there is not one can pick an arbitrary Bondi coordinate system and decompose the Cauchy data for the ingoing states in a similar manner. Now consider one of the $\mathcal{I}^+$ states $p_{\omega lm}$ propagating backwards through this space-time into the collapsing body and out again onto $\mathcal{I}^-$. Take a null geodesic generator γ of the event horizon and extend it backwards beyond its past end-point to intersect $\mathcal{I}^-$ at a point y on a null geodesic generator λ of $\mathcal{I}^-$. Choose a pair of null vectors $(l^a, \hat{n}^a)$ at y with l^a tangent to γ and $\hat{n}^a$ tangent to λ. Parallelly propagate $l^a, \hat{n}^a$ along γ to a point x in the region of space-time where the metric is almost that of the final Schwarzchild solution. At x $\hat{n}^a$ will be some linear combination of l^a and the radial inward directed null vector n^a. This means that the vector $-\varepsilon\hat{n}^a$ will connect x to a surface of phase $-\omega/\kappa$ $(\log\varepsilon - \log E)$ of the solution $p_{\omega lm}$ where E is some constant. As before, by the geometric optics approximation, the vector $-\varepsilon\hat{n}^a$ at y will connect y to a surface of phase $-\omega/\kappa$ $(\log\varepsilon - \log E)$ of $p^{(2)}_{\omega lm}$ where $p^{(2)}_{\omega lm}$ is the part of $p_{\omega lm}$ which enters the collapsing body. Thus on the null geodesic generator λ of $\mathcal{I}^-$ the phase of $p^{(2)}_{\omega lm}$ will be

$$-\frac{i\omega}{\kappa}(\log(v_0 - v) - \log H) \tag{2.30}$$

where v is an affine parameter on λ with value v_0 at y and H is a constant. By the geometrical optics approximation, the value of $p^{(2)}_{\omega lm}$ on λ will be

$$L \exp\left\{-\frac{i\omega}{\kappa}[\log(v_0 - v) - \log H]\right\} \tag{2.31}$$

for $v_0 - v$ small and positive and zero for $v > v_0$ where L is a constant. On each null geodesic generator of $\mathcal{I}^-$ $p^{(2)}_{\omega lm}$ will have the form (2.31) with different values of L, v_0, and H. The lack of spherical symmetry during the collapse will cause $p^{(2)}_{\omega lm}$ on $\mathcal{I}^-$ to contain components of spherical harmonics with indices (l', m') different from (l, m). This means that one now has to express $p^{(2)}_{\omega lm}$ in the form

$$p^{(2)}_{\omega lm} = \sum_{l'm'} \int_0^\infty \{\alpha^{(2)}_{\omega lm\omega'l'm'} f_{\omega'l'm'} + \beta^{(2)}_{\omega lm\omega'l'm'} \bar{f}_{\omega'l'm'}\} d\omega'. \tag{2.32}$$

Because of (2.31), the coefficients $\alpha^{(2)}$ and $\beta^{(2)}$ will have the same ω' dependence as in (2.19) and (2.20). Thus one still has the same relation as (2.21):

$$|\alpha^{(2)}_{\omega lm\omega'l'm'}| = \exp(\pi\omega\kappa^{-1})|\beta^{(2)}_{\omega lm\omega'l'm'}|. \tag{2.33}$$

As before, for each (l, m), one can make up wave packets p_{jnlm}. The number of particles emitted in such a wave packet mode is

$$\sum_{l'm'} \int_0^\infty |\beta_{jnlm\omega'l'm'}|^2 d\omega'. \tag{2.34}$$

Similarly, the fraction Γ_{jnlm} of the wave packet that enters the collapsing body is

$$\Gamma_{jnlm} = \sum_{l',m'} \int_0^\infty \{|\alpha^{(2)}_{jnlm\omega' l' m'}|^2 - |\beta^{(2)}_{jnlm\omega' l' m'}|^2\} d\omega' . \tag{2.35}$$

Again, Γ_{jnlm} is equal to the fraction of a similar wave packet coming from $\mathscr{I}^-$ that would have been absorbed by the black hole. Thus, using (2.33), one finds that the emission is just that of a body of temperature $\kappa/2\pi$: the emission at late retarded times depends only on the final quasi-stationary state of the black hole and not on the details of the gravitational collapse.

3. Angular Momentum and Charge

If the collapsing body was rotating or electrically charged, the resulting black hole would settle down to a stationary state which was described, not by the Schwarzchild solution, but by a charged Kerr solution characterised by the mass M, the angular momentum J, and the charge Q. As these solutions are stationary and axisymmetric, one can separate solutions of the wave equations in them into a factor $e^{i\omega u}$ or $e^{i\omega v}$ times $e^{-im\phi}$ times a function of r and θ. In the case of the scalar wave equation one can separate this last expression into a function of r times a function of θ [26]. One can also completely separate any wave equation in the non-rotating charged case and Teukolsky [27] has obtained completely separable wave equations for neutrino, electromagnetic and linearised gravitational fields in the uncharged rotating case.

Consider a wave packet of a classical field of charge e with frequency ω and axial quantum number m incident from infinity on a Kerr black hole. The change in mass dM of the black hole caused by the partial absorption of the wave packet will be related to the change in area, angular momentum and charge by the classical first law of black holes:

$$dM = \frac{\kappa}{8\pi} dA + \Omega dJ + \Phi dQ \tag{3.1}$$

where Ω and Φ are the angular frequency and electrostatic potential respectively of the black hole [13]. The fluxes of energy, angular momentum and charge in the wave packet will be in the ratio $\omega:m:e$. Thus the changes in the mass, angular momentum and charge of the black hole will also be in this ratio. Therefore

$$dM(1 - \Omega m\omega^{-1} - e\Phi\omega^{-1}) = \frac{\kappa}{8\pi} dA . \tag{3.2}$$

A wave packet of a classical Boson field will obey the weak energy condition: the local energy density for any observer is non-negative. It follows from this [7, 12] that the change in area dA induced by the wave-packet will be non-negative. Thus if

$$\omega < m\Omega + e\Phi \tag{3.3}$$

the change in mass dM of the black hole must be negative. In other words, the black hole will lose energy to the wave packet which will therefore be scattered with the same frequency but increased amplitude. This is the phenomenon known as "superradiance".

For classical fields of half-integer spin, detailed calculations [28] show that there is no superradiance. The reason for this is that the scalar product for half-integer spin fields is positive definite unlike that for integer spins. This means that the probability flux across the event horizon is positive and therefore, by conservation of probability, the probability flux in the scattered wave packet must be less than that in the incident wave packet. The reason that the above argument based on the first law breaks down is that the energy-momentum tensor for a classical half-integer spin field does not obey the weak energy condition. On a quantum, particle level one can understand the absence of superradiance for fermion fields as a consequence of the fact that the Exclusion Principle does not allow more than one particle in each outgoing wave packet mode and therefore does not allow the scattered wave-packet to be stronger than the incident wave-packet.

Passing now to the quantum theory, consider first the case of an unchanged, rotating black hole. One can as before pick an arbitrary Bondi coordinate frame on $\mathscr{I}^-$ and decompose the operator ϕ in terms of a family $\{f_{\omega lm}\}$ of incoming solutions where the indices ω, l, and m refer to the advanced time and angular dependence of f on $\mathscr{I}^-$ in the given coordinate system. On $\mathscr{I}^+$ the final quasi-stationary state of the black hole defines a preferred Bondi coordinate system using which one can define a family $\{p_{\omega lm}\}$ of outgoing solutions. The index l in this case labels the spheroidal harmonics in terms of which the wave equation is separable. One proceeds as before to calculate the asymptotic form of $p_{\omega lm}^{(2)}$ on $\mathscr{I}^-$. The only difference is that because the horizon is rotating with angular velocity Ω with respect to $\mathscr{I}^+$, the effective frequency near a generator of the event horizon is not ω but $\omega - m\Omega$. This means that the number of particles emitted in the wave-packet mode p_{jnlm} is

$$\{\exp(2\pi\kappa^{-1}(\omega - m\Omega)) \mp 1\}^{-1}\Gamma_{jnlm}\,. \tag{3.4}$$

The effect of this is to cause the rate of emission of particles with positive angular momentum m to be higher than that of particles with the same frequency ω and quantum number l but with negative angular momentum $-m$. Thus the particle emission tends to carry away the angular momentum. For Boson fields, the factor in curly brackets in (3.4) is negative for $\omega < m\Omega$. However the fraction Γ_{jnlm} of the wave-packet that would have been absorbed by the black hole is also negative in this case because $\omega < m\Omega$ is the condition for superradiance. In the limit that the temperature $\kappa/2\pi$ is very low, the only particle emission occurs is an amount $\mp\Gamma_{jnlm}$ in the modes for which $\omega < m\Omega$. This amount of particle creation is equal to that calculated by Starobinski [16] and Unruh [29], who considered only the final stationary Kerr solution and ignored the gravitational collapse.

One can treat a charged non-rotating black hole in a rather similar way. The behaviour of fields like the electromagnetic or gravitational fields which do not carry an electric charge will be the same as before except that the charge on the black will reduce the surface gravity k and hence the temperature of the black hole. Consider now the simple case of a massless charged scalar field ϕ which obeys the minimally coupled wave equation

$$g^{ab}(\Gamma_a - ieA_a)(\Gamma_b - ieA_b)\phi = 0\,. \tag{3.5}$$

The phase of a solution p_ω of the wave equation (3.5) is not gauge-invariant but the propagation vector $ik_a = \nabla_a(\log p_\omega) - ieA_a$ is. In the geometric optics or WKB limit the vector k_a is null and propagates according to

$$k_{a;b}k^b = -eF_{ab}k^b. \qquad (3.6)$$

An infinitessimal vector z^a will connect points with a "guage invariant" phase difference of ik_az^a. If z^a is propagated along the integral curves of k^a according to

$$z^a_{;b}k^b = -eF^a_b z^b \qquad (3.7)$$

z^a will connect surfaces of constant guage invariant phase difference.

In the final stationary region one can choose a guage such that the electromagnetic potential A_a is stationary and vanishes on $\mathcal{I}^+$. In this guage the field equation (3.5) is separable and has solutions p_ω with retarded time dependence $e^{i\omega u}$. Let x be a point on the event horizon in the final stationary region and let l^a and n^a be a pair of null vectors at x. As before, the vector $-\varepsilon n^a$ will connect the event horizon with the surface of actual phase $-\omega/\kappa \, (\log\varepsilon - \log C)$ of the solution p_ω. However the guage invariant phase will be $-\kappa^{-1}(\omega - e\Phi)(\log\varepsilon - \log C)$ where $\Phi = K^a A_a$ is the electrostatic potential on the horizon and K^a is the time-translation Killing vector. Now propagate l^a like k^a in Eq. (3.6) back until it intersects a generator λ of $\mathcal{I}^-$ at a point y and propagate n^a like z^a in Eq. (3.7) along the integral curve of l^a. With this propagation law, the vector $-\varepsilon n^a$ will connect surfaces of constant guage invariant phase. Near $\mathcal{I}^-$ one can use a different electromagnetic guage such that A^a is zero on $\mathcal{I}^-$. In this guage the phase of $p_\omega^{(2)}$ along each generator of $\mathcal{I}^-$ will have the form

$$-(\omega - e\phi)\kappa^{-1}\{\log(v_0 - v) - \log H\} \qquad (3.8)$$

where H is a constant along each generator. This phase dependence gives the same thermal emission as before but with ω replaced by $\omega - e\Phi$. Similar remarks apply about charge loss and superradiance. In the case that the black hole is both rotating and charged one can simply combine the above results.

4. The Back-Reaction on the Metric

I now come to the difficult problem of the back-reaction of the particle creation on the metric and the consequent slow decrease of the mass of the black hole. At first sight it might seem that since all the time dependence of the metric in Fig. 4 is in the collapsing phase, all the particle creation must take place in the collapsing body just before the formation of the event horizon, and that an infinite number of created particles would hover just outside the event horizon, escaping to $\mathcal{I}^+$ at a steady rate. This does not seem reasonable because it would involve the collapsing body knowing just when it was about to fall through the event horizon whereas the position of the event horizon is determined by the whole future history of the black hole and may be someway outside the apparent horizon, which is the only thing that can be determined locally [7].

Consider an observer falling through the horizon at some time after the collapse. He can set up a local inertial coordinate patch of radius $\sim M$ centred on the point where he crosses the horizon. He can pick a complete family $\{h_\omega\}$ of solutions of the wave equations which obey the condition:

$$\tfrac{1}{2}i \int_S (h_{\omega_1}\bar{h}_{\omega_2;a} - \bar{h}_{\omega_2}h_{\omega_1;a})d\Sigma^a = \delta(\omega_1 - \omega_2) \tag{4.1}$$

(where S is a Cauchy surface) and which have the approximate coordinate dependence $e^{i\omega t}$ in the coordinate patch. This last condition determines the splitting into positive and negative frequencies and hence the annihilation and creation operators fairly well for modes h_ω with $\omega > M$ but not for those with $\omega < M$. Because the $\{h_\omega\}$, unlike the $\{p_\omega\}$, are continuous across the event horizon, they will also be continuous on $\mathscr{I}^-$. It is the discontinuity in the $\{p_\omega\}$ on $\mathscr{I}^-$ at $v = v_0$ which is responsible for creating an infinite total number of particles in each mode. p_ω by producing an $(\omega')^{-1}$ tail in the Fourier transforms of the $\{p_\omega\}$ at large negative frequencies ω'. On the other hand, the $\{h_\omega\}$ for $\omega > M$ will have very small negative frequency components on $\mathscr{I}^-$. This means that the observer at the event horizon will see few particles with $\omega > M$. He will not be able to detect particles with $\omega < M$ because they will have a wavelength bigger than his particle detector which must be smaller than M. As described in the introduction, there will be an indeterminacy in the energy density of order M^{-4} corresponding to the indeterminacy in the particle number for these modes.

The above discussion shows that the particle creation is really a global process and is not localised in the collapse: an observer falling through the event horizon would not see an infinite number of particles coming out from the collapsing body. Because it is a non-local process, it is probably not reasonable to expect to be able to form a local energy-momentum tensor to describe the back-reaction of the particle creation on the metric. Rather, the negative energy density needed to account for the decrease in the area of the horizon, should be thought of as arising from the indeterminacy of order of M^{-4} of the local energy density at the horizon. Equivalently, one can think of the area decrease as resulting from the fact that quantum fluctuations of the metric will cause the position and the very concept of the event horizon to be somewhat indeterminate.

Although it is probably not meaningful to talk about the local energy-momentum of the created particles, one may still be able to define the total energy flux over a suitably large surface. The problem is rather analogous to that of defining gravitational energy in classical general relativity: there are a number of different energy-momentum pseudo-tensors, none of which have any invariant local significance, but which all agree when integrated over a sufficiently large surface. In the particle case there are similarly a number of different expressions one can use for the renormalised energy-momentum tensor. The energy-momentum tensor for a classical field ϕ is

$$T_{ab} = \phi_{;a}\phi_{;b} - \tfrac{1}{2}g_{ab}g^{cd}\phi_{;c}\phi_{;d} . \tag{4.2}$$

If one takes this expression over into the quantum theory and regards the ϕ's as operators one obtains a divergent result because there is a creation operator for each mode to the right of an annihilation operator. One therefore has to subtract out the divergence in some way. Various methods have been proposed for this (e.g. [30]) but they all seem a bit ad hoc. However, on the analogy of the pseudo-tensor, one would hope that the different renormalisations would all give the same integrated fluxes. This is indeed the case in the final quasi-stationary region: all renormalised energy-momentum operators T_{ab} which obey the conservation equations $T^{ab}_{;b} = 0$, which are stationary i.e. which have zero Lie derivative with respect to the time translation Killing vector K^a and which agree near $\mathscr{I}^+$ will give the same fluxes of energy and angular momentum over any surface of constant r outside the event horizon. It is therefore sufficient to evaluate the energy flux near $\mathscr{I}^+$: by the conservation equations this will be equal to the energy flux out from the event horizon. Near $\mathscr{I}^+$ the obvious way to renormalise the energy-momentum operator is to normal order the expression (4.2) with respect to positive and negative frequencies defined by the time-translation Killing vector K^a of the final quasi-stationary state. Near the event horizon normal ordering with respect to K^a cannot be the correct way to renormalise the energy-momentum operator since the normal-ordered operator diverges at the horizon. However it still gives the same energy outflow across any surface of constant r. A renormalised operator which was regular at the horizon would have to violate the weak energy condition by having negative energy density. This negative energy density is not observable locally.

In order to evaluate the normal ordered operator one wants to choose the $\{q_i\}$ which describe waves crossing the event horizon, to be positive frequency with respect to the time parameter defined by K^a along the generators of the horizon in the final quasi-stationary state. The condition on the $\{q_i\}$ in the time-dependent collapse phase is not determined but this should not affect wave packets on the horizon at late times. If one makes up wave-packets $\{q_{jn}\}$ like the $\{p_{jn}\}$, one finds that a fraction Γ_{jn} penetrates through the potential barrier around the black hole and gets out to $\mathscr{I}^-$ with the same frequency ω that it had on the horizon. This produces a $\delta(\omega - \omega')$ behaviour in $\gamma_{jn\omega'}$. The remaining fraction $1 - \Gamma_{jn}$ of the wave-packet is reflected back by the potential barrier and passes through the collapsing body and out onto $\mathscr{I}^-$. Here it will have a similar form to $p_{jn}^{(2)}$. Thus for large ω',

$$|\gamma^{(2)}_{jn\omega'}| = \exp(\pi\omega\kappa^{-1})|\eta^{(2)}_{jn\omega'}|. \tag{4.3}$$

By a similar argument to that used in Section (2) one would conclude that the number of particles crossing the event horizon in a wave-packet mode peaked at late times would be

$$(1 - \Gamma_{jn})\{\exp(2\pi\omega\kappa^{-1}) - 1\}^{-1}. \tag{4.4}$$

For a given frequency ω, i.e. a given value of j, the absorption fraction Γ_{jn} goes to zero as the angular quantum number l increases because of the centrifugal barrier. Thus at first sight it might seem that each wave-packet mode of high l value would contain

$$\{\exp(2\pi\omega\kappa^{-1}) - 1\}^{-1}$$

particles and that the total rate of particles and energy crossing the event horizon would be infinite. This calculation would, of course, be inconsistent with the result obtained above that an observer crossing the event horizon would see only a finite small energy density of order M^{-4}. The reason for this discrepancy seems to be that the wave-packets $\{p_{jn}\}$ and $\{q_{jn}\}$ provide a complete basis for solutions of the wave equation only in the region outside the event horizon and not actually on the event horizon itself. In order to calculate the particle flux over the horizon one therefore has to calculate the flux over some surface just outside the horizon and take the limit as the surface approaches the horizon.

To perform this calculation it is convenient to define new wave-packets $x_{jn} = p_{jn}^{(2)} + q_{jn}^{(2)}$ which represent the part of p_{jn} and q_{jn} which passes through the collapsing body and $y_{jn} = p_{jn}^{(1)} + q_{jn}^{(1)}$ which represents the part of p_{jn} and q_{jn} which propagates out to $\mathscr{I}^-$ through the quasi-stationary metric of the final black hole. In the initial vacuum state the $\{y_{jn}\}$ modes will not contain any particles but each x_{jn} mode will contain $\{\exp(2\pi\omega\kappa^{-1}) - 1\}^{-1}$ particles. These particles will appear to leave the collapsing body just outside the event horizon and will propagate radially outwards. A fraction Γ_{jn} will penetrate through the potential barrier peaked at $r = 3M$ and will escape to $\mathscr{I}^+$ where they will constitute the thermal emission of the black hole. The remaining fraction $1-\Gamma_{jn}$ will be reflected back by the potential barrier and will cross the event horizon. Thus the net particle flux across a surface of constant r just outside the horizon will be Γ_{jn} directed outwards.

I shall now show that using the normal ordered energy momentum operator, the average energy flux across a surface of constant r between retarded times u_1 and u_2

$$(u_2 - u_1)^{-1} \int_{u_1}^{u_2} \langle 0_- | T_{ab} | 0_- \rangle K^a d\Sigma^b \tag{4.5}$$

is directed outwards and is equal to the energy flux for the thermal emission from a hot body. Because the $\{y_{jn}\}$ contain no negative frequencies on $\mathscr{I}^-$, they will not make any contribution to the expectation value (4.5) of the normal ordered energy-momentum operator. Let

$$x_{jn} = \int_0^\infty (\zeta_{jn\omega'} f_{\omega'} + \xi_{jn\omega'} \bar{f}_{\omega'}) d\omega' . \tag{4.6}$$

Near $\mathscr{I}^+$

$$x_{jn} = (\Gamma_{jn})^{\frac{1}{2}} p_{jn} . \tag{4.7}$$

Thus

$$(4.5) = (u_2 - u_1)^{-1} \operatorname{Re} \left\{ \sum_{j,n} \sum_{j'',n''} \int_0^\infty \int_{u_1}^{u_2} \omega\omega'' \Gamma_{jn}^{\frac{1}{2}} p_{jn} \bar{\xi}_{jn\omega'} \right.$$
$$\left. \cdot (\bar{\Gamma}_{j''n''}^{\frac{1}{2}} \bar{p}_{j''n''} \xi_{j''n''\omega'} - \Gamma_{j''n''}^{\frac{1}{2}} p_{j''n''} \zeta_{j''n''\omega'}) d\omega' du \right\} \tag{4.8}$$

where ω and ω'' are the frequencies of the wave-packets p_{jn} and $p_{j''n''}$ respectively. In the limit $u_2 - u_1$ tends to infinity, the second term in the integrand in (4.8) will integrate out and the first term will contribute only for $(j'', n'') = (j, n)$. By arguments similar to those used in Section 2,

$$\int_0^\infty |\xi_{jn\omega'}|^2 d\omega' = \{\exp(2\pi\omega\kappa^{-1}) - 1\}^{-1} . \tag{4.9}$$

Therefore

$$(4.5) = \int_0^\infty \Gamma_\omega \omega \{\exp(2\pi\omega\kappa^{-1}) - 1\}^{-1} d\omega \tag{4.10}$$

where $\Gamma_\omega = \lim_{n \to \infty} \Gamma_{jn}$ is the fraction of wave-packet of frequency that would be absorbed by the black hole. The energy flux (4.10) corresponds exactly to the rate of thermal emission calculated in Section 2. Any renormalised energy momentum

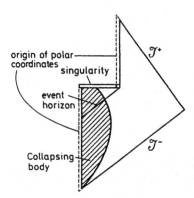

Fig. 5. The Penrose diagram for a gravitational collapse followed by the slow evaporation and eventual disappearance of the black hole, leaving empty space with no singularity at the origin

operator which agrees with the normal ordered operator near $\mathscr{I}^+$, which obeys the conservation equations, and which is stationary in the final quasi-stationary region will give the same energy flux over any surface of constant r. Thus it will give positive energy flux out across the event horizon or, equivalently, a negative energy flux in across the event horizon.

This negative energy flux will cause the area of the event horizon to decrease and so the black hole will not, in fact, be in a stationary state. However, as long as the mass of the black hole is large compared to the Planck mass 10^{-5} g, the rate of evolution of the black hole will be very slow compared to the characteristic time for light to cross the Schwarzschild radius. Thus it is a reasonable approximation to describe the black hole by a sequence of stationary solutions and to calculate the rate of particle emission in each solution. Eventually, when the mass of the black hole is reduced to 10^{-5} g, the quasi-stationary approximation will break down. At this point, one cannot continue to use the concept of a classical metric. However, the total mass or energy remaining in the system is very small. Thus, provided the black hole does not evolve into a negative mass naked singularity there is not much it can do except disappear altogether. The baryons or leptons that formed the original collapsing body cannot reappear because all their rest mass energy has been carried away by the thermal radiation. It is tempting to speculate that this might be the reason why the universe now contains so few baryons compared to photons: the universe might have started out with baryons only, and no radiation. Most of the baryons might have fallen into small black holes which then evaporated giving back the rest mass energy of baryons in the form of radiation, but not the baryons themselves.

The Penrose diagram of a black hole which evaporates and leaves only empty space is shown in Fig. 5. The horizontal line marked "singularity" is really a region where the radius of curvature is of the order the Planck length. The matter that runs into this region might reemerge in another universe or it might even reemerge in our universe through the upper vertical line thus creating a naked singularity of negative mass.

References

1. Isham, C. J.: Preprint (1973)
2. Ashtekar, A., Geroch, R. P.: Quantum theory of gravity (preprint 1973)
3. Penrose, R.: Phys. Rev. Lett. 14, 57—59 (1965)
4. Hawking, S. W.: Proc. Roy. Soc. Lond. A 300, 187—20 (1967)
5. Hawking, S. W., Penrose, R.: Proc. Roy. Soc. Lond. A 314, 529—548 (1970)
6. Hawking, S. W., Ellis, G. F. R.: The large scale structure of space-time. London: Cambridge University Press 1973
7. Hawking, S. W.: The event horizon. In: Black holes. Ed. C. M. DeWitt, B. S. DeWitt. New York: Gordon and Breach 1973
8. Bardeen, J. M., Carter, B., Hawking, S. W.: Commun. math. Phys. 31, 161—170 (1973)
9. Hawking, S. W.: Mon, Not. Roy. astr. Soc. 152, 75—78 (1971)
10. Carr, B. J., Hawking, S. W.: Monthly Notices Roy. Astron. Soc. 168, 399—415 (1974)
11. Hagedorn, R.: Astron. Astrophys. 5, 184 (1970)
12. Hawking, S. W.: Commun. math. Phys. 25, 152—166 (1972)
13. Carter, B.: Black hole equilibrium states. In: Black holes. Ed. C. M. DeWitt, B. S. DeWitt. New York: Gordon and Breach 1973
14. Misner, C. W.: Bull. Amer. Phys. Soc. 17, 472 (1972)
15. Press, W. M., Teukolsky, S. A.: Nature 238, 211 (1972)
16. Starobinsky, A. A.: Zh. E.T.F. 64, 48 (1973)
17. Starobinsky, A. A., Churilov, S. M.: Zh. E.T.F. 65, 3 (1973)
18. Bjorken, T. D., Drell, S. D.: Relativistic quantum mechanics. New York: McGraw Hill 1965
19. Beckenstein, J. D.: Phys. Rev. D. 7, 2333—2346 (1973)
20. Beckenstein, J. D.: Phys. Rev. D. 9,
21. Penrose, R.: Phys. Rev. Lett. 10, 66—68 (1963)
22. Sachs, R. K.: Proc. Roy. Soc. Lond. A 270, 103 (1962)
23. Eardley, D., Sachs, R. K.: J. Math. Phys. 14 (1973)
24. Schmidt, B. G.: Commun. Math. Phys. 36, 73—90 (1974)
25. Bondi, H., van der Burg, M. G. J., Metzner, A. W. K.: Proc. Roy. Soc. Lond. A 269, 21 (1962)
26. Carter, B.: Commun. math. Phys. 10, 280—310 (1968)
27. Teukolsky, S. A.: Ap. J. 185, 635—647 (1973)
28. Unruh, W.: Phys. Rev. Lett. 31, 1265 (1973)
29. Unruh, W.: Phys. Rev. D. 10, 3194—3205 (1974)
30. Zeldovich, Ya. B., Starobinsky, A. A.: Zh. E.T.F. 61, 2161 (1971), JETP 34, 1159 (1972)

Communicated by J. Ehlers

S. W. Hawking
California Institute of Technology
W. K. Kellogg Radiation Lab. 106-38
Pasadena, California 91125, USA

PHYSICAL REVIEW D   VOLUME 15, NUMBER 10   15 MAY 1977

Action integrals and partition functions in quantum gravity

G. W. Gibbons* and S. W. Hawking

Department of Applied Mathematics and Theoretical Physics, University of Cambridge, England

(Received 4 October 1976)

One can evaluate the action for a gravitational field on a section of the complexified spacetime which avoids the singularities. In this manner we obtain finite, purely imaginary values for the actions of the Kerr-Newman solutions and de Sitter space. One interpretation of these values is that they give the probabilities for finding such metrics in the vacuum state. Another interpretation is that they give the contribution of that metric to the partition function for a grand canonical ensemble at a certain temperature, angular momentum, and charge. We use this approach to evaluate the entropy of these metrics and find that it is always equal to one quarter the area of the event horizon in fundamental units. This agrees with previous derivations by completely different methods. In the case of a stationary system such as a star with no event horizon, the gravitational field has no entropy.

I. INTRODUCTION

In the path-integral approach to the quantization of gravity one considers expressions of the form

$$Z = \int d[g] d[\phi] \exp\{iI[g, \phi]\}, \qquad (1.1)$$

where $d[g]$ is a measure on the space of metrics g, $d[\phi]$ is a measure on the space of matter fields ϕ, and $I[g, \phi]$ is the action. In this integral one must include not only metrics which can be continuously deformed into the flat-space metric but also homotopically disconnected metrics such as those of black holes; the formation and evaporation of macroscopic black holes gives rise to effects such as baryon nonconservation and entropy production.[1-4] One would therefore expect similar phenomena to occur on the elementary-particle level. However, there is a problem in evaluating the action I for a black-hole metric because of the spacetime singularities that it necessarily contains.[5-7] In this paper we shall show how one can overcome this difficulty by complexifying the metric and evaluating the action on a real four-dimensional section (really a contour) which avoids the singularities. In Sec. II we apply this procedure to evaluating the action for a number of stationary exact solutions of the Einstein equations. For a black hole of mass M, angular momentum J, and charge Q we obtain

$$I = i\pi\kappa^{-1}(M - Q\Phi), \qquad (1.2)$$

where

$$\kappa = (r_+ - r_-)\, 2^{-1}(r_+{}^2 + J^2 M^{-2})^{-1},$$

$$\Phi = Q r_+ (r_+{}^2 + J^2 M^{-2})^{-1},$$

$$r_\pm = M \pm (M^2 - J^2 M^{-2} - Q^2)^{1/2}$$

in units such that

$$G = c = \hbar = k = 1.$$

One interpretation of this result is that it gives a probability, in an appropriate sense, of the occurrence in the vacuum state of a black hole with these parameters. This aspect will be discussed further in another paper. Another interpretation which will be discussed in Sec. III of this paper is that the action gives the contribution of the gravitational field to the logarithm of the partition function for a system at a certain temperature and angular velocity. From the partition function one can calculate the entropy by standard thermodynamic arguments. It turns out that this entropy is zero for stationary gravitational fields such as those of stars which contain no event horizons. However, both for black holes and de Sitter space[8] it turns out that the entropy is equal to one quarter of the area of the event horizon. This is in agreement with results obtained by completely different methods.[1,4,8]

II. THE ACTION

The action for the gravitational field is usually taken to be

$$(16\pi)^{-1}\int R(-g)^{1/2}\, d^4x.$$

However, the curvature scalar R contains terms which are linear in second derivatives of the metric. In order to obtain an action which depends only on the first derivatives of the metric, as is required by the path-integral approach, the second derivatives have to be removed by integration by parts. The action for the metric g over a region Y with boundary ∂Y has the form

$$I = (16\pi)^{-1}\int_Y R(-g)^{1/2} d^4x + \int_{\partial Y} B(-h)^{1/2} d^3x. \quad (2.1)$$

The surface term B is to be chosen so that for metrics g which satisfy the Einstein equations the action I is an extremum under variations of the metric which vanish on the boundary ∂Y but which may have nonzero normal derivatives. This will be satisfied if $B = (8\pi)^{-1} K + C$, where K is the trace of the second fundamental form of the boundary ∂Y in the metric g and C is a term which depends only on the induced metric h, on ∂Y. The term C gives rise to a term in the action which is independent of the metric g. This can be absorbed into the normalization of the measure on the space of all metrics. However, in the case of asymptotically flat metrics, where the boundary ∂Y can be taken to be the product of the time axis with a two-sphere of large radius, it is natural to choose C so that $I = 0$ for the flat-space metric η. Then $B = (8\pi)^{-1} [K]$, where $[K]$ is the difference in the trace of the second fundamental form of ∂Y in the metric g and the metric η.

We shall illustrate the procedure for evaluating the action on a nonsingular section of a complexified spacetime by the example of the Schwarzschild solution. This is normally given in the form

$$ds^2 = -(1 - 2Mr^{-1})dt^2 + (1 - 2Mr^{-1})^{-1} dr^2 + r^2 d\Omega^2 . \tag{2.2}$$

This has singularities at $r = 0$ and at $r = 2M$. As is now well known, the singularity at $r = 2M$ can be removed by transforming to Kruskal coordinates in which the metric has the form

$$ds^2 = 32M^3 r^{-1} \exp[-r(2M)^{-1}](-dz^2 + dy^2) + r^2 d\Omega^2 , \tag{2.3}$$

where

$$-z^2 + y^2 = [r(2M)^{-1} - 1]\exp[r(2M)^{-1}], \tag{2.4}$$

$$(y + z)(y - z)^{-1} = \exp[t(2M)^{-1}]. \tag{2.5}$$

The singularity at $r = 0$ now lies on the surface $z^2 - y^2 = 1$. It is a curvature singularity and cannot be removed by coordinate changes. However, it can be avoided by defining a new coordinate $\zeta = iz$. The metric now takes the positive-definite or Euclidean form

$$ds^2 = 32M^3 r^{-1}\exp[-r(2M)^{-1}](d\zeta^2 + dy^2) + r^2 d\Omega^2 , \tag{2.6}$$

where r is now defined by

$$\zeta^2 + y^2 = [r(2M)^{-1} - 1] \exp[r(2M)^{-1}]. \tag{2.7}$$

On the section on which ζ and y are real (the Euclidean section), r will be real and greater than or equal to $2M$. Define the imaginary time by $\tau = it$. It follows from Eq. (2.5) that τ is periodic

with period $8\pi M$. On the Euclidean section τ has the character of an angular coordinate about the "axis" $r = 2M$. Since the Euclidean section is nonsingular we can evaluate the action (2.1) on a region Y of it bounded by the surface $r = r_0$. The boundary ∂Y has topology $S^1 \times S^2$ and so is compact.

The scalar curvature R vanishes so the action is given by the surface term

$$I = (8\pi)^{-1} \int [K]d\Sigma . \tag{2.8}$$

But

$$\int K d\Sigma = \frac{\partial}{\partial n} \int d\Sigma , \tag{2.9}$$

where $(\partial / \partial n) \int d\Sigma$ is the derivative of the area $\int d\Sigma$ of ∂Y as each point of ∂Y is moved an equal distance along the outward unit normal n. Thus in the Schwarzschild solution

$$\int K d\Sigma = -32\pi^2 M(1 - 2Mr^{-1})^{1/2}$$
$$\times \frac{d}{dr}[ir^2(1 - 2Mr^{-1})^{1/2}]$$
$$= -32\pi^2 iM(2r - 3M). \tag{2.10}$$

The factor $-i$ arises from the $(-h)^{1/2}$ in the surface element $d\Sigma$. For flat space $K = 2r^{-1}$. Thus

$$\int K d\Sigma = -32\pi^2 iM(1 - 2Mr^{-1})^{1/2} 2r . \tag{2.11}$$

Therefore

$$I = (8\pi)^{-1}\int [K]d\Sigma$$
$$= 4\pi iM^2 + O(M^2 r_0^{-1})$$
$$= \pi iM\kappa^{-1} + O(M^2 r_0^{-1}), \tag{2.12}$$

where $\kappa = (4M)^{-1}$ is the surface gravity of the Schwarzschild solution.

The procedure is similar for the Reissner-Nordström solution except that now one has to add on the action for the electromagnetic field F_{ab}. This is

$$-(16\pi)^{-1}\int F_{ab}F^{ab}(-g)^{1/2}d^4 x . \tag{2.13}$$

For a solution of the Maxwell equations, $F^{ab}_{;b} = 0$ so the integrand of (2.13) can be written as a divergence

$$F_{ab}F_{cd}g^{ac}g^{bd} = (2F^{ab}A_a)_{;b}. \tag{2.14}$$

Thus the value of the action is

$$-(8\pi)^{-1}\int F^{ab}A_a d\Sigma_b . \tag{2.15}$$

The electromagnetic vector potential A_a for the Reissner-Nordström solution is normally taken to be

$$A_a = Qr^{-1} t_{;a}. \tag{2.16}$$

However, this is singular on the horizon as t is not defined there. To obtain a regular potential one has to make a gauge transformation

$$A'_a = (Qr^{-1} - \Phi)t_{;a}, \tag{2.17}$$

where $\Phi = Q(r_+)^{-1}$ is the potential of the horizon of the black hole. The combined gravitational and electromagnetic actions are

$$I = i\pi\kappa^{-1}(M - Q\Phi). \tag{2.18}$$

We have evaluated the action on a section in the complexified spacetime on which the induced metric is real and positive-definite. However, because R, F_{ab}, and K are holomorphic functions on the complexified spacetime except at the singularities, the action integral is really a contour integral and will have the same value on any section of the complexified spacetime which is homologous to the Euclidean section even though the induced metric on this section may be complex. This allows us to extend the procedure to other spacetimes which do not necessarily have a real Euclidean section. A particularly important example of such a metric is that of the Kerr-Newman solution. In this one can introduce Kruskal coordinates y and z and, by setting $\zeta = iz$, one can define a nonsingular section as in the Schwarzschild case. We shall call this the "quasi-Euclidean section." The metric on this section is complex and it is asymptotically flat in a coordinate system rotating with angular velocity Ω, where $\Omega = JM^{-1}(r_+^2 + J^2M^{-2})^{-1}$ is the angular velocity of the black hole. The regularity of the metric at the horizon requires that the point (t, r, θ, ϕ) be identified with the point $(t + i2\pi\kappa^{-1}, r, \theta, \phi + i2\pi\Omega\kappa^{-1})$. The rotation does not affect the evaluation of the $\int [K]d\Sigma$ so the action is still given by Eq. (2.18). One can also evaluate the gravitational contribution to the action for a stationary axisymmetric solution containing a black hole surrounded by a perfect fluid rigidly rotating at some different angular velocity. The action is

$$I = i2\pi\kappa^{-1}\left[(16\pi)^{-1}\int_\Sigma R K^a d\Sigma_a + 2^{-1}M\right], \tag{2.19}$$

where $K^a \partial/\partial x_a = \partial/\partial t$ is the time-translation Killing vector and Σ is a surface in the quasi-Euclidean section which connects the boundary at $r = r_0$ with the "axis" or bifurcation surface of the horizon $r = r_+$. The total mass, M, can be expressed as

$$M = M_H + 2\int_\Sigma (T_{ab} - \tfrac{1}{2}g_{ab}T)K^a d\Sigma^b, \tag{2.20}$$

where

$$M_H = (4\pi)^{-1}\kappa A + 2\Omega_H J_H. \tag{2.21}$$

M_H is the mass of the black hole, A is the area of the event horizon, and Ω_H and J_H are respectively the angular velocity and angular momentum of the black hole.[9] The energy-momentum tensor of the fluid has the form

$$T_{ab} = (p + \rho)u_a u_b + p g_{ab}, \tag{2.22}$$

where ρ is the energy density and p is the pressure of the fluid. The 4-velocity u_a can be expressed as

$$\lambda u^a = K^a + \Omega_m m^a, \tag{2.23}$$

where Ω_m is the angular velocity of the fluid, m^a is the axial Killing vector, and λ is a normalization factor. Substituting (2.21) and (2.22) in (2.20) one finds that

$$M = (4\pi)^{-1}\kappa A + 2\Omega_H J_H + 2\Omega_m J_m$$
$$- \int (\rho + 3p)K^a d\Sigma_a, \tag{2.24}$$

where

$$J_m = -\int T_{ab} m^a d\Sigma^b \tag{2.25}$$

is the angular momentum of the fluid. By the field equations, $R = 8\pi(\rho - 3p)$, so this action is

$$I = 2\pi i\kappa^{-1}\left[M - \Omega_H J_H - \Omega_m J_m - \kappa A(8\pi)^{-1} + \int \rho K^a d\Sigma_a\right]. \tag{2.26}$$

One can also apply (2.26) to a situation such as a rotating star where there is no black hole present. In this case the regularity of the metric does not require any particular periodicity of the time coordinate and $2\pi\kappa^{-1}$ can be replaced by an arbitrary periodicity β. The significance of such a periodicity will be discussed in the next section.

We conclude this section by evaluating the action for de Sitter space. This is given by

$$I = (16\pi)^{-1}\int_Y (R - 2\Lambda)(-g)^{1/2}d^4x$$
$$+ (8\pi)^{-1}\int_{\partial Y} [K]d\Sigma, \tag{2.27}$$

where Λ is the cosmological constant. By the field equations $R = 4\Lambda$. If one were to take Y to be the ordinary real de Sitter space, i.e., the section on which the metric was real and Lorentzian, the volume integral in (2.27) would be infinite. However, the complexified de Sitter space contains a section on which the metric is the real positive-definite metric of a 4-sphere of radius $3^{1/2}\Lambda^{-1/2}$. This Euclidean section has no boundary so that the value of this action on it is

$$I = -12\pi i\Lambda^{-1}, \qquad (2.28)$$

where the factor of $-i$ comes from the $(-g)^{1/2}$.

III. THE PARTITION FUNCTION

In the path-integral approach to the quantization of a field ϕ one expresses the amplitude to go from a field configuration ϕ_1 at a time t_1 to a field configuration ϕ_2 at time t_2 as

$$\langle \phi_2, t_2 | \phi_1, t_1 \rangle = \int d[\phi]\exp(iI[\phi]), \qquad (3.1)$$

where the path integral is over all field configurations ϕ which take the values ϕ_1 at time t_1 and ϕ_2 at time t_2. But

$$\langle \phi_2, t_2 | \phi_1, b_1 \rangle = \langle \phi_2 | \exp[-iH(t_2 - t_1)] | \phi_1 \rangle, \quad (3.2)$$

where H is the Hamiltonian. If one sets $t_2 - t_1 = -i\beta$ and $\phi_1 = \phi_2$ and the sums over all ϕ_1 one obtains

$$\text{Tr}\exp(-\beta H) = \int d[\phi]\exp(iI[\phi]), \qquad (3.3)$$

where the path integral is now taken over all fields which are periodic with period β in imaginary time. The left-hand side of (3.3) is just the partition function Z for the canonical ensemble consisting of the field ϕ at temperature $T = \beta^{-1}$. Thus one can express the partition function for the system in terms of a path integral over periodic fields.[10] When there are gauge fields, such as the electromagnetic or gravitational fields, one must include the Faddeev-Popov ghost contributions to the path integral.[11-13]

One can also consider grand canonical ensembles in which one has chemical potentials μ_i associated with conserved quantities C_i. In this case the partition function is

$$Z = \text{Tr}\exp\left[-\beta\left(H - \sum_i \mu_i C_i\right)\right]. \qquad (3.4)$$

For example, one could consider a system at a temperature $T = \beta^{-1}$ with a given angular momentum J and electric charge Q. The corresponding chemical potentials are then Ω, the angular velocity, and Φ, the electrostatic potential. The partition function will be given by a path integral over all fields ϕ whose value at the point $(t+i\beta, r, \theta, \phi +i\beta\Omega)$ is $\exp(q\beta\Phi)$ times the value at (t, r, θ, ϕ), where q is the charge on the field.

The dominant contribution to the path integral will come from metrics g and matter fields ϕ which are near background fields g_0 and ϕ_0 which have the correct periodicities and which extremize the action, i.e., are solutions of the classical field equations. One can express g and ϕ as

$$g = g_0 + \bar{g}, \quad \phi = \phi_0 + \bar{\phi} \qquad (3.5)$$

and expand the action in a Taylor series about the background fields

$$I[g, \phi] = I[g_0, \phi_0] + I_2[\bar{g}] + I_2[\bar{\phi}]$$
$$+ \text{higher-order terms}, \qquad (3.6)$$

where $I_2[\bar{g}]$ and $I_2[\bar{\phi}]$ are quadratic in the fluctuations $\bar{g}$ and $\bar{\phi}$. If one neglects higher-order terms, the partition function is given by

$$\ln Z = iI[g_0, \phi_0] + \ln\int d[\bar{g}]\exp(iI_2[\bar{g}])$$
$$+ \ln\int d[\bar{\phi}]\exp(iI_2[\bar{\phi}]). \qquad (3.7)$$

But the normal thermodynamic argument

$$\ln Z = -WT^{-1}, \qquad (3.8)$$

where $W = M - TS - \sum_i \mu_i C_i$ is the "thermodynamic potential" of the system. One can therefore regard $iI[g_0, \phi_0]$ as the contribution of the background to $-WT^{-1}$ and the second and third terms in (3.7) as the contributions arising from thermal gravitons and matter quanta with the appropriate chemical potentials. A method for evaluating these latter terms will be given in another paper.

One can apply the above analysis to the Kerr-Newman solutions because in them the points (t, r, θ, ϕ) and $(t + 2\pi i\kappa^{-1}, r, \theta, \phi + 2\pi i\Omega\kappa^{-1})$ are identified (the charge q of the graviton and photon are zero). It follows that the temperature T of the background field is $\kappa(2\pi)^{-1}$ and the thermodynamic potential is

$$W = \tfrac{1}{2}(M - \Phi Q), \qquad (3.9)$$

but

$$W = M - TS - \Phi Q - \Omega J. \qquad (3.10)$$

Therefore

$$\tfrac{1}{2}M = TS + \tfrac{1}{2}\Phi Q + \Omega J, \qquad (3.11)$$

but by the generalized Smarr formula[9,14]

$$\tfrac{1}{2}M = \kappa(8\pi)^{-1}A + \tfrac{1}{2}\Phi Q + \Omega J. \qquad (3.12)$$

Therefore

$$S = \tfrac{1}{4}A, \qquad (3.13)$$

in complete agreement with previous results.

For de Sitter space

$$WT^{-1} = -12\pi\Lambda^{-1}, \qquad (3.14)$$

but in this case $W = -TS$, since $M = J = Q = 0$ because this space is closed. Therefore

$$S = 12\pi\Lambda^{-1}, \qquad (3.15)$$

which again agrees with previous results. Note that the temperature T of de Sitter space cancels out the period. This is what one would expect since the temperature is observer dependent and related to the normalization of the timelike Killing vector.

Finally we consider the case of a rotating star in equilibrium at some temperature T with no event horizons. In this case we must include the contribution from the path integral over the matter fields as it is these which are producing the gravitational field. For matter quanta in thermal equilibrium at a temperature T volume $V \gg T^{-3}$ of flat space the thermodynamic potential is given by

$$WT^{-1} = -i\int p(-\eta)^{1/2} d^4 x = -pVT^{-1}. \qquad (3.16)$$

In situations in which the characteristic wavelengths, T^{-1}, are small compared to the gravitational length scales it is reasonable to use this fluid approximation for the density of thermodynamic potential; thus the matter contributing to the thermodynamic potential will be given by

$$W_m T^{-1} = -i\int p(-g)^{1/2} d^4 x = T^{-1}\int pK^a d\Sigma_a \quad (3.17)$$

(because of the signature of our metric $K^a d\Sigma_a$ is negative), but by Eq. (2.26) the gravitational contribution to the total thermodynamic potential is

$$W_g = M - \Omega_m J_m + \int_\Sigma \rho K^a d\Sigma_a. \qquad (3.18)$$

Therefore the total thermodynamic potential is

$$W = M - \Omega_m J_m + \int_\Sigma (p + \rho)K^a d\Sigma_a, \qquad (3.19)$$

but

$$p + \rho = \overline{T}s + \sum_i \overline{\mu}_i n_i, \qquad (3.20)$$

where $\overline{T}$ is the local temperature, s is the entropy density of the fluid, $\overline{\mu}_i$ is the local chemical potentials, and n_i is the number densities of the ith species of particles making up the fluid. Therefore

$$W = M - \Omega_m J_m + \int_\Sigma \left(\overline{T}s + \sum_i \overline{\mu}_i n_i\right)K^a d\Sigma_a. \quad (3.21)$$

In thermal equilibrium

$$\overline{T} = T\lambda^{-1}, \qquad (3.22)$$

$$\overline{\mu}_i = \mu_i \lambda^{-1}, \qquad (3.23)$$

where T and μ_i are the values of $\overline{T}$ and $\overline{\mu}_i$ at infinity.[9] Thus the entropy is

$$S = -\int su^a d\Sigma_a. \qquad (3.24)$$

This is just the entropy of the matter. In the absence of the event horizon the gravitational field has no entropy.

---

*Present address: Max-Planck-Institute für Physik und Astrophysik, 8 München 40, Postfach 401212, West Germany. Telephone: 327001.

[1]S. W. Hawking, Commun. Math. Phys. **43**, 199 (1975).

[2]R. M. Wald, Commun. Math. Phys. **45**, 9 (1975).

[3]S. W. Hawking, Phys. Rev. D **14**, 2460 (1976).

[4]S. W. Hawking, Phys. Rev. D **13**, 191 (1976).

[5]R. Penrose, Phys. Rev. Lett. **14**, 57 (1965).

[6]S. W. Hawking and R. Penrose, Proc. R. Soc. London **A314**, 529 (1970).

[7]S. W. Hawking and G. F. R. Ellis, *The Large Scale Structure of Spacetime* (Cambridge Univ. Press, Cambridge, England, 1973).

[8]G. W. Gibbons and S. W. Hawking, preceding paper, Phys. Rev. D **15**, 2738 (1977).

[9]J. Bardeen, B. Carter, and S. W. Hawking, Commun. Math. Phys. **31**, 161 (1973).

[10]R. P. Feynman and Hibbs, *Quantum Mechanics and Path Integrals* (McGraw-Hill, New York, 1965).

[11]C. W. Bernard, Phys. Rev. D **9**, 3312 (1974).

[12]L. Dolan and R. Jackiw, Phys. Rev. D **9**, 3320 (1974).

[13]L. D. Faddeev and V. N. Popov, Phys. Lett. **25B**, 29 (1967).

[14]L. Smarr, Phys. Rev. Lett. **30**, 71 (1973); **30**, 521(E) (1973).

PHYSICAL REVIEW D   VOLUME 14, NUMBER 10   15 NOVEMBER 1976

Breakdown of predictability in gravitational collapse*

S. W. Hawking[†]

*Department of Applied Mathematics and Theoretical Physics, University of Cambridge, Cambridge, England
and California Institute of Technology, Pasadena, California 91125*

(Received 25 August 1975)

The principle of equivalence, which says that gravity couples to the energy-momentum tensor of matter, and the quantum-mechanical requirement that energy should be positive imply that gravity is always attractive. This leads to singularities in any reasonable theory of gravitation. A singularity is a place where the classical concepts of space and time break down as do all the known laws of physics because they are all formulated on a classical space-time background. In this paper it is claimed that this breakdown is not merely a result of our ignorance of the correct theory but that it represents a fundamental limitation to our ability to predict the future, a limitation that is analogous but additional to the limitation imposed by the normal quantum-mechanical uncertainty principle. The new limitation arises because general relativity allows the causal structure of space-time to be very different from that of Minkowski space. The interaction region can be bounded not only by an initial surface on which data are given and a final surface on which measurements are made but also a "hidden surface" about which the observer has only limited information such as the mass, angular momentum, and charge. Concerning this hidden surface one has a "principle of ignorance": The surface emits with equal probability all configurations of particles compatible with the observers limited knowledge. It is shown that the ignorance principle holds for the quantum-mechanical evaporation of black holes: The black hole creates particles in pairs, with one particle always falling into the hole and the other possibly escaping to infinity. Because part of the information about the state of the system is lost down the hole, the final situation is represented by a density matrix rather than a pure quantum state. This means there is no S matrix for the process of black-hole formation and evaporation. Instead one has to introduce a new operator, called the superscattering operator, which maps density matrices describing the initial situation to density matrices describing the final situation.

I. INTRODUCTION

Gravity is by far the weakest interaction known to physics: The ratio of the gravitational to electrical forces between two electrons is about one part in 10^{43}. In fact, gravity is so weak that it would not be observable at all were it not distinguished from all other interactions by having the property known as the principle of universality or equivalence: Gravity affects the trajectories of all freely moving particles in the same way. This has been verified experimentally to an accuracy of about 10^{-11} by Roll, Krotkov, and Dicke[1] and by Braginsky and Panov.[2] Mathematically, the principle of equivalence is expressed as saying that gravity couples to the energy-momentum tensor of matter. This result and the usual requirement from quantum theory that the local energy density should be positive imply that gravity is always attractive. The gravitational fields of all the particles in large concentrations of matter therefore add up and can dominate over all other forces. As predicted by general relativity and verified experimentally, the universality of gravity extends to light. A sufficiently high concentration of mass can therefore produce such a strong gravitational field that no light can escape. By the principle of special relativity, nothing else can escape either since nothing can travel faster than light. One thus has

a situation in which a certain amount of matter is trapped in a region whose boundary shrinks to zero in a finite time. Something obviously goes badly wrong. In fact, as was shown in a series of papers by Penrose and this author,[3-6] a space-time singularity is inevitable in such circumstances provided that general relativity is correct and that the energy-momentum tensor of matter satisfies a certain positive-definite inequality.

Singularities are predicted to occur in two areas. The first is in the past at the beginning of the present expansion of the universe. This is thought to be the "big bang" and is generally regarded as the beginning of the universe. The second area in which singularities are predicted is the collapse of isolated regions of high-mass concentration such as burnt-out stars.

A singularity can be regarded as a place where there is a breakdown of the classical concept of space-time as a manifold with a pseudo-Reimannian metric. Because all known laws of physics are formulated on a classical space-time background, they will all break down at a singularity. This is a great crisis for physics because it means that one cannot predict the future: One does not know what will come out of a singularity.

Many physicists are very unwilling to believe that physics breaks down at singularities. The following attempts were therefore made in order to try to avoid this conclusion.

1. General relativity does not predict singularities. This was widely believed at one time (e.g., Lifshitz and Khalatnikov[8]). It was, however, abandoned after the singularity theorems mentioned above and it is now generally accepted that the classical theory of general relativity does indeed predict singularities (Lifshitz and Khalatnikov[9]).

2. Modify general relativity. In order to prevent singularities the modifications have to be such as to make gravity repulsive in some situations. The simplest viable modification is probably the Brans-Dicke theory,[10]. In this, however, gravity is always attractive so that the theory predicts singularities just as in general relativity.[7] The Einstein-Cartan theory[11] contains a spin-spin interaction which can be repulsive. This might prevent singularities in some cases but there are situations (such as a purely gravitational and electromagnetic fields) in which singularities will still occur. Most other modifications of general relativity appear either to be in conflict with observations or to have undesirable features like negative energy or fourth-order equations.

3. The "cosmic censorship" hypothesis: Nature abhors a naked singularity. In other words, if one starts out with an initially nonsingular asymptotically flat situation, any singularities which subsequently develop due to gravitational collapse will be hidden from the view of an observer at infinity by an event horizon. This hypothesis, though unproved, is probably true for the classical theory of general relativity with an appropriate definition of nontrivial singularities to rule out such cases as the world lines of pressure-free matter intersecting on caustics. If the cosmic censorship hypothesis held, one might argue that one could ignore the breakdown of physics at space-time singularities because this could never cause any detectable effect for observers careful enough not to fall into a black hole. This is a rather selfish attitude because it ignores the question of what happens to an observer who does fall through an event horizon. It also does not solve the problem of the big-bang singularity which definitely is naked. The final blow to this attempt to evade the issue of breakdown at singularities, however, has been the discovery by this author[12,13] that black holes create and emit particles at a steady rate with a thermal spectrum. Because this radiation carries away energy, the black holes must presumably lose mass and eventually disappear. If one tries to describe this process of black-hole evaporation by a classical space-time metric, there is inevitably a naked singularity when the black hole disappears. Even if the black hole does not evaporate completely one can regard the emitted particles as having come

from the singularity inside the black hole and having tunnelled out through the event horizon on spacelike trajectories. Thus even an observer at infinity cannot avoid seeing what happens at a singularity.

4. Quantize general relativity. One would expect quantum gravitational effects to be important in the very strong fields near a singularity. A number of people have hoped, therefore, that these quantum effects might prevent the singularity from occurring or might smear it out in some way such as to maintain complete predictability within the limits set by the uncertainty principle. However, serious difficulties have arisen in trying to treat quantum gravity like quantum electrodynamics by using perturbation theory about some background metric (usually flat space). Usually in electrodynamics one makes a perturbation expansion in powers of the small parameter $e^2/\hbar c$, the charge squared. Because of the principle of equivalence, the quantity in general relativity that corresponds to charge in electrodynamics is the energy of a particle. The perturbation expansion is therefore really a series in powers of the various energies involved divided by the Planck mass $\hbar^{1/2}c^{1/2}G^{-1/2} \simeq 10^{-5}$ g.

This works well for low-energy tree-approximation diagrams but it breaks down for diagrams with closed loops where one has to integrate over all energies. At energies of the Planck mass, all diagrams become equally important and the series diverges. This is the basic reason why general relativity is not renormalizable.[14,15]

Each additional closed loop appears to involve a new infinite subtraction. There appears to be an infinite sequence of finite remainders or renormalization parameters which are not determined by the theory. One therefore cannot, as was hoped, construct an S matrix which would make definite predictions. The trouble with perturbation theory is that it uses the light cones of a fixed background space. It therefore cannot describe situations in which horizons or worm holes develop by vacuum fluctuations. This is not to say that one cannot quantize gravity, but that one needs a new approach.

One possible view of the failure of the above attempts to avoid the breakdown of predictability would be that we have not yet discovered the correct theory. The aim of this paper, however, is to show this cannot be the case if one accepts that quantum effects will cause a black hole to radiate. In this case there is a basic limitation on our ability to predict which is similar but additional to the usual quantum-mechanical uncertainty principle. This extra limitation arises because general relativity allows the causal structure of space-time to be very different from that of Minkowski space. For example, in the case of gravitational collapse which produces a black hole there is an event horizon which prevents observers at infinity from measuring the internal state of the black hole apart from its mass, angular momentum, and charge. This means that measurements at future infinity are insufficient to determine completely the state of the system at past infinity: One also needs data on the event horizon describing what fell into the black hole. One might think that one could have observers stationed just outside the event horizon who would signal to the observers at future infinity every time a particle fell into the black hole. However, this is not possible, just as one cannot have observers who will measure both the position and the velocity of a particle. To signal accurately the time at which a particle crossed the event horizon would require a photon of the same wavelength and therefore the same energy as that of the infalling particle. If this were done for every particle which underwent gravitational collapse to form the black hole, the total energy required to signal would be equal to that of the collapsing body and there would be no energy left over to form the black hole. It therefore follows that when a black hole forms, one cannot determine the results of measurements at past infinity from observations at future infinity. This might not seem so terrible because one is normally more concerned with prediction than postdiction. However, although in such a situation one could classically determine future infinity from knowledge of past infinity, one cannot do this if quantum effects are taken into account. For example, quantum mechanics allows particles to tunnel on spacelike or past-directed world lines. It is therefore possible for a particle to tunnel out of the black hole through the event horizon and escape to future infinity. One can interpret such a happening as being the spontaneous creation in the gravitational field of the black hole of a pair of particles, one with negative and one with positive energy with respect to infinity. The particle with negative energy would fall into the black hole where there are particle states with negative energy with respect to infinity. The particles with positive energy can escape to infinity where they constitute the recently predicted thermal emission from black holes. Because these particles come from the interior of the black hole about which an external observer has no knowledge, he cannot predict the amplitudes for them to be emitted but only the *probabilities* without the phases.

In Secs. III and IV of this paper it is shown that the quantum emission from a black hole is completely random and uncorrelated. Similar results have been found by Wald[16] and Parker.[17] The black hole emits with equal probability every configuration of particles compatible with conservation of energy, angular momentum, and charge (not every configuration escapes to infinity with equal probability because there is a potential barrier around the black hole which depends on the angular momentum of the particles and which may reflect some of the particles back into the black holes). This result can be regarded as a quantum version of the "no hair" theorems because it implies that an observer at infinity cannot predict the internal state of the black hole apart from its mass, angular momentum, and charge: If the black hole emitted some configuration of particles with greater probability than others, the observer would have some *a priori* information about the internal state. Of course, if the observer measures the wave functions of all the particles that are emitted in a particular case he can then *a posteriori* determine the internal state of the black hole but it will have disappeared by that time.

A gravitational collapse which produces an event horizon is an example of a situation in which the interaction region is bounded by an initial surface on which data are prescribed, a final surface on which measurements are made, and, in addition, a third "hidden" surface about which the observer can have only limited information such as the flux of energy, angular momentum, or charge. Such hidden surfaces can surround either singularities (as in the Schwarzschild solution) or "wormholes" leading to other space-time regions about which the observer has no knowledge (as in the Reissner-Nordström or other solutions). About this surface one has the *principle of ignorance*.

All data on a "hidden" surface compatible with the observer's limited information are equally probable.

So far the discussion has been in terms of quantized matter fields on a fixed classical background metric (the semiclassical approximation). However, one can extend the principle to treatments in which the gravitational field is also quantized by means of the Feynman sum over histories. In this one performs an integration (with an as yet undetermined measure) over all configuration of both matter and gravitational fields. The classical example of black-hole event horizons shows that in this integral one has to include metrics in which the interaction region (i.e., the region over which the action is evaluated) is bounded, not only by the initial and final surfaces, but by a hidden surface as well. Indeed, in any quantum gravitational situation there is the possibility of "virtual" black holes which arise from vacuum fluctuations and which appear out of nothing and then disappear again. One therefore has to include in the sum over histories metrics containing transient holes, leading either to singularities or to other space-time regions about which one has no knowledge. One therefore has to introduce a hidden surface around each of these holes and apply the principle of ignorance to say that all field configurations on these hidden surfaces are equally probable provided they are compatible with the conservation of mass, angular momentum, etc. which can be measured by surface integrals at a distance from the hole.

Let H_1 be the Hilbert space of all possible data on the initial surface, H_2 be the Hilbert space of all possible data on the hidden surface, and H_3 be the Hilbert space of all possible data on the final surface. The basic assumption of quantum theory is that there is some tensor S_{ABC} whose three indices refer to H_3, H_2, and H_1, respectively, such that if

$$\xi_C \in H_1, \quad \zeta_B \in H_2, \quad \chi_A \in H_3,$$

then

$$\sum \sum \sum S_{ABC} \chi_A \zeta_B \xi_C$$

is the amplitude to have the initial state ξ_C, the final state χ_A, and the state ζ_B on the hidden surface. Given only the initial state ξ_C one cannot determine the final state but only the element $\sum S_{ABC} \xi_C$ of the tensor product $H_2 \otimes H_3$. Because one is ignorant of the state on the hidden surface one cannot find the amplitude for measurements on the final surface to give the answer χ_A but one can calculate the probability for this outcome to be $\sum \sum \sum \rho_{CD} \bar{\chi}_C \chi_D$, where

$$\rho_{CD} = \sum \sum \sum \bar{S}_{CBE} S_{DBF} \bar{\xi}_E \xi_F$$

is the density matrix which completely describes observations made only on the future surface and not on the hidden surface. Note that one gets this density matrix from $\sum S_{ABC} \xi_C$ by summing with equal weight over all the unobserved states on the "hidden" surface.

One can see from the above that there will not be an S matrix or operator which maps initial states to final states, because the observed final situation is described, not by a pure quantum state, but by a density matrix. In fact, the initial situation in general will also be described not by a pure state but by a density matrix because of the hidden surface occurring at earlier times. Instead of an S matrix one will have a new operator called the superscattering operator S, which maps density matrices describing the initial situation to density matrices describing the final situation. This operator can be regarded as a 4-index tensor

S_{ABCD} where the first two indices operate on the final space $H_3 \otimes H_3$ and the last two indices operate on the space $H_1 \otimes H_1$. It is related to the 3-index tensor S_{ABC} by

$$S_{ABCD} = \tfrac{1}{2} \sum (\bar{S}_{AEC} S_{BED} + \bar{S}_{BEC} S_{AED}).$$

The final density matrix ρ_{2AB} is given in terms of the initial density matrix ρ_{1CD} by

$$\rho_{2AB} = \sum \sum S_{ABCD} \rho_{1CD}.$$

The superscattering operator is discussed further in Sec. V.

The fact that in gravitational interactions the final situation at infinity is described by a density matrix and not a pure state indicates that quantum gravity cannot, as was hoped, be renormalized to give a well-defined S matrix with only a finite number of undetermined parameters. It seems reasonable to conjecture that there is a close connection between the infinite sequence of renormalization constants that occur in perturbation theory and the loss of predictability which arises from hidden surfaces.

One can also appeal to the principle of ignorance to provide a possible explanation of the observations of the microwave background and of the abundances of helium and deuterium which indicate that the early universe was very nearly spatially homogeneous and isotropic and in thermal equilibrium. One could regard a surface very close to the initial big-bang singularity (say, at the Planck time 10^{-43} sec) as being a "hidden surface" in the sense that we have no *a priori* information about it. The initial surface would thus emit all configurations of particles with equal probability. To obtain a thermal distribution one would need to impose some constraint on the total energy of the configurations where the total energy is the rest-mass energy of the particles plus their kinetic energy of expansion minus their gravitational potential energy. Observationally this energy is very nearly, if not exactly, zero and this can be understood as a necessary condition for our existence: If the total energy were large and positive, the universe would expand too rapidly for galaxies to form, and if the total energy were large and negative, the universe would collapse before intelligent life had time to develop. We therefore do have some limited knowledge of the data on the initial surface from the fact of our own existence. If one assumes that the initial surface emitted with equal probability all configurations of particles with total energy (with some appropriate definition) nearly equal to zero, then an approximately thermal distribution is the most probable macrostate since it corresponds to the largest number of microstates. Any significant departure from homogeneity or isotropy could be regarded as the presence in some long-wavelength modes of a very large number of gravitons, a number greatly in excess of that for a thermal distribution and therefore highly improbable. It should be pointed out that this view of the generality of isotropic expansion is the opposite of that adopted by Collins and Hawking.[18] The difference arises from considering microscopic rather than macroscopic configurations.

One might also think to explain the observed net baryon number of the universe by saying that we, as observers, could result only from initial configurations that had a net baryon number. An alternative explanation might be that CP violations in the highly T-nonsymmetric early universe caused expanding configurations in which baryons predominated to have lower energies than similar expanding configurations in which antibaryons predominated. This would mean that for a given energy density there would be more configurations with a positive baryon number than with a negative baryon number, thus the expectation value of the baryon number would be positive. Alternatively, there might be a sort of spontaneous symmetry breaking which resulted in regions of pure baryons or pure antibaryons having lower energy densities than regions containing a mixture of baryons and antibaryons. In this case, as suggested by Omnès,[19] one would get a phase transition in which regions of pure baryons were separated from regions of pure antibaryons. Unlike the case considered by Omnès, there is no reason why the separation should not be over length scales larger than the particle horizon. Such a greater separation would overcome most of the difficulties of the Omnès model.

There is a close connection between the above proposed explanation for the isotropy of the universe and the suggestion by Zel'dovich[20] that it is caused by particle creation in anisotropic regions. In Zel'dovich's work, however, in order to define particle creation, one has to pretend that the universe was time-independent at early times (which is obviously not the case). The present approach avoids the difficulty of talking about early times; one merely has to count the configurations at some convenient late time.

The conclusion of this paper is that gravitation introduces a new level of uncertainty or randomness into physics over and above the uncertainty usually associated with quantum mechanics. Einstein was very unhappy about the unpredictability of quantum mechanics because he felt that "God does not play dice." However, the results given here indicate that "God not only plays dice, He sometimes throws the dice where they cannot be seen."

II. QUANTUM THEORY IN CURVED SPACE-TIME

In this section a brief outline is given of the formalism of quantum theory on a given space-time background which was used by Hawking[13] to derive the quantum-mechanical emission from black holes. This formalism will be used in Sec. III to show that the radiation which escapes to infinity is completely thermal and uncorrelated. In Sec. IV a specific choice of states for particles going into the black hole is used to calculate explicitly both the ingoing and the emitted particles. This shows that the particles are created in pairs with one member of the pair always falling into the hole and the other member either falling in or escaping to infinity. Section V contains a discussion of the superscattering operator S which maps density matrices describing the initial situation to density matrices describing the final situation.

For simplicity only a massless Hermitian scalar field ϕ and an uncharged nonrotating black hole will be considered. The extension to charged massive fields of higher spin and charged rotating black holes is straightforward along the lines indicated in Ref. 13. Throughout the paper units will be used in which $G = c = \hbar = k = 1$.

Figure 1 is a diagram of the situation under consideration: A gravitational collapse creates a black hole which slowly evaporates and eventually disappears by the quantum-mechanical creation and emission of particles. Except in the final stages of the evaporation, when the black hole gets down to the Planck mass, the back reaction on the gravitational field is very small and it can be treated as an unquantized external field. The metric at late times can be approximated by a sequence of time-independent Schwarzschild solutions and the gravitational collapse can be taken to be spherically symmetric (it was shown in Ref. 13 that departures from spherical symmetry made no essential difference).

The scalar field operator ϕ satisfies wave equation

$$\Box \phi = 0 \qquad (2.1)$$

in this metric and the commutation relations

$$[\phi(x), \phi(y)] = iG(x, y), \qquad (2.2)$$

where $G(x, y)$ is the half-retarded minus half-advanced Green's function. One can express the operator ϕ as

$$\phi = \sum (f_i a_i + \bar{f}_i a_i^\dagger), \qquad (2.3)$$

where the $\{f_i\}$ are a complete orthonormal family

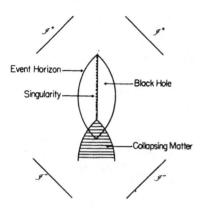

FIG. 1. A gravitational collapse produces a black hole which slowly evaporates by the emission of radiation to future null infinity $\mathcal{I}^+$. Because of the loss of energy, the black hole decreases in size and eventually disappears.

of complex-valued solutions of the wave equation $\Box f_i = 0$ which contain only positive frequencies at past null infinity $\mathcal{I}^-$. The operators a_i are position independent and obey the commutation relations

$$[a_i, a_j] = 0, \tag{2.4}$$

$$[a_i, a_j^\dagger] = \delta_{ij}. \tag{2.5}$$

The operators a_i and $a_i^\dagger$ are respectively the annihilation and creation operators for particles in the ith mode at past infinity. The initial vacuum state for scalar particles $|0_-\rangle$, i.e., the state which contains no scalar particles at past infinity, is defined by

$$a_i |0_-\rangle = 0 \quad \text{for all } i. \tag{2.6}$$

One can also express ϕ in the form

$$\phi = \sum_i (p_i b_i + \bar{p}_i b_i^\dagger + q_i c_i + \bar{q}_i c_i^\dagger). \tag{2.7}$$

Here the $\{p_i\}$ are a complete orthonormal family of solutions of the wave equation which contain only positive frequencies at future null infinity $\mathcal{I}^+$ and which are purely outgoing, i.e., they have zero Cauchy data on the event horizon H. The $\{q_i\}$ are a complete orthonormal set of solutions of the wave equation which contain no outgoing component. The position-independent operators b_i and c_i obey the commutation relations

$$[b_i, b_j] = [c_i, c_j] = 0, \tag{2.8}$$

$$[b_i, c_j] = [b_i, c_j^\dagger] = 0, \tag{2.9}$$

$$[b_i, b_j^\dagger] = [c_i, c_j^\dagger] = \delta_{ij}. \tag{2.10}$$

The operators b_i and $b_i^\dagger$ are respectively the annihilation and creation operators for outgoing parti-

cles at future infinity. By analogy one could regard the operators c_i and $c_i^\dagger$ as the annihilation and creation operators for particles falling into the black hole. However, because one cannot uniquely define positive frequency for the $\{q_i\}$, the division into annihilation and creation parts is not unique and so one should not attach too much physical significance to this interpretation. The nonuniqueness of the $\{c_i\}$ and the $\{c_i^\dagger\}$ does not affect any observable at future infinity. In Sec. IV a particular choice of the $\{q_i\}$ will be made which will allow an explicit calculation of the particles going into the black hole. The final scalar-particle vacuum state $|0_+\rangle$, i.e., the state which contains no outgoing particles at future infinity or particles going into the black hole, is defined by

$$b_i |0_+\rangle = c_i |0_+\rangle = 0. \tag{2.11}$$

It can be represented as $|0_I\rangle |0_H\rangle$, where the b and c operators act on $|0_I\rangle$ and $|0_H\rangle$, respectively, which are the vacua for outgoing particles and for particles falling into the hole. $|0_I\rangle$ is uniquely defined by the positive-frequency condition on the $\{p_i\}$ but the ambiguity in the choice of the $\{q_i\}$ means that $|0_H\rangle$ is not unique.

Because massless fields are completely determined by their data on $\mathcal{I}^-$ one can express $\{p_i\}$ and $\{q_i\}_*$ as linear combinations of the $\{f_i\}$ and $\{\bar{f}_i\}$:

$$p_i = \sum_j (\alpha_{ij} f_j + \beta_{ij} \bar{f}_j), \tag{2.12}$$

$$q_i = \sum_j (\gamma_{ij} f_j + \eta_{ij} \bar{f}_j). \tag{2.13}$$

These relations lead to corresponding relations between the operators:

$$b_i = \sum_j (\bar{\alpha}_{ij} a_j - \bar{\beta}_{ij} a_j^\dagger), \tag{2.14}$$

$$c_i = \sum_j (\bar{\gamma}_{ij} a_j - \bar{\eta}_{ij} a_j^\dagger). \tag{2.15}$$

In the situation under consideration the metric is spherically symmetric. This means the angular dependence of the $\{f_i\}$, $\{p_i\}$, and $\{q_i\}$ can be taken to be that of spherical harmonics Y_{lm}. The relations (2.12) and (2.13) will connect only solutions with the same values of l and $|m|$. (This is not true if the collapse is not exactly spherically symmetric but it was shown in Ref. 13 that this makes no essential difference.) For computational purposes it is convenient to use f and p solutions which have time dependence of the form $e^{i\omega' v}$ and $e^{i\omega u}$, respectively, where v and u are advanced and retarded times. The solutions will be denoted by

$\{f_{\omega'}\}$ and $\{p_\omega\}$ and will have continuum normalization. They can be superposed to form wave-packet solutions of finite normalization. The summations in Eqs. (2.3), (2.7), and (2.12) are replaced by integrations over frequency. The operators a_ω, b_ω, etc. obey similar commutation relations involving δ functions in the frequency.

The advantage of using Fourier components with respect to time is that one can calculate the coefficients $\alpha_{\omega\omega'}$ and $\beta_{\omega\omega'}$ in the approximation that the mass of the black hole is changing only slowly. One considers a solution p_ω propagating backwards in time from future infinity. A part $p_\omega^{(1)}$ is reflected by the static Schwarzschild metric and reaches past infinity with the same frequency. This gives a term $r_\omega\delta(\omega-\omega')$ in $\alpha_{\omega\omega'}$, where r_ω is the reflection coefficient of the Schwarzschild metric for the frequency ω and the given angular mode. More interesting is the behavior of the part $p_\omega^{(2)}$ which propagates through the collapsing body and out to past infinity with a very large blue-shift. This gives contributions to $\alpha_{\omega\omega'}$ and $\beta_{\omega\omega'}$ of the form

$$\alpha_{\omega\omega'}^{(2)} \simeq t_\omega (2\pi)^{-1} e^{i(\omega-\omega')v_0}\left(\frac{\omega'}{\omega}\right)^{1/2}$$

$$\times\Gamma\left(1-\frac{i\omega}{\kappa}\right)(-i\omega')^{-1+i\omega/\kappa}, \qquad (2.16)$$

$$\beta_{\omega\omega'}^{(2)} \simeq -i\alpha_{\omega(-\omega')}^{(2)}, \qquad (2.17)$$

where $\kappa = (4M)^{-1}$ is the surface gravity of the black hole and where t_ω is the transmission coefficient for the given Schwarzschild metric, i.e.,

$$|t_\omega|^2 = \Gamma_\omega$$

is the fraction of a wave with frequency ω and the given angular dependence which penetrates through the potential barrier into the hole,

$$|t_\omega|^2 + |r_\omega|^2 = 1.$$

III. THE OUTGOING RADIATION

One assumes that there are no scalar particles present in the infinite past, i.e., the system is in the initial scalar-particle vacuum state $|0_-\rangle$. (It is not a complete vacuum because it contains the matter that will give rise to the black hole.) The state $|0_-\rangle$ will not coincide with the final scalar-particle vacuum state $|0_+\rangle$ because there is particle creation. One can express $|0_-\rangle$ as a linear combination of states with different numbers of particles going out to infinity and into the horizon:

$$|0_-\rangle = \sum\sum \lambda_{AB}|A_I\rangle|B_H\rangle, \qquad (3.1)$$

where $|A_I\rangle$ is the outgoing state with n_{ja} particles in the jth outgoing mode and $|B_H\rangle$ is the horizon

state with n_{kb} particles in the kth mode going into the hole. In other words,

$$|A_I\rangle = \prod_j (n_{ja}!)^{-1/2}(b_j^\dagger)^{n_{ja}}|0_I\rangle, \qquad (3.2)$$

$$|B_H\rangle = \prod_k (n_{kb}!)^{-1/2}(c_k^\dagger)^{n_{kb}}|0_H\rangle. \qquad (3.3)$$

An operator Q which corresponds to an observable at future infinity will be composed only of the $\{b_j\}$ and the $\{b_j^\dagger\}$ and will operate only on the vectors $|A_I\rangle$. Thus the expectation value of this operator will be

$$\langle 0_-|Q|0_-\rangle = \sum\sum \rho_{AC}Q_{CA} \qquad (3.4)$$

where $Q_{CA} = \langle C_I|Q|A_I\rangle$ in the matrix element of the operator Q on the Hilbert space of outgoing states and $\rho_{AC} = \sum\lambda_{AB}\bar\lambda_{CB}$ is the density matrix which completely describes all observations which are made only at future infinity and do not measure what went into the hole. The components of ρ_{AC} can be completely determined from the expectation values of polynomials in the operators $\{b_j\}$ and $\{b_j^\dagger\}$. Thus the density matrix is independent of the ambiguity in the choice of the $\{q_j\}$ which describes particles going into the hole.

As an example of such a polynomial consider $b_j^\dagger b_j$, which is the number operator for the jth outgoing mode. Then

$$\langle n_j\rangle = \sum n_{ja}\rho_{AA}$$

$$= \langle 0_-|b_j^\dagger b_j|0_-\rangle$$

$$= \sum_k |\beta_{jk}|^2. \qquad (3.5)$$

In order to calculate this last expression one expands the finite-normalization wave-packet mode p_j in terms of continuum-normalization modes p_ω,

$$p_j(u) = \int^\infty \tilde p_j(\omega)p_\omega(u)d\omega, \qquad (3.6)$$

where

$$\int^\infty \bar{\tilde p}_j\tilde p_l d\omega = \delta_{jl}, \qquad (3.7)$$

then

$$\langle n_j\rangle = \iiint \bar{\tilde p}_j(\omega_1)\tilde p_j(\omega_2),$$

$$\times\beta_{\omega_1\omega'}\bar\beta_{\omega_2\omega'}d\omega_1 d\omega_2 d\omega'.$$

If the wave packet is sharply peaked around frequency ω, one can use Eq. (2.14) to show that

$$\int \beta_{\omega_1 \omega'} \bar{\beta}_{\omega_2 \omega'} d\omega' = (2\pi)^{-2} |t_\omega|^2 |\Gamma(1 - i\omega\kappa^{-1})|^2$$

$$\times e^{i(\omega_1 - \omega_2)v_0 \omega^{-1}} e^{-\pi\omega\kappa^{-1}}$$

$$\times \int_{-\infty}^{\infty} e^{iy\kappa^{-1}(\omega_1 - \omega_2)} dy, \qquad (3.8)$$

where $y = \ln(-\omega')$. The factor $e^{-\pi\omega\kappa^{-1}}$ arises from the analytic continuation of ω' to negative values in the expression (2.15) for $\beta_{\omega\omega'}$,

$$\text{Eq. (3.8)} = |t_\omega|^2 (e^{2\pi\omega\kappa^{-1}} - 1)^{-1} \delta(\omega_1 - \omega_2), \qquad (3.9)$$

therefore

$$\langle n_j \rangle = |t_\omega|^2 (e^{2\pi\omega\kappa^{-1}} - 1)^{-1}. \qquad (3.10)$$

This is precisely the expectation value for a body emitting thermal radiation with a temperature $T = \kappa/2\pi$. To show that the probabilities of emitting different numbers of particles in the jth mode and not just the average number are in agreement with thermal radiation, one can calculate the expectation values of n_j^2, n_j^3, and so on. For example,

$$\langle n_j^2 \rangle = \langle 0_- | b_j^\dagger b_j b_j^\dagger b_j | 0_- \rangle$$

$$= \langle n_j \rangle + \langle 0_- | (b_j^\dagger)^2 (b_j)^2 | 0_- \rangle. \qquad (3.11)$$

One can evaluate the second term on the right-hand side of (3.11) using Eqs. (2.14) and (2.15) as above. The terms $\alpha_{\omega\omega'}^{(2)}$ give rise to expressions involving functions like $\delta(\omega_1 + \omega_2)$ which do not contribute, since ω_1 and ω_2 are both positive. The terms in $\alpha_{\omega\omega'}^{(1)}$ give rise to expressions involving functions like $\int \bar{p}_j^2(\omega) d\omega$ which vanish because for wave packets at late times the phase of $\bar{p}_j(\omega)$ varies very rapidly with ω. Thus,

$$\langle n_j^2 \rangle = \frac{x\Gamma[1 + (2\Gamma - 1)x]}{(1 - x)^2}, \qquad (3.12)$$

where $x = e^{-\omega T^{-1}}$ and $\Gamma = |t_\omega|^2$. Proceeding inductively one can calculate the higher moments $\langle n_j^3 \rangle$, etc. These are all consistent with the probability distribution for n particles in the jth mode,

$$P(n_j) = \frac{(1 - x)(x\Gamma)^n}{[1 - (1 - \Gamma)x]^{n+1}}. \qquad (3.13)$$

This is just the combination of the thermal probability $(1 - x)x^m$ to emit m particles in the given mode with the probability Γ that a given emitted particle will escape to infinity and not be reflected back into the hole by the potential barrier.

One can also investigate whether there is any correlation between the phases for emitting different numbers of particles in the same mode by examining the expectation values of operators like $b_j b_j$ which connect components of the density matrix with different numbers of particles in the jth mode. These expectation values are all zero. To see whether there are any correlations between different modes one can consider the expectation values of operators like $b_j^\dagger b_l$ which relate to other nondiagonal components of the density matrix. These are also all zero. Thus the density matrix is completely diagonal in a basis of states with definite particle numbers in modes which are sharply peaked in frequency. One can express the density matrix explicitly as

$$\rho_{AC} = \prod_j \delta_{n_{ja} n_{jc}} P(n_{ja}). \qquad (3.14)$$

The density matrix (3.14) is exactly what one would expect for a body emitting thermal radiation.

As the black hole emits radiation its mass will go down and its temperature will go up. This variation will be slow except when the mass of the black hole has gone down to nearly the Planck mass. Thus to a good approximation the probability of n_j particles being emitted in the jth wave-packet mode will be given by Eq. (3.13) where the temperature corresponds to the mass of the black hole at the retarded time around which the jth mode is peaked. After the black hole has completely evaporated and disappeared, the only possible states $|A_I\rangle$ for the radiation at future infinity will be those for which the total energy of the particles is equal to the initial mass M_0 of the black hole. The probability of such a state occurring will be

$$P(A) = \rho_{AA}$$

$$= \prod_j P(n_{ja}). \qquad (3.15)$$

If Γ were 1 for all modes,

$$\ln[P(A)] = \sum_j \ln(1 - x_{ja}) - \sum 8\pi n_{ja} \omega_j M_{ja}, \qquad (3.16)$$

where M_{ja} is the mass to which the black hole has been reduced by the retarded time of the jth mode by emission of particles in configuration A. By conservation of energy $\sum n_{ja} \omega_j = M_0$ for all possible configurations A of the emitted particles. Because M_{ja} is only a slowly varying function of the mode number j, the last term in Eq. (3.16) will be nearly the same for all configurations A. Thus the black hole emits all configurations with equal probability. The probabilities of different configurations at future infinity are not equal because the Γ factors are different for different modes.

IV. THE INGOING PARTICLES

In this section a specific choice will be made of the ingoing solutions $\{q_i\}$ which will allow an explicit calculation to be made of the coefficients λ_{AB} so that the state of the system can be expressed in terms of particles falling into the black hole and particles escaping to infinity. The outgoing solutions $\{p_j\}$ are chosen to be purely positive frequency along the orbits of the approximate time-translation Killing vector K in the quasistationary region outside the black hole at late times. They therefore, correspond to particle modes that would be measured by an observer with a detector moving along a world-line at constant distance from the black hole. They do not correspond to what would be detected by nonstationary observers, in particular observers falling into the black hole, because they are not purely positive frequency along the world lines of such observers.

A stationary observer outside the black hole could regard a particle he detected in a mode $\{p_j\}$ as being one member of a pair of particles created by the gravitational yield of the collapse, the other member having negative energy and having fallen into the black hole. The horizon states $\{q_j\}$ will be chosen so that some of them describe those negative-energy particles which the stationary observer considers to exist inside the black hole. The remaining $\{q_j\}$ will describe those positive-energy particles which are reflected back by the potential barrier around the black hole and which fall through the event horizon. It should be emphasized that this choice of $\{q_j\}$ does not correspond to anything that an infalling observer would measure since they are not positive frequency along his world line. However, given the $\{p_j\}$, the choice of the $\{q_j\}$ that will be used is minimal in the sense that any other choice would describe the creation of extra pairs of particles, both of which fell into the black hole.

To calculate the coefficients α and β which relate the $\{p_i\}$ to the $\{f_i\}$ and $\{\bar{f}_i\}$ one decomposes the $\{p_i\}$ into Fourier components $\{p_\omega\}$ with time dependence of the form $e^{i\omega u}$, where $u = t - r - 2M \ln(r - 2M)$ is the retarded time coordinate in the Schwarzschild solution. Because u tends to $+\infty$ in the exterior region as one approaches the future horizon, the surfaces of constant phase of p_ω pile up just outside the future horizon (Fig. 2). In other words, p_ω is blue-shifted to a very high frequency near the future horizon. This means that it propagates by geometric optics back through the collapsing body and out to past null infinity $\mathcal{I}^-$ where it has time dependence of the form

$$e^{-i\omega\kappa^{-1}\ln(v_0-v)} \quad \text{for} \quad v < v_0$$

and   (4.1)

$$0 \quad \text{for} \quad v > v_0,$$

where $v = t + r + 2M \ln(r - 2M)$ is the advanced time

coordinate and v_0 is the last advanced time before which a null geodesic could leave $\mathcal{I}^-$, pass through the center of the collapsing object, and escape to $\mathcal{I}^+$. Similarly, to calculate the coefficients γ and η which express the $\{q_i\}$ in terms of the $\{f_i\}$ and the $\{\bar{f}_i\}$ one decomposes the $\{q_i\}$ into Fourier components $\{q_\omega\}$. In the quasistationary region the part $\{q_\omega^{(3)}\}$ that crosses the future horizon in the quasistationary region will have time dependence of the form $e^{i\omega v}$. The part $q_\omega^{(4)}$ which crosses the horizon just after its formation will have time dependence of the form $e^{-i\omega u}$ (the minus sign is because in the interior region the direction of increase of u is reversed). The surfaces of constant phase of $\{q_\omega^{(4)}\}$ pile up just inside the horizon (Fig. 2). One can therefore propagate them backwards also by geometric optics through the collapsing body and out to $\mathcal{I}^-$, where they will have time dependence of the form

$$e^{i\omega\kappa^{-1}\ln(v-v_0)} \quad \text{for} \quad v > v_0$$

and   (4.2)

$$0 \quad \text{for} \quad v < v_0.$$

In order to calculate the coefficients α, β, γ, and η one can decompose (4.1) and (4.2) into positive- and negative-frequency components of the form $e^{i\omega v}$ and $e^{-i\omega v}$ in terms of the advanced time v at $\mathcal{I}^-$. However, one can obtain the same results

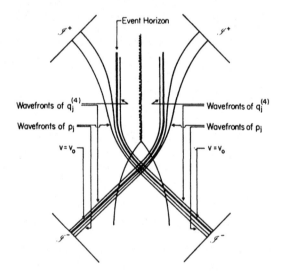

FIG. 2. The wave fronts or surfaces of constant phase of the solutions p_j pile up just outside the event horizon because of the large blue-shift. They propagate by geometric optics through the collapsing body and out to past null infinity $\mathcal{I}^-$ just before the advanced time $v = v_0$. Similarly the wave fronts of $q_j^{(4)}$ will pile up just inside the horizon and will propagate through the collapsing body out to $\mathcal{I}^-$ just after the advanced time $v = v_0$.

if one leaves out the collapsing body and analytically extends back to the past horizon the Schwarzschild solution that represents the quasistationary region. Instead of propagating p_ω and q_ω back through the collapsing body to $\mathscr{I}^-$ and analyzing them there into positive- and negative-frequency components with respect to the advanced time v, one propagates them back to the past horizon H^- and analyzes them into positive- and negative-frequency components with respect to an affine parameter U along the generators of H^-. (A similar construction has been used by Unruh.[21]) One can then discuss the creation of particles in terms of the Penrose diagram (Fig. 3) of the analytically extended Schwarzschild solution. The initial vacuum state $|0_-\rangle$ is now defined as the state which on $\mathscr{I}^-$ has no positive-frequency components with respect to the advanced time v and which on the past horizon H^- has no positive-frequency components with respect to affine parameter U. In other words, one can express the operator ϕ in the form

$$\phi = \int_0^\infty (a_\omega^{(1)} f_\omega^{(1)} + a_\omega^{(2)} f_\omega^{(2)} + \text{H.c.})d\omega, \qquad (4.3)$$

where $\{f_\omega^{(1)}\}$ are a family of solutions of the wave equation in the analytically extended Schwarzschild solution with continuum normalization which have zero Cauchy data on the past horizon and have time dependence of the form $e^{i\omega v}$ on $\mathscr{I}^-$, and $\{f_\omega^{(2)}\}$ are a family of solutions with continuum normalization which have zero Cauchy data on $\mathscr{I}^-$ and have time dependence of the form $e^{i\omega U}$ on the past horizon. The initial vacuum state is then defined by

$$a_\omega^{(1)}|0_-\rangle = a_\omega^{(2)}|0_-\rangle = 0. \qquad (4.4)$$

This definition of the vacuum state is different from that used by Boulware[22] for the analytically extended Schwarzschild solution. The above definition, however, reproduces the results on particle creation by a black hole which was formed by a collapse.

The affine parameter U on the past horizon is related to the retarded time u by

$$u = -\kappa^{-1}\ln(-U), \qquad (4.5)$$

where $-\infty < u < \infty$, $U < 0$. One can analytically continue (4.5) past the logarithmetic singularity at $U = 0$. In doing so, one picks up an imaginary part of $\pm\pi\kappa^{-1}$ depending on whether one passes above or below the singularity, respectively. Define the two analytic continuations u_+ and u_- by

$$u_+ = u_- = -\kappa^{-1}\ln(-U) \quad \text{for } U < 0,$$
$$u_\pm = -\kappa^{-1}\ln U \pm i\pi\kappa^{-1} \quad \text{for } U > 0. \qquad (4.6)$$

Because u_+ is holomorphic in the upper half U plane, the functions $e^{i\omega u_+}$ and $e^{-i\omega u_+}$ defined all the way up the past horizon from $U = -\infty$ to $U = +\infty$ both contain only positive frequencies with respect to U. This means that one can replace the family of solutions $\{f_\omega^{(2)}\}$, which have zero Cauchy data on $\mathscr{I}^-$ and only positive frequencies with respect to U on the past horizon, by two orthogonal families of solutions $\{f_\omega^{(3)}\}$ and $\{f_\omega^{(4)}\}$, with continuum normalization which have zero Cauchy data on $\mathscr{I}^-$, and which have time dependence on the past horizon of the form $e^{i\omega u_+}$ and $e^{-i\omega u_+}$, respectively. One can then express ϕ as

$$\phi = \int (a_\omega^{(1)} f_\omega^{(1)} + a_\omega^{(3)} f_\omega^{(3)} + a_\omega^{(4)} f_\omega^{(4)} + \text{H.c.})d\omega.$$

$$(4.7)$$

Equation (4.4) then becomes

$$a_\omega^{(1)}|0_-\rangle = a_\omega^{(3)}|0_-\rangle = a_\omega^{(4)}|0_-\rangle = 0. \qquad (4.8)$$

Equation (4.8) says that there are no scalar particles in the modes $\{f_\omega^{(3)}\}$ and $\{f_\omega^{(4)}\}$. However, these modes extend across both the interior and exterior regions of the analytically continued Schwarzschild solution. An observer at future null infinity $\mathscr{I}^+$ cannot measure these modes but only the part of them outside the future horizon. To correspond with what an observer sees, define a new basis consisting of three orthogonal families $\{w_\omega\}$, $\{y_\omega\}$, and $\{z_\omega\}$ of solutions with continuum normalization with the following properties:

$\{w_\omega\}$ have zero Cauchy data on $\mathscr{I}^-$ and on the past horizon for $U < 0$. On the past horizon for $U > 0$ they have time dependence of the form $e^{-i\omega u_+}$. (The minus sign is necessary in order for the $\{w_\omega\}$ to have positive Klein-Gordon norm and thus for the associated annihilation and creation operators to have the right commutation relations.)

$\{y_\omega\}$ have zero Cauchy data on $\mathscr{I}^-$ and the past horizon for $U > 0$. On the past horizon for $U < 0$ they have time dependence of the form $e^{i\omega u_+}$.

$\{z_\omega\}$ have zero Cauchy data on the past horizon and on $\mathscr{I}^-$ they have time dependence of the form $e^{i\omega v}$.

The modes $\{z_\omega\}$ represent particles which come in from $\mathscr{I}^-$ and pass through the future horizon with probability $|t_\omega|^2$ or are reflected back to $\mathscr{I}^+$ with probability $|r_\omega|^2$. The modes $\{y_\omega\}$ represent particles which, in the analytically extended Schwarzschild space, appear to come from the past horizon and which escape to $\mathscr{I}^+$ with probability $|t_\omega|^2$ or are reflected back to the future horizon with probability $|r_\omega|^2$. In the spacetime which includes the collapsing body, the outgoing and incoming solutions $\{p_\omega\}$ and $\{q_\omega\}$ in the quasistationary region outside the horizon correspond to linear combinations of the $\{y_\omega\}$ and the $\{z_\omega\}$:

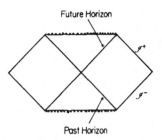

FIG. 3. The Penrose diagram of the r-t plane of the analytically extended Schwarzschild space. Null lines are at ±45° and a conformal transformation has been made to bring infinity, represented by $\mathcal{J}^+$ and $\mathcal{J}^-$, to a finite distance. Each point in this diagram represents a sphere of area $4\pi r^2$.

$$p_\omega = t_\omega y_\omega + r_\omega z_\omega,$$
$$q_\omega = \overline{r}_\omega y_\omega - \overline{t}_\omega z_\omega. \qquad (4.9)$$

The modes $\{w_\omega\}$ represent particles which, in the analytically extended Schwarzschild space, are always inside the future horizon and which do not enter the exterior region. In the real space-time with the collapsing body they correspond to particles which cross the event horizon just after its formation.

The modes $\{z_\omega\}$ have the same Cauchy data as the $\{f_\omega^{(1)}\}$, therefore they are the same everywhere, i.e.,

$$z_\omega = f_\omega^{(1)} \qquad (4.10)$$

on the past horizon for $U < 0$,

$$y_\omega = (1-x)^{1/2} f_\omega^{(3)}$$
$$= x^{-1/2}(1-x)^{1/2} \overline{f}_\omega^{(4)}, \qquad (4.11)$$

where $x = e^{-2\pi\omega\kappa^{-1}}$. The factors $(1-x)^{1/2}$ and $x^{-1/2}(1-x)^{1/2}$ appear because of the normalization. On the past horizon for $U > 0$

$$w_\omega = x^{-1/2}(1-x)^{1/2} \overline{f}_\omega^{(3)}$$
$$= (1-x)^{1/2} f_\omega^{(4)}. \qquad (4.12)$$

This implies that $(1-x)^{-1/2}(y_\omega + x^{1/2}\overline{w}_\omega)$ has the same Cauchy data as $f_\omega^{(3)}$ and therefore is the same everywhere, i.e.,

$$\overline{f}_\omega^{(3)} = (1-x)^{-1/2}(y_\omega + x^{1/2}\overline{w}_\omega). \qquad (4.13)$$

Similarly,

$$f_\omega^{(4)} = (1-x)^{-1/2}(w_\omega + x^{1/2}\overline{y}_\omega). \qquad (4.14)$$

One can express the operator ϕ in terms of the basis $\{w_\omega, y_\omega, z_\omega\}$:

$$\phi = \int (g_\omega w_\omega + h_\omega y_\omega + j_\omega z_\omega + \text{H.c.})d\omega, \qquad (4.15)$$

where the $\{g_\omega\}$ and the $\{g_\omega^\dagger\}$, etc., are the annihilation and creation operators for particles in the modes $\{w_\omega\}$, etc. Comparing (4.15) with (4.7) and using (4.13) and (4.14) one sees that

$$a_\omega^{(3)} = (1-x)^{-1/2}(h_\omega - x^{1/2} g_\omega^\dagger),$$
$$a_\omega^{(4)} = (1-x)^{-1/2}(g_\omega - x^{1/2} h_\omega^\dagger), \qquad (4.16)$$
$$a_\omega^{(1)} = j_\omega.$$

One can superimpose the continuum-normalization solutions $\{f_\omega^{(1)}\}$, etc., $\{w_\omega\}$, etc. to form families of orthonormal wave-packet solutions $\{f_j^{(1)}\}$, $\{f_j^{(3)}\}$, $\{f_j^{(4)}\}$, $\{w_j\}$, $\{y_j\}$, $\{z_j\}$. If the wave packets are sharply peaked around frequency ω, the corresponding operators $a_j^{(1)}$, etc., g_j, etc. will be related by Eq. (4.16), where the suffix ω is replaced by j and modes with the same suffix j are taken to be made up from continuum modes in the same way, i.e., they have the same Fourier transforms.

One can define a future vacuum state $|0_+\rangle$ by

$$g_j |0_+\rangle = h_j |0_+\rangle = j_j |0_+\rangle = 0. \qquad (4.17)$$

One can then define states $|A; B; C\rangle$ which contain n_{1a} particles in the mode w_1, n_{2a} particles in the mode w_2, etc., n_{1b} particles in the mode y_1, etc., and n_{1c} particles in the mode z_1, etc. by

$$|A; B; C\rangle = \left[\prod (n_{ja}!)^{-1/2}(g_j^\dagger)^{n_{ja}}\right]$$
$$\times \left[\prod (n_{jb}!)^{-1/2}(h_j^\dagger)^{n_{jb}}\right]$$
$$\times \left[\prod (n_{jc}!)^{-1/2}(j_j^\dagger)^{n_{jc}}\right] |0_+\rangle. \qquad (4.18)$$

The initial vacuum state $|0_-\rangle$ can be expressed as a linear combination of these states:

$$|0_-\rangle = \sum \mu(A; B; C)|A; B; C\rangle. \qquad (4.19)$$

The coefficients $\mu(A; B; C)$ may be found by using Eqs. (4.8) and (4.16) which give

$$(g_k - x^{1/2} h_k^\dagger)|0_-\rangle = 0, \qquad (4.20)$$

$$(h_k - x^{1/2} g_k^\dagger)|0_-\rangle = 0, \qquad (4.21)$$

$$j_k |0_-\rangle = 0. \qquad (4.22)$$

Equation (4.22) implies that the coefficients μ will be nonzero only for states with no particles in the $\{z_j\}$ modes, i.e., states for which $n_{jc} = 0$ for all j. Equation (4.20) connects the coefficients μ for states with m particles in the w_k mode and s particles in the y_k mode, with the coefficients μ for states with $m-1$ particles in the w_k mode and $s-1$ particles in the y_k mode, i.e.,

$$(m_k)^{1/2} \mu(A[m_k]; B[s_k]; 0)$$

$$- x^{1/2}(s_k)^{1/2} \mu(A[m - 1_k]; B[(s - 1)_k]; 0) = 0,$$

$$(4.23)$$

where $\mu(A[m_k]; B[s_k]; 0)$ is the coefficient for the state $\{n_{1a}, n_{2a}, \ldots; n_{1b}, n_{2b}, \ldots; 0\}$, where $n_{ka} = m$ and $n_{kb} = s$. By induction on (4.23) one sees that

$$\mu(A[m_k]; B[s_k]; 0) = \delta_{ms} x^{m/2} \mu(A[0_k]; B[0_k]; 0).$$

$$(4.24)$$

In other words, if one compares states with the same numbers of particles in all modes except the w_k mode and the y_k mode, the relative probabilities of having m and s particles, respectively, in these modes is zero unless $m = s$, in which case it is proportional to x^m. One can interpret this as saying that the particles are created in pairs in the corresponding w and y modes. The particle in the w mode enters the black hole shortly after its formation. The particle in the y mode is emitted from the black hole and will escape to infinity with probability $|t_\omega|^2$ or be reflected back into the black hole with probability $|r_\omega|^2$. The relative probabilities of different numbers of particles being emitted in the y modes correspond exactly to the probability distribution for thermal radiation.

By applying (4.24) to each value of k one obtains

$$\mu(A; B; 0) = \exp\left(-\pi\kappa^{-1} \sum n_{ja}\omega_j\right) \mu(0; 0; 0)$$

$$(4.25)$$

if $\{n_{1a}, n_{2a}, \ldots\} = \{n_{1b}, n_{2b}, \ldots\}$, $\mu(A; B; 0) = 0$ otherwise. Strictly speaking, $\mu(0; 0; 0)$ is zero because in the approximation that has been used the back reaction of the created particles has been ignored and the space-time has been represented by a Schwarzschild solution of constant mass. This means that the black hole goes on emitting at a steady rate for an infinite time and therefore the probability of emitting any given finite number of particles is vanishingly small. However, if one considers the emission only over some finite period of time in which the mass of the black hole does not change significantly, Eq. (4.25) gives the correct relative probabilities of emitting different configurations of particles. Again one sees that the probabilities of emitting all configurations with some given energy are equal.

If one puts in the angular dependence Y_{lm} of the modes, one finds that because (4.13) and (4.14) connect w_ω and $\bar{y}_\omega$, they connect modes with the opposite angular momenta, (l, m) and $(l, -m)$. This means that the particles are created in pairs in the w and y modes with opposite angular momenta. Because the w modes have time dependence

of the form $e^{-i\omega u}$ while the y modes have time dependence of the form $e^{i\omega u}$, there is also a sense in which they have opposite signs of energy: The y particles have positive energy and can escape to infinity while the w particles have negative energy and reduce the mass of the black hole.

The particle creation that is observed at infinity comes about because an observer at infinity divides the modes of the scalar field in a manner which is discontinuous at the event horizon and loses all information about modes inside the horizon. An observer who was falling into the black hole would not make such a discontinuous division. Instead, he would analyze the field into modes which were continuous against the event horizon. When propagated back to the past horizon, these modes would merely be blue-shifted by some constant factor and therefore would still be purely positive frequency with respect to the affine parameter U on the past horizon. Thus the observer falling into the black hole would not see any created particles.

V. THE SUPERSCATTERING OPERATOR $

It was shown in Sec. III that observations at future infinity had to be described in terms of a density operator or matrix rather than a pure quantum state. The reason for this was that part of the information about the quantum state of the system was lost down the black hole. One might think that this information might reemerge during the final stages of the evaporation and disappearance of the black hole so that what one would be left with at future infinity would be a pure quantum state after all. However, this cannot be the case; there must be nonconservation of information in black-hole formation and evaporation just as there must be a nonconservation of baryon number. A large black hole formed by the collapse of a star consisting mainly of baryons will have a very low temperature. It will therefore emit most of its rest-mass energy in the form of particles of zero rest mass. By the time it becomes hot enough to emit baryons it will have lost all but a small fraction of its original mass and there will be insufficient energy available to emit the number of baryons that went into forming the black hole. Thus, if the black hole disappears completely, there will be nonconservation of baryon number. The situation with regard to information nonconservation is similar. The black hole is formed by the collapse of some well-ordered body with low entropy. During the quasistationary emission phase the black hole sends out random thermal radiation with a large amount of entropy. In order to end up in a pure quantum state the black hole would have to emit a similar amount of negative entropy or information in the final stages of the evaporation. However, information like baryon number requires energy and there is simply not enough energy available in the final stages of the evaporation. To carry the large amount of information needed would require the emission in the final stages of about the same number of particles as had already been emitted in the quasistationary phase.

Because one ends up with a density operator rather than pure quantum space, the process of black-hole formation and evaporation cannot be described by an S matrix. In general, the initial situation will not be a pure quantum state either because of the evaporation of black holes at earlier times. What one has therefore is an operator, which will be called the superscattering operator $\mathcal{S}$, that maps density operators describing the initial situation to density operators describing the final situation. By the superposition principle this mapping must be linear. Thus if one regards the initial and final density operators ρ_1 and ρ_2 as second-rank tensors or matrices ρ_{1AB} and ρ_{2CD} on the initial and final Hilbert spaces, respectively, the superscattering operator will be a 4-index tensor $\mathcal{S}_{ABCD}$ such that

$$\rho_{2CD} = \sum \sum \mathcal{S}_{CDAB}\rho_{1AB}. \tag{5.1}$$

When the initial situation is a pure quantum state ξ_A the initial density operator will be

$$\rho_{1AB} = \xi_A \overline{\xi}_B. \tag{5.2}$$

If the initial state is such as to have a very small probability of forming a black hole, the final situation will also be a pure quantum state ζ_C which is related to the initial state by the S matrix:

$$\zeta_C = \sum S_{CA}\xi_A. \tag{5.3}$$

The final density operator will be

$$\rho_{2CD} = \zeta_C \overline{\zeta}_D. \tag{5.4}$$

Thus the components of the $\mathcal{S}$ operator on these states can be expressed as the product of two S matrices:

$$\mathcal{S}_{CDAB} = \tfrac{1}{2}(S_{CA}S_{BD}^{-1} + S_{AD}^{-1}S_{CB}). \tag{5.5}$$

However, for initial states that have a significant probability of forming a black hole, there is no S matrix and so one cannot represent $\mathcal{S}$ in the form (5.5).

Consider, for example, the scattering of two gravitons. In this case the initial situation is a pure quantum state and, if the energy is low, the final situation will be also a nearly pure state. This can be recognized by computing the entropy of the final situation which can be defined as

$$S_2 = -\sum \sum \rho_{2CD}\ln(\rho_{2CD}). \tag{5.6}$$

In this expression the logarithm is to be understood as the inverse of the exponential of a matrix. It can be computed by transforming to a basis in which ρ_{2CD} is diagonal. For energies for which there is a low probability of forming a black hole, the entropy S_2 will be nearly zero. However, as the center-of-mass energy of the gravitons is increased to the Planck mass, there will be a significant probability of a black hole forming and evaporating and the entropy S_2 will be nonzero.

The tensor $\mathcal{S}_{CDAB}$ is Hermitian in the first and second pairs of indices. Any density matrix has unit trace because, in a basis in which it is diagonal, the diagonal entries are the probabilities of being in the different states of the basis. Since ρ_{2CD} must have unit trace for any initial density matrix ρ_{1AB},

$$\sum \mathcal{S}_{CCAB} = \delta_{AB}. \tag{5.7}$$

One can regard this as saying that, starting from any initial state, the probabilities of ending up in different final states must sum to unity. The corresponding relation

$$\sum \mathcal{S}_{CDAA} = \delta_{CD} \tag{5.8}$$

would imply that for any given final state, the probabilities of it arising from different initial states should sum to unity. Two arguments will be given for Eq. (5.8). The first is a thermodynamic argument based on the impossibility of constructing perpetual-motion machines. The second is based on CPT invariance.

Because the mass measured from infinity is conserved, the superscattering operator $\mathcal{S}$ will connect only initial and final states with the same energy. Thus (5.7) will hold when the initial and final state indices are restricted to states with some given energy E. Similarly, if (5.8) holds, it should also hold when restricted to initial and final states of energy E. For convenience, in order to make the number of states finite, consider states between energy E and $E + \Delta E$ contained in a very large box with perfectly reflecting walls. Define ψ_{CD} to be $\sum \mathcal{S}_{CDAA}$, where the summation is over the finite number of states specified above. Suppose that

$$\psi_{CD} \neq \delta_{CD}. \tag{5.9}$$

By (5.7) restricted to the same states, $\sum \psi_{CC} = N$, where N is the number of states. By transforming to a basis in which ψ_{CD} is diagonal, one can see that (5.9) would imply that there was some state ξ_C such that

$$\sum \sum \psi_{CD} \xi_C \overline{\xi}_D = \sum \sum \sum \mathcal{S}_{CDAA} \xi_C \overline{\xi}_D > 1.$$

(5.10)

This would imply that the sum of the probabilities of arriving at the final state ξ_C from all the different possible initial states was greater than unity. If one now left the energy E in the box for a very long time, the system would evolve to various different configurations. For most of the time the box would contain particles in approximately thermal distribution. Occasionally, a large number of particles would get together in a small region and would create a black hole which would then evaporate again. To a good approximation one could regard the time development of the density matrix of the system as being given by successive applications of the $\mathcal{S}$ operator restricted to the finite number of states. On the normal assumptions of thermal equilibrium and ergodicity one would expect that after a long time the probability of finding the system in any given state would be N^{-1} and the entropy would be $\ln N$. However, if (5.10) held, the probability of the system being in the state ξ_C would be greater than N^{-1} and so the entropy would be less than $\ln N$. One could therefore extract useful energy and run a perpetual-motion machine by periodically allowing the system to relax to entropy $\ln N$. If one assumes that this is impossible, (5.8) must hold.

The second argument for Eq. (5.8) is based on CPT invariance. Because the Einstein equations are separately invariant under C, P, and T, pure quantum gravity will also be invariant under these operations if the boundary conditions at hidden surfaces are similarly invariant. The matter fields are not necessarily locally invariant under C, P, and T separately, but they are locally invariant under CPT because their Lagrangian density is a scalar under local proper Lorentz transformations. Thus the quantum theory of coupled gravitational and matter fields will be invariant under CPT provided that the boundary conditions at hidden surfaces are invariant under CPT. That the boundary conditions at hidden surfaces should be invariant under CPT would seem a very reasonable assumption. In fact, the assumption of CPT for quantum gravity and the assumption that one cannot build a perpetual-motion machine are equivalent in that each of them implies the other. With CPT invariance, Eq. (5.8) follows from (5.7). Because black holes can form when there was no black hole present beforehand, CPT implies that they must also be able to evaporate completely; they cannot stabilize at the Planck mass, as has been suggested by some authors. CPT invariance also implies that for an observer at infinity there is no operational distinction between a black hole and a white hole: The formation and evaporation of a black hole can be regarded equally well in the reverse direction of time as the formation and evaporation of a white hole.[23] An observer who falls into a hole will always think that it is a black hole but he will not be able to communicate his measurements to an observer at infinity.

ACKNOWLEDGMENTS

The author is very grateful for helpful discussions with a number of colleagues, in particular, B. J. Carr, W. Israel, D. N. Page, R. Penrose, and R. M. Wald.

*Work supported in part by the National Science Foundation under Grant No. MPS75-01398 at the California Institute of Technology.

† Sherman Fairchild Distinguished Scholar at the California Institute of Technology.

[1] P. J. Roll, R. Krotkov, and R. H. Dicke, Ann. Phys. (N.Y.) 26, 442 (1964).

[2] V. B. Braginsky and V. I. Panov, Zh. Eksp. Teor. Fiz. 61, 873 (1971) [Sov. Phys.—JETP 34, 464 (1972)].

[3] R. Penrose, Phys. Rev. Lett. 14, 57 (1965).

[4] S. W. Hawking, Proc. R. Soc. London A294, 511 (1966).

[5] S. W. Hawking, Proc. R. Soc. London A295, 460 (1966).

[6] S. W. Hawking, Proc. R. Soc. London A300, 187 (1967).

[7] S. W. Hawking and R. Penrose, Proc. R. Soc. London A314, 529 (1970).

[8] E. M. Lifshitz and I. M. Khalatnikov, Adv. Phys. 12, 185 (1963).

[9] E. M. Lifshitz and I. M. Khalatnikov, Phys. Rev. Lett. 24, 76 (1970).

[10] C. Brans and R. H. Dicke, Phys. Rev. 124, 925 (1961).

[11] A. Trautman, Colloques, CNRS Report No. 220, 1973 (unpublished).

[12] S. W. Hawking, Nature 248, 30 (1974).

[13] S. W. Hawking, Commun. Math. Phys. 43, 199 (1975).

[14] G. 't Hooft and M. Veltman, Ann. Inst. H. Poincaré 20A, 69 (1974).

[15] S. Deser and P. van Nieuwenhuizen, Phys. Rev. D 10, 401 (1974); 10, 411 (1974).

[16] R. M. Wald, Commun. Math. Phys. 45, 9 (1975).

[17] L. Parker, Phys. Rev. D 12, 1519 (1975).

[18] C. B. Collins and S. W. Hawking, Astrophys. J. 180, 317 (1973).

[19] R. Omnés, Phys. Rep. 3C, 1 (1972).

[20] Ya. B. Zel'dovich and A. Starobinsky, Zh. Eksp. Teor. Fiz. 61, 2161 (1971) [Sov. Phys.—JETP 34, 1159 (1972)].

[21] W. G. Unruh, Phys. Rev. D 14, 870 (1976).

[22] D. G. Boulware, Phys. Rev. D 12, 350 (1975).

[23] S. W. Hawking, Phys. Rev. D 13, 191 (1976).

VOLUME 69, NUMBER 3   PHYSICAL REVIEW LETTERS   20 JULY 1992

Evaporation of Two-Dimensional Black Holes

S. W. Hawking

California Institute of Technology, Pasadena, California 91125
and Department of Applied Mathematics and Theoretical Physics, University of Cambridge,
Silver Street, Cambridge CB3 9EW, United Kingdom
(Received 23 March 1992)

An interesting two-dimensional model theory has been proposed that allows one to consider black-hole evaporation in the semiclassical approximation. The semiclassical equations will give a singularity where the dilaton field reaches a certain critical value. This singularity will be hidden behind a horizon. As the evaporation proceeds, the dilaton field on the horizon will approach the critical value but the temperature and rate of emission will remain finite. These results indicate either that there is a naked singularity, or (more likely) that the semiclassical approximation breaks down.

PACS numbers: 97.60.Lf, 04.20.Cv, 04.60.+n

Callan, Giddings, Harvey, and Strominger (CGHS) [1] have suggested an interesting two-dimensional theory with a metric coupled to a dilaton field and N minimal scalar fields. The Lagrangian is

$$L = \frac{1}{2\pi}\sqrt{-g}\left[e^{-2\phi}[R + 4(\nabla\phi)^2 + 4\lambda^2] - \frac{1}{2}\sum_{i=1}^{N}(\nabla f_i)^2 \right].$$

If one writes the metric in the form

$$ds^2 = e^{2\rho}dx_+ dx_- ,$$

the classical field equations are

$$\partial_+\partial_- f_i = 0 ,$$

$$2\partial_+\partial_-\phi - 2\partial_+\phi\,\partial_-\phi - \tfrac{1}{2}\lambda^2 e^{2\rho} = \partial_+\partial_-\rho ,$$

$$\partial_+\partial_-\phi - 2\partial_+\phi\,\partial_-\phi - \tfrac{1}{2}\lambda^2 e^{2\rho} = 0 .$$

These equations have a solution

$$\phi = -b\ln(-x_+ x_-) - c - \ln\lambda ,$$

$$\rho = -\tfrac{1}{2}\ln(-x_+ x_-) + \ln(2b/\lambda) ,$$

where b and c are constants and b can be taken to be positive without loss of generality. A change of coordinates

$$u\pm = \pm(2b/\lambda)\ln(\pm x_\pm) \pm (1/\lambda)(c + \ln\lambda)$$

gives a flat metric and a linear dilaton field

$$\rho = 0 ,$$

$$\phi = -\tfrac{1}{2}\lambda(u_+ - u_-) .$$

This solution is known as the linear dilaton. The solution is independent of the constants b and c which correspond to freedom in the choice of coordinates. Normally b is taken to have the value $\tfrac{1}{2}$.

These equations also admit a solution

$$\phi = \rho - c = -\tfrac{1}{2}\ln(M\lambda^{-1} - \lambda^2 e^{2c} x_+ x_-) .$$

This represents a two-dimensional black hole with horizons at $x_\pm = 0$ and singularities at $x_+ x_- = M\lambda^{-2}e^{-2c}$. Note that there is still freedom to shift the ρ field on the

horizon by a constant and compensate by rescaling the coordinates $x_\pm$, but there is nothing corresponding to the freedom to choose the constant b. In terms of the coordinates $u_\pm$ defined as before with $b = \tfrac{1}{2}$,

$$\rho = -\tfrac{1}{2}\ln(1 - M\lambda^{-1}e^{-\lambda(u_+ - u_-)}) ,$$

$$\phi = -\tfrac{1}{2}\lambda(u_+ - u_-) - \tfrac{1}{2}\ln(1 - M\lambda^{-1}e^{-\lambda(u_+ - u_-)}) .$$

This black-hole solution is periodic in the imaginary time with period $2\pi\lambda^{-1}$. One would therefore expect it to have a temperature

$$T = \lambda/2\pi$$

and to emit thermal radiation [2]. This is confirmed by CGHS. They considered a black hole formed by sending in a thin shock wave of one of the f_i fields from the weak-coupling region (large negative ϕ) of the linear dilaton. One can calculate the energy-momentum tensors of the f_i fields, using the conservation and trace anomaly equations. If one imposes the boundary condition that there is no incoming energy momentum apart from the shock wave, one finds that at late retarded times u_- there is a steady flow of energy in each f_i field at the mass-independent rate

$$\lambda^2/48 .$$

If this radiation continued indefinitely, the black hole would radiate an infinite amount of energy, which seems absurd. One might therefore expect that the backreaction would modify the emission and cause it to stop when the black hole had radiated away its initial mass. A fully quantum treatment of the backreaction seems very difficult even in this two-dimensional theory. But CGHS suggested that in the limit of a large number N of scalar fields f_i, one could neglect the quantum fluctuations of the dilaton and the metric and treat the backreaction of the radiation in the f_i fields semiclassically by adding to the action a trace anomaly term

$$\tfrac{1}{12}N\partial_+\partial_-\rho .$$

The evolution equations that result from this action are

$$\partial_+\partial_-\phi = (1 - \tfrac{1}{24}Ne^{2\phi})\partial_+\partial_-\rho\,,$$

$$2(1 - \tfrac{1}{12}Ne^{2\phi})\partial_+\partial_-\phi = (1 - \tfrac{1}{24}Ne^{2\phi})$$

$$\times (4\partial_+\phi\,\partial_-\phi + \lambda^2 e^{2\rho})\,.$$

In addition, there are two equations that can be regarded as constraints on the data on characteristic surfaces of constant $x_\pm$,

$$\partial_+^2\phi - 2\partial_+\rho\,\partial_+\phi = \tfrac{1}{24}Ne^{2\phi}[\partial_+^2\rho - \partial_+\rho\,\partial_+\rho - t_+(x^+)]\,,$$

$$\partial_-^2\phi - 2\partial_-\rho\,\partial_-\phi = \tfrac{1}{24}Ne^{2\phi}[\partial_-^2\rho - \partial_-\rho\,\partial_-\rho - t_-(x^-)]\,,$$

where $t_\pm(x^\pm)$ are determined by the boundary conditions in a manner that will be explained later.

Even these semiclassical equations seem too difficult to solve in closed form. CGHS suggested that a black hole formed from an f wave would evaporate completely without there being any singularity. The solution would approach the linear dilaton at late retarded times u_- and there would be no horizons. They therefore claimed that there would be no loss of quantum coherence in the formation and evaporation of a two-dimensional black hole: The radiation would be in a pure quantum state, rather than in a mixed state.

In [3,4] it was shown that this scenario could not be correct. The solution would develop a singularity on the incoming f wave at the point where the dilaton field reached the critical value

$$\phi_0 = -\tfrac{1}{2}\ln(N/12)\,.$$

This singularity will be spacelike near the f wave [4]. Thus at least part of the final quantum state will end up on the singularity, which implies that the radiation at infinity in the weak-coupling region will not be in a pure quantum state.

The outstanding question is: How does the spacetime evolve to the future of the f wave? There seem to be two main possibilities: (1) The singularity remains hidden behind an event horizon. One can continue an infinite distance into the future on a line of constant $\phi < \phi_0$ without ever seeing the singularity. If this were the case, the rate of radiation would have to go to zero. (2) The singularity is naked. That is, it is visible from a line of constant ϕ at a finite time to the future of the f wave. Any evolution of the solution after this would not be uniquely determined by the semiclassical equations and the initial data. Indeed, it is likely that the point at which the singularity became visible was itself singular and that the solution could not be evolved to the future for more than a finite time.

In what follows I shall present evidence that suggests the semiclassical equations lead to possibility (2). This probably indicates that the semiclassical approximation breaks down as the dilaton field on the horizon approaches the critical value.

Static black holes.—If the solution were to evolve without a naked singularity, it would presumably approach a static state in which a singularity was hidden behind an event horizon. This motivates a study of static black-hole solutions of the semiclassical equations. One could look for solutions in which ϕ and ρ were independent of the "time" coordinate $\tau = x_+ + \bar{x}_-$ and depended only on a "radial" variable $\sigma = x_+ - x_-$ but this has the disadvantage that the Killing vector $\partial/\partial\tau$ is timelike everywhere. This means the black-hole horizon is at $\sigma = -\infty$. Instead it seems better to choose the Killing vector to be that corresponding to boosts in the background two-dimensional Minkowski space. Then the past and future horizons will be the null lines $x_\pm = 0$ intersecting at the origin. One can define a radial coordinate that is left invariant by the boost as

$$r^2 = -x_+x_-\,.$$

It is straightforward to verify that r is regular on a spacelike surface through the origin and has nonzero gradient there if one chooses the positive square root on one side of the intersection of the horizons at $r = 0$ and the negative root on the other. In the r coordinate the field equations for a static solution are

$$\phi'' + \frac{1}{r}\phi' = \left(1 - \frac{N}{24}e^{2\phi}\right)\left(\rho'' + \frac{1}{r}\rho'\right),$$

$$\left(1 - \frac{N}{12}e^{2\phi}\right)\left(\phi'' + \frac{1}{r}\phi'\right) = 2\left(1 - \frac{N}{24}e^{2\phi}\right)[(\phi')^2 - \lambda^2 e^{2\rho}]\,.$$

The boundary conditions for a regular horizon are

$$\phi' = \rho' = 0\,.$$

A static black-hole solution is therefore determined by the values of ϕ and ρ on the horizon. The value of ρ, however, can be changed by a constant by rescaling the coordinates $x_\pm$. The physical distinct static solutions with a horizon are therefore characterized simply by ϕ_h, the value of the dilaton on the horizon.

If $\phi_h > \phi_0$, ϕ would increase away from the horizon and would always be greater than its horizon value. This shows that to get a static black-hole solution that is asymptotic to the weak-coupling region of the linear dilaton, ϕ_h must be less than the critical value ϕ_0. One can then show that both ϕ and ρ must decrease with increasing r. This means the backreaction terms proportional to N will become unimportant. For large r one can therefore approximate by putting $N = 0$. This gives

$$\phi = \rho - (2b-1)\ln r - c\,,$$

$$\phi'' + (1/r)\phi' = 2\{[\rho' - (2b-1)r^{-1}]^2 - \lambda^2 e^{2\rho}\}\,.$$

Asymptotically these have the solution

$$\rho = -\ln r + \ln\frac{2b}{\lambda} - \frac{K + L\ln r}{r^{4b}} + \cdots\,,$$

where b, c, K, L are parameters that determine the solution. The parameters b and c correspond to the coordinate freedom in the linear dilaton that the solution approaches at large r. If $L=0$, the parameter K can be related to the Arnowitt-Deser-Misner (ADM) mass M of the solution. However, if $L \neq 0$, the ADM mass will be infinite. This is what one would expect for a static black hole in equilibrium with radiation at a nonzero temperature because there will be incoming and outgoing radiation all the way to infinity. Of course a solution formed by sending in an f wave to the linear dilaton will have a finite mass. But one might hope that it would settle down to a static black-hole solution which has finite mass because there is no incoming radiation (by boundary conditions) and no outgoing radiation (because the rate of radiation has gone to zero). Indeed this is what would have to happen if the singularity were to remain hidden for all time.

For $\phi_h \ll \phi_0$, the backreaction terms will be small at all values of r and the solutions of the semiclassical equations will be almost the same as the classical black holes. So

$$\phi_h = -\tfrac{1}{2}\ln(M/\lambda),$$

where M is the mass at a finite distance from the black hole.

Consider a situation in which a black hole of large mass ($M \gg N\lambda/12$) is created by sending in an f wave. One could approximate the subsequent evolution by a sequence of static black-hole solutions with a steadily increasing value of ϕ on the horizon. However, when the value of ϕ on the horizon approaches the critical value ϕ_0, the backreaction will become important and will change the black-hole solutions significantly. Let

$$\phi = \phi_0 + \bar{\phi}, \quad \rho = \ln\lambda + \bar{\rho}.$$

Then N and λ disappear and the equations for static black holes become

$$\phi'' + \frac{1}{r}\bar{\phi}' = \tfrac{1}{2}(2 - e^{2\bar{\phi}})\left(\bar{\rho}'' + \frac{1}{r}\bar{\rho}'\right),$$

$$(1 - e^{2\bar{\phi}})\left(\bar{\phi}'' + \frac{1}{r}\bar{\phi}'\right) = (2 - e^{2\bar{\phi}})[(\bar{\phi}')^2 - e^{2\bar{\rho}}].$$

As the dilaton field on the horizon approaches the critical value ϕ_0, the term $1 - e^{2\bar{\phi}}$ will approach 2ϵ, where $\epsilon = \phi_0 - \phi_h$. This will cause the second derivative of $\bar{\phi}$ to be very large until $\bar{\phi}'$ approaches $-e^{\bar{\rho}_h}$ in a coordinate distance Δr of order 4ϵ. By the above equations, ρ' approaches $-2e^{\bar{\rho}_h}$ in the same distance. A power series solution and numerical calculations carried out by Jonathan Brenchley confirm that in the limit as ϵ tends to zero, the solution tends to a limiting form $\bar{\phi}_c, \bar{\rho}_c$.

The limiting black hole is regular everywhere outside the horizon, but has a fairly mild singularity on the horizon with R diverging like r^{-1}. At large values of r, the

solution will tend to the linear dilaton in the manner of the asymptotic expansion given before. One or both of the constants K and L must be nonzero, because the solution is not exactly the linear dilaton. Fitting to the asymptotic expansion gives a value

$$b_c \approx 0.4.$$

If the singularity inside the black hole were to remain hidden at all times, as in possibility (1) above, one might expect that the temperature and rate of evolution of the black hole would approach zero as the dilaton field on the horizon approached the critical value. However, this is not what happens. The fact that the black holes tend to the limiting solution $\bar{\phi}_c, \bar{\rho}_c$ means that the period in imaginary time will tend to $4\pi b_c/\lambda$. Thus the temperature will be

$$T_c = \lambda/4\pi b_c.$$

The energy-momentum tensor of one of the f_i fields can be calculated from the conservation equations. In the $x_{\pm}$ coordinates, they are

$$\langle T^f_{++}\rangle = -\tfrac{1}{12}[\partial_+\bar{\rho}\,\partial_+\bar{\rho} - \partial_+^2\bar{\rho} + t_+(x_+)],$$

$$\langle T^f_{--}\rangle = -\tfrac{1}{12}[\partial_-\bar{\rho}\,\partial_-\bar{\rho} - \partial_-^2\bar{\rho} + t_-(x_-)],$$

where $t_{\pm}(x_{\pm})$ are chosen to satisfy the boundary conditions on the energy-momentum tensor. In the case of a black hole formed by sending in an f wave, the boundary condition is that the incoming flux $\langle T^f_{++}\rangle$ should be zero at large r. This would imply that

$$t_+ = 1/4x_+^2.$$

The energy-momentum tensor would not be regular on the past horizon, but this does not matter as the physical spacetime would not have a past horizon but would be different before the f wave.

On the other hand, the energy-momentum tensor should be regular on the future horizon. This would imply that $t_-(x_-)$ should be regular at $x_- = 0$. Converting to the coordinates $u_{\pm}$, one then would obtain a steady rate

$$\lambda^2/192 b_c^2$$

of energy outflow in each f field at late retarded times u_-.

In conclusion, the fact that the temperature and rate of emission of the limiting black hole do not go to zero establishes a contradiction with the idea that the black hole settles down to a stable state. Of course, this does not tell us what the semiclassical equations will predict, but it makes it very plausible that they will lead either to a naked singularity or to a singularity that spreads out to infinity at some finite retarded time.

The semiclassical evolution of these two-dimensional black holes is very similar to that of charged black holes in four dimensions with a dilaton field [5]. If one supposes that there are no fields in the theory that can carry away the charge, the steady loss of mass would suggest that the black hole would approach an extreme state. However, unlike the case of the Reissner-Nordström solutions, the extreme black holes with a dilaton have a finite temperature and rate of emission. So one obtains a similar contradiction. If the solution were to evolve to a state of lower mass but the same charge, the singularity would become naked.

There seems to be no way of avoiding a naked singularity in the context of the semiclassical theory. If spacetime is described by a semiclassical Lorentz metric, a black hole cannot disappear completely without there being some sort of naked singularity. But there seem to be zero-temperature nonradiating black holes only in a few cases, for example, charged black holes with no dilaton field and no fields to carry away the charge.

What seems to be happening is that the semiclassical approximation is breaking down in the strong-coupling regime. In conventional general relativity, this breakdown occurs only when the black hole gets down to the Planck mass. But in the two- and four-dimensional dilatonic theories, it can occur for macroscopic black holes when the dilaton field on the horizon approaches the critical value. When the coupling becomes strong, the semiclassical approximation will break down. Quantum fluctuations of the metric and the dilaton could no longer be neglected. One could imagine that this might lead to a tremendous explosion in which the remaining mass energy of the black hole was released. Such explosions might be detected as gamma-ray bursts.

Even though the semiclassical equations seem to lead to a naked singularity, one would hope that this would not happen in a full quantum treatment. Exactly what it means not to have naked singularities in a quantum theory of gravity is not immediately obvious. One possible interpretation is the no boundary condition [6]: Spacetime is nonsingular and without boundary in the Euclidean regime. If this proposal is correct, some sort of Euclidean wormhole would have to occur, which would carry away the particles that went in to form the black hole, and bring in the particles to be emitted. These wormholes could be in a coherent state described by alpha parameters [7]. These parameters might be determined by the minimization of the effective gravitational constant G [7–9]. In this case, there would be no loss of quantum coherence if a black hole were to evaporate and disappear completely or the alpha parameters might be different moments of a quantum field α on superspace [10]. In this case there would be effective loss of quantum coherence, but it might be possible to measure all the alpha parameters involved in the evaporation of a black hole of a given mass. In that case, there would be no further loss of quantum coherence when black holes of up to that mass evaporated.

I was greatly helped by talking to S. B. Giddings and A. Strominger who were working along similar lines. I also had useful discussions with G. Hayward, G. T. Horowitz, and J. Preskill. This work was carried out during a visit to Caltech as a Sherman Fairchild Distinguished Scholar. This work was supported in part by the U.S. Department of Energy under Contract No. DEAC-03-81ER40050.

[1] C. G. Callan, S. B. Giddings, J. A. Harvey, and A. Strominger, Phys. Rev. D **45**, R1005 (1992).

[2] S. W. Hawking, Commun. Math. Phys. **43**, 199 (1975).

[3] T. Banks, A. Dabholkar, M. R. Douglas, and M. O'Loughlin, Phys. Rev. D **45**, 3607 (1992).

[4] J. G. Russo, L. Susskind, and L. Thorlacius, Report No. SU-ITP-92-4 (unpublished).

[5] D. Garfinkle, G. T. Horowitz, and A. Strominger, Phys. Rev. D **43**, 3140 (1991).

[6] J. B. Hartle and S. W. Hawking, Phys. Rev. D **28**, 2960 (1983).

[7] S. Coleman, Nucl. Phys. **B310**, 643 (1988).

[8] J. Preskill, Nucl. Phys. **B323**, 141 (1989).

[9] S. W. Hawking, Nucl. Phys. **B335**, 155 (1990).

[10] S. W. Hawking, Nucl. Phys. **B363**, 117 (1991).

PHYSICAL REVIEW D   VOLUME 15, NUMBER 10   15 MAY 1977

Cosmological event horizons, thermodynamics, and particle creation

G. W. Gibbons[*] and S. W. Hawking

D.A.M.T.P., University of Cambridge, Silver Street, Cambridge, United Kingdom

(Received 4 March 1976)

It is shown that the close connection between event horizons and thermodynamics which has been found in the case of black holes can be extended to cosmological models with a repulsive cosmological constant. An observer in these models will have an event horizon whose area can be interpreted as the entropy or lack of information of the observer about the regions which he cannot see. Associated with the event horizon is a surface gravity κ which enters a classical "first law of event horizons" in a manner similar to that in which temperature occurs in the first law of thermodynamics. It is shown that this similarity is more than an analogy: An observer with a particle detector will indeed observe a background of thermal radiation coming apparently from the cosmological event horizon. If the observer absorbs some of this radiation, he will gain energy and entropy at the expense of the region beyond his ken and the event horizon will shrink. The derivation of these results involves abandoning the idea that particles should be defined in an observer-independent manner. They also suggest that one has to use something like the Everett-Wheeler interpretation of quantum mechanics because the back reaction and hence the spacetime metric itself appear to be observer-dependent, if one assumes, as seems reasonable, that the detection of a particle is accompanied by a change in the gravitational field.

I. INTRODUCTION

The aim of this paper is to extend to cosmological event horizons some of the ideas of thermodynamics and particle creation which have recently been successfully applied to black-hole event horizons. In a black hole the inward-directed gravitational field produced by a collapsing body is so strong that light emitted from the body is dragged back and does not reach an observer at a large distance. There is thus a region of spacetime which is not visible to an external observer. The boundary of the region is called the event horizon of the black hole. Event horizons of a different kind occur in cosmological models with a repulsive Λ term. The effect of this term is to cause the universe to expand so rapidly that for each observer there are regions from which light can never reach him. We shall call the boundary of this region the cosmological event horizon of the observer.

The "no hair" theorems (Israel,[1] Muller zum Hagen et al.,[2] Carter,[3] Hawking,[4] Robinson[5,6]) imply that a black hole formed in a gravitational collapse will rapidly settle down to a quasistationary state characterized by only three parameters, the mass M_H, the angular momentum J_H, and the charge Q_H. A black hole of a given M_H, J_H, Q_H therefore has a large number of possible unobservable internal configurations which reflect the different possible initial configurations of the body that collapsed to produce the hole. In purely classical theory this number of internal configurations would be infinite because one could make a given black hole out of an infinitely large number of

particles of indefinitely small mass. However, when quantum mechanics is taken into account, one would expect that in order to obtain gravitational collapse the energies of the particle would have to be restricted by the requirement that their wavelength be less than the size of the black hole. It would therefore seem reasonable to postulate that the number of internal configurations is finite. In this case one could associate with the black hole an entropy S_H which would be the logarithm of this number of possible internal configurations.[7,8,9] For this to be consistent the black hole would have to emit thermal radiation like a body with a temperature

$$T_H = G^2 \left[\left(\frac{\partial S}{\partial M} \right)_{J,Q} \right]^{-1}.$$

The mechanism by which this thermal radiation arises can be understood in terms of pair creation in the gravitational potential well of the black hole. Inside the black hole there are particle states which have negative energy with respect to an external stationary observer. It is therefore energetically possible for a pair of particles to be spontaneously created near the event horizon. One particle has positive energy and escapes to infinity, the other particle has negative energy and falls into the black hole, thereby reducing its mass. The existence of the event horizon would prevent this happening classically but it is possible quantum-mechanically because one or other of the particles can tunnel through the event horizon. An equivalent way of looking at the pair creation is to regard the positive- and negative-energy particles as being the same particle which tunnels out from the black hole on a spacelike or past-directed timelike world line and is scattered onto a future-directed world line (Hartle and Hawking[10]). When one calculates the rate of particle emission by this process it turns out to be exactly what one would expect from a body with a temperature $T_H = \hbar(2\pi kc)^{-1}\kappa_H$, where κ_H is the surface gravity of the black hole and is related to M_H, J_H, and Q_H by the formulas

$$\kappa_H = (r_+ - r_-)c^2 r_0^{-2},$$

$$r_\pm = c^{-2}[GM \pm (G^2M^2 - J^2M^{-2}c^2 - GQ^2)^{1/2}],$$

$$r_0^2 = r_+^2 + G^{-2}J^2M^{-2}c^2,$$

$$A_H = 4\pi r_0^2.$$

A_H is the area of the event horizon of the black hole.

Combining this quantum-mechanical argument with the thermodynamic argument above, one finds that the total number of internal configurations is indeed finite and that the entropy is given by

$$S_H = (4G\hbar)^{-1}kc^3 A_H.$$

Cosmological models with a repulsive Λ term which expand forever approach de Sitter space asymptotically at large times. In de Sitter space future infinity is spacelike.[11,12] This means that for each observer moving on a timelike world line there is an event horizon separating the region of spacetime which the observer can never see from the region that he can see if he waits long enough. In other words, the event horizon is the boundary of the past of the observer's world line. Such a cosmological event horizon has many formal similarities with a black-hole event horizon. As we shall show in Sec. III it obeys laws very similar to the zeroth, first, and second laws of black-hole mechanics in the classical theory.[13] It also bounds the region in which particles can have negative energy with respect to the observer. One might therefore expect that particle creation with a thermal spectrum would also occur in these cosmological models. In Secs. IV and V we shall show that this is indeed the case: An observer will detect thermal radiation with a characteristic wavelength of the order of the Hubble radius. This would correspond to a temperature of less than 10^{-28} °K so that it is not of much practical significance. It is, however, important conceptually because it shows that thermodynamic arguments can be applied to the universe as a whole and that the close relationship between event horizons, gravitational fields, and thermodynamics that was found for black holes has a wider validity.

One can regard the area of the cosmological

event horizon as a measure of one's lack of knowledge about the rest of the universe beyond one's ken. If one absorbs the thermal radiation, one gains energy and entropy at the expense of this region and so, by the first law mentioned above, the area of the horizon will go down. As the area decreases, the temperature of the cosmological radiation goes down (unlike the black-hole case), so the cosmological event horizon is stable. On the other hand, if the observer chooses not to absorb any radiation, there is no change in area of the horizon. This is another illustration of the fact that the concept of particle production and the back reaction associated with it seem not to be uniquely defined but to be dependent upon the measurements that one wishes to consider.[14-16]

The plan of the paper is as follows. In Sec. II we describe the black-hole asymptotically de Sitter solutions found by Carter.[20] In Sec. III we derive the classical laws governing both cosmological and black-hole event horizons. In Sec. IV we discuss particle creation in de Sitter space. We abandon the concept of particles as being observer-independent and consider instead what an observer moving on a timelike geodesic and equipped with a particle detector would actually measure. We find that he would detect an isotropic background of thermal radiation with a temperature $(2\pi)^{-1}\kappa_C$ where $\kappa_C = \Lambda^{1/2}3^{-1/2}$ is the surface gravity of the cosmological event horizon of the observer. Any other observer moving on a timelike geodesic will also see isotropic radiation with the same temperature even though he is moving relative to the first observer. This shows that they are not observing the same particles: Particles are observer-dependent. In Sec. V we extend these results to asymptotically de Sitter spaces containing black holes. The implications are considered in Sec. VI. It seems necessary to adopt something like the Everett-Wheeler interpretation of quantum mechanics because the back reaction and hence the spacetime metric will be observer-dependent, if one assumes, as seems reasonable, that the detection of a particle is accompanied by a change in the gravitational field.

We shall adopt units in which $G = \hbar = k = c = 1$. We shall use a metric with signature +2 and our conventions for the Riemann and the Ricci tensors are

$$v_{a;[b;c]} = \tfrac{1}{2} R^d{}_{abc} v_d,$$

$$R_{ab} = R_a{}^c{}_{bc}.$$

II. EXACT SOLUTIONS WITH COSMOLOGICAL EVENT HORIZONS

In this section we shall give some examples of event horizons in exact solutions of the Einstein equations

$$R_{ab} - \tfrac{1}{2} g_{ab} R + \Lambda g_{ab} = 8\pi T_{ab} \,. \tag{2.1}$$

We shall consider only the case of Λ positive (corresponding to repulsion). Models with negative Λ do not, in general, have event horizons.

The simplest example is de Sitter space which is a solution of the field equations with $T_{ab} = 0$. One can write the metric in the static form

$$ds^2 = -(1 - \Lambda r^2 3^{-1})dt^2 + dr^2(1 - \Lambda r^2 3^{-1})^{-1}$$
$$+ r^2(d\theta^2 + \sin^2\theta \, d\phi^2) \,. \tag{2.2}$$

This metric has an apparent singularity at $r = 3^{1/2}\Lambda^{-1/2}$. This singularity caused considerable discussion when the metric was first discovered.[17,18] However, it was soon realized that it arose simply from a bad choice of coordinates and that there are other coordinate systems in which the metric can be analytically extended to a geodesically complete space of constant curvature with topology $R^1 \times S^3$. For a detailed description of these coordinate systems the reader is referred to Refs. 12 and 19. For our purposes it will be convenient to express the de Sitter metric in "Kruskal coordinates":

$$ds^2 = 3\Lambda^{-1}(UV - 1)^{-2}$$
$$\times [-4dU\,dV + (UV + 1)^2(d\theta^2 + \sin^2\theta \, d\phi^2)]$$
$$\tag{2.3}$$

where

$$r = 3^{1/2}\Lambda^{-1/2}(UV + 1)(1 - UV) \,. \tag{2.4}$$

$$\exp(2\Lambda^{1/2}3^{-1/2}t) = -VU^{-1} \,. \tag{2.5}$$

The structure of this space is shown in Fig. 1. In this diagram radial null geodesics are at $\pm 45°$ to the vertical. The dashed curves $UV = -1$ are timelike and represent the origin of polar coordinates and the antipodal point on a three-sphere. The solid curves $UV = +1$ are spacelike and represent past and future infinity $\mathcal{I}^-$ and $\mathcal{I}^+$, respectively.

In region I ($U < 0$, $V > 0$, $UV > -1$) the Killing vector $K = \partial/\partial t$ is timelike and future-directed. However, in region IV ($U > 0$, $V < 0$, $UV > -1$, K is still timelike but past-directed, while in regions II and III ($0 < UV < 1$) K is spacelike. The Killing vector K is null on the two surfaces $U = 0$, $V = 0$. These are respectively the future and past event horizons for any observer whose world line remains in region I; in particular for any observer moving along a curve of constant r in region I.

By applying a suitable conformal transformation one can make the Kruskal diagram finite and convert it to the Penrose-Carter form (Fig. 2). Radial null geodesics are still $\pm 45°$ to the vertical but the freedom of the conformal factor has been used

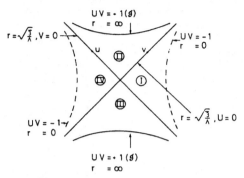

FIG. 1. Kruskal diagram of the (r,t) plane of de Sitter space. In this figure null geodesics are at $\pm 45°$ to the vertical. The dashed curves $r = 0$ are the antipodal origins of polar coordinates on a three-sphere. The solid curves $r = \infty$ are past and future infinity $\mathcal{I}^-$ and $\mathcal{I}^+$, respectively. The lines $r = 3^{1/2}\Lambda^{-1/2}$ are the past and future event horizons of observers at the origin.

to make the origin of polar coordinates, $r = 0$, and future and past infinity, $\mathcal{I}^+$ and $\mathcal{I}^-$, straight lines. Also shown are some orbits of the Killing vector $K = \partial/\partial t$. Because de Sitter space is invariant under the ten-parameter de Sitter group, $SO(4, 1)$, K will not be unique. Any timelike geodesic can be chosen as the origin of polar coordinates and the surfaces $U = 0$ and $V = 0$ in such coordinates will be the past and future event horizons of an observer moving on this geodesic. If one normalizes K to have unit magnitude at the origin, one can define a "surface gravity" for the horizon by

$$K_{a;b}K^b = \kappa_C K_a \tag{2.6}$$

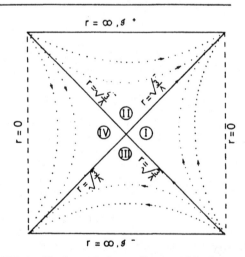

FIG. 2. The Penrose-Carter diagram of de Sitter space. The dotted curves are orbits of the Killing vector.

on the horizon. This gives

$$\kappa_C = \Lambda^{1/2} 3^{-1/2}. \tag{2.7}$$

The area of the cosmological horizon is

$$A_C = 12\pi\Lambda^{-1}. \tag{2.8}$$

One can also construct solutions which generalize the Kerr-Newman family to the case when Λ is nonzero.[20,21] The simplest of these is the Schwarzschild–de Sitter metric. When $\Lambda = 0$ the unique spherically symmetric vacuum spacetime is the Schwarzschild solution. The metric of this can be written in static form:

$$ds^2 = -(1 - 2Mr^{-1})dt^2 + dr^2(1 - 2Mr^{-1})^{-1}$$
$$+ r^2(d\theta^2 + \sin^2\theta\, d\phi^2). \tag{2.9}$$

As is now well known, the apparent singularities at $r = 2M$ correspond to a horizon and can be removed by changing to Kruskal coordinates in which the metric has the form

$$ds^2 = -32M^3 r^{-1}\exp(-2^{-1}M^{-1}r)dU\,dV$$
$$+ r^2(d\theta^2 + \sin^2\theta\, d\phi^2), \tag{2.10}$$

where

$$UV = (1 - 2^{-1}M^{-1}r)\exp(2^{-1}M^{-1}r) \tag{2.11}$$

and

$$UV^{-1} = -\exp(-2^{-1}M^{-1}t). \tag{2.12}$$

The Penrose-Carter diagram of the Schwarzschild solution is shown in Fig. 3. The wavy lines marked $r = 0$ are the past and future singularities. Region I is asymptotically flat and is bounded on the right by past and future null infinity $\mathscr{I}^-$ and $\mathscr{I}^+$. It is bounded on the left by the surfaces $U = 0$ and $V = 0$, $r = 2M$. These are future and past event horizons for observers who remain outside $r = 2M$. On the

left-hand side of the diagram there is another a asymptotically flat region IV. The Killing vector $K = \partial/\partial t$ is now uniquely defined by the condition that it be timelike and of unit magnitude near $\mathscr{I}^+$ and $\mathscr{I}^-$. It is timelike and future-directed in region I, timelike and past-directed in region IV, and spacelike in regions II and III. The Killing vector K is null on the horizons which have area $A_H = 16\pi M^2$. The surface gravity, defined by (2.6), is $\kappa_H = (4M)^{-1}$.

The Schwarzschild solution is usually interpreted as a black hole of mass M in an asymptotically flat space. There is a straightforward generalization to the case of nonzero Λ which represents a black hole in asymptotically de Sitter space. The metric can be written in the static form

$$ds^2 = -(1 - 2Mr^{-1} - \Lambda r^2 3^{-1})dt^2$$
$$+ dr^2(1 - 2Mr^{-1} - \Lambda r^2 3^{-1})^{-1}$$
$$+ r^2(d\theta^2 + \sin^2\theta\, d\phi^2). \tag{2.13}$$

If $\Lambda > 0$ and $9\Lambda M^2 < 1$, the factor $(1 - 2Mr^{-1} - \Lambda r^2 3^{-1})$ is zero at two positive values of r. The smaller of of these values, which we shall denote by r_+, can be regarded as the position of the black-hole event horizon, while the larger value r_{++} represents the position of the cosmological event horizon for observers on world lines of constant r between r_+ and r_{++}. By using Kruskal coordinates as above one can remove the apparent singularities in the metric at r_+ and r_{++}. One has to employ separate coordinate patches at r_+ and r_{++}. We shall not give the expressions in full because they are rather messy; however, the general structure can be seen from the Penrose-Carter diagram shown in Fig. 4. Instead of having two regions (I and IV) in which the Killing vector $K = \partial/\partial t$ is timelike, there are now an infinite sequence of such regions, also labeled I and IV depending upon whether K is future- or past-directed. There are also infinite sequences of $r = 0$ singularities and spacelike infinities $\mathscr{I}^+$ and $\mathscr{I}^-$. The surfaces $r = r_+$ and $r = r_{++}$ are black-hole and cosmological event horizons for observers moving on world lines of constant

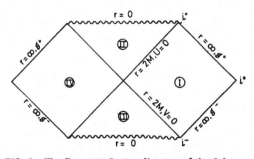

FIG. 3. The Penrose-Carter diagram of the Schwarzschild solution. The wavy lines and the top and bottom are the future and past singularities. The diagonal lines bounding the diagram on the right-hand side are the past and future null infinity of asymptotically flat space. The region IV on the left-hand-side is another asymptotically flat space.

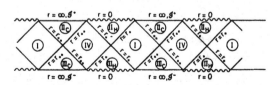

FIG. 4. The Penrose-Carter diagram for Schwarzschild–de Sitter space. There is an infinite sequence of singularities $r = 0$ and spacelike infinities $r = \infty$. The Killing vector $K = \partial/\partial t$ is timelike and future-directed in regions I, timelike and past-directed in regions IV and spacelike in the others.

r between r_+ and r_{++}.

The Killing vector $K = \partial/\partial t$ is uniquely defined by the conditions that it be null on both the black-hole and the cosmological horizons and that its magnitude should tend to $\Lambda^{1/2}3^{-1/2}r$ as r tends to infinity. One can define black-hole and cosmological surface gravities κ_H and κ_C by

$$K_{a;b}K^b = \kappa K_a \qquad (2.14)$$

on the horizons. These are given by

$$\kappa_H = \Lambda 6^{-1}r_+{}^{-1}(r_{++} - r_+)(r_+ - r_{--}), \qquad (2.15a)$$

$$\kappa_C = \Lambda 6^{-1}r_{++}{}^{-1}(r_{++} - r_+)(r_{++} - r_{--}), \qquad (2.15b)$$

where $r = r_{--}$ is the negative root of

$$3r - 6M - \Lambda r^3 = 0. \qquad (2.16)$$

The areas of the two horizons are

$$A_H = 4\pi r_+{}^2 \qquad (2.17)$$

and

$$A_C = 4\pi r_{++}{}^2. \qquad (2.18)$$

If one keeps Λ constant and increases M, r_+ will increase and r_{++} will decrease. One can understand this in the following way. When $M = 0$ the gravitational potential $g(\partial/\partial t, \partial/\partial t)$ is $1 - \Lambda r^2 3^{-1}$. The introduction of a mass M at the origin produces an additional potential of $-2Mr^{-1}$. Horizons occur at the two values of r at which $g(\partial/\partial t, \partial/\partial t)$ vanishes. Thus as M increases, the black-hole horizon r_+ increases and the cosmological horizon r_{++} decreases. When $9\Lambda M^2 = 1$ the two horizons coincide. The surface gravity K can be thought of as the gravitational field or gradient of the potential at the horizons. As M increases both κ_H and κ_C decrease.

The Kerr–Newman–de Sitter space can be expressed in Boyer-Lindquist-type coordinates as[20,21]

$$ds^2 = \rho^2(\Delta_r{}^{-1}dr^2 + \Delta_\theta{}^{-1}d\theta^2)$$
$$+ \rho^{-2}\Xi^{-2}\Delta_\theta[adt - (r^2 + a^2)d\phi]^2$$
$$- \Delta_r\Xi^{-2}\rho^{-2}(dt - a\sin^2\theta d\phi)^2, \qquad (2.19)$$

where

$$\rho^2 = r^2 + a^2\cos^2\theta, \qquad (2.20)$$

$$\Delta_r = (r^2 + a^2)(1 - \Lambda r^2 3^{-1}) - 2Mr + Q^2, \qquad (2.21)$$

$$\Delta_\theta = 1 + \Lambda a^2 3^{-1}\cos^2\theta, \qquad (2.22)$$

$$\Xi = 1 + \Lambda a^2 3^{-1}. \qquad (2.23)$$

The electromagnetic vector potential A_a is given by

$$A_a = Qr\rho^{-2}\Xi^{-1}(\delta_a^t - a\sin^2\theta\delta_a^\phi). \qquad (2.24)$$

Note that our Λ has the opposite sign to that in Ref. 21.

There are apparent singularities in the metric at the values of r for which $\Delta_r = 0$. As before, these correspond to horizons and can be removed by using appropriate coordinate patches. The Penrose-Carter diagram of the symmetry axis $(\theta = 0)$ of these spaces is shown in Fig. 5 for the case that Δ_r has 4 distinct roots: r_{--}, r_-, r_+, and r_{++}. As before, r_{++} and r_+ can be regarded as the cosmological and black-hole event horizons, respectively. In addition, however, there is now an inner black-hole horizon at $r = r_-$. Passing through this, one comes to the ring singularity at $r = 0$, on the other side of which there is another cosmological horizon at $r = r_{--}$ and another infinity. The diagram shown is the simplest one to draw but it is not simply connected; one can take covering spaces. Alternatively one can identify regions in this diagram.

The Killing vector $\vec{K} = \partial/\partial\phi$ is uniquely defined by the condition that its orbits should be closed curves with parameter length 2π. The other Kill-

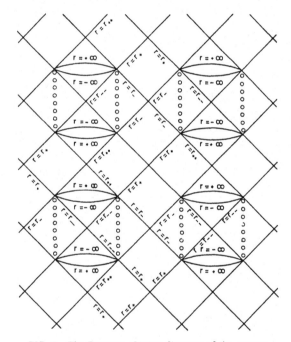

FIG. 5. The Penrose-Carter diagram of the symmetry axis of the Kerr–Newman–de Sitter solution for the case that Δ_r has four distinct real roots. The infinities $r = +\infty$ and $r = -\infty$ are not joined together. The external cosmological horizon occurs at $r = r_{++}$ the exterior black-hole horizon at $r = r_+$, the inner black-hole horizon at $r = r_-$. The open circles mark where the ring singularity occurs, although this is not on the symmetry axis. On the other side of the ring at negative values of r there is another cosmological horizon at $r = r_{--}$ and another infinity.

ing vector $K = \partial/\partial t$ is not so specially picked out. One can add different constants multiples of $\bar{K}$ to K to obtain Killing vectors which are null on the different horizons and one can then define surface gravities as before. We shall be interested only in those for the r_+, r_{++} horizons. They are

$$\kappa_H = \Lambda 6^{-1} \Xi^{-1} (r_+ - r_{--})(r_+ - r_-)(r_{++} - r_+)(r_+{}^2 + a^2)^{-1},$$
$$(2.25)$$

$$\kappa_C = \Lambda 6^{-1} \Xi^{-1} (r_{++} - r_+)(r_{++} - r_-)(r_{++} - r_-)(r_+{}^2 + a^2)^{-1}.$$
$$(2.26)$$

The areas of these horizons are

$$A_H = 4\pi(r_+{}^2 + a^2), \qquad (2.27)$$

$$A_C = 4\pi(r_{++}{}^2 + a^2). \qquad (2.28)$$

III. CLASSICAL PROPERTIES OF EVENT HORIZONS

In this section we shall generalize a number of results about black-hole event horizons in the classical theory to spacetimes which are not asymptotically flat and may have a nonzero cosmological constant, and to event horizons which are not black-hole horizons. The event horizon of a black hole in asymptotically flat spacetimes is normally defined as the boundary of the region from which one can reach future null infinity, $\mathscr{I}^+$, along a future-directed timelike or null curve. In other words it is $\dot{J}^-(\mathscr{I}^+)$ [or equivalently $\dot{I}^-(\mathscr{I}^+)$], where an overdot indicates the boundary and J^- is the causal past (I^- is the chronological past). However, one can also define the black-hole horizon as $\dot{I}^-(\lambda)$, the boundary of the past of a timelike curve λ which has a future end point at future timelike infinity, i^+ in Fig. 3. One can think of λ as the world line of an observer who remains outside the black hole and who does not accelerate away to infinity. The event horizon is the boundary of the region of spacetime that he can see if he waits long enough. It is this definition of event horizon that we shall extend to more general spacetimes which are not asymptotically flat.

Let λ be a future inextensible timelike curve representing an observer's world line. For our considerations of particle creation in the next section we shall require that the observer have an indefinitely long time in which to detect particles. We shall therefore assume that λ has infinite proper length in the future direction. This means that it does not run into a singularity. The past of λ, $I^-(\lambda)$, is a terminal indecomposable past set, or TIP in the language of Geroch, Kronheimer, and Penrose.[22] It represents all the events that the observer can ever see. We shall assume that what the observer sees at late times can be predicted (classically at least) from a spacelike surface $\mathscr{S}$,

i.e., $I^-(\lambda) \cap J^+(\mathscr{S})$ is contained in the future Cauchy development $D^+(\mathscr{S})$.[12] We shall also assume that $\dot{I}^-(\lambda) \cap J^+(\mathscr{S})$, the portion of the event horizon to the future of $\mathscr{S}$, is contained in $D^+(\mathscr{S})$. Such an event horizon will be said to be predictable. The event horizon will be generated by null geodesic segments which have no future end points but which have past end points if and where they intersect other generators.[12] In another paper[23] it is shown that the generators of a predictable event horizon cannot be converging if the Einstein equations hold (with or without cosmological constant), provided that the energy-momentum tensor satisfies the strong energy condition $T_{ab}u^a u^b \geq \frac{1}{2} T_a{}^a u^b u_b$ for any timelike vector u_a, i.e., provided that $\mu + P_i \geq 0$, $\mu + \sum_{i=1}^{i=3} P_i \geq 0$, where μ is the energy density and P_i are the principal pressures. This gives immediately the following result, which, because of the very suggestive analogy with thermodynamics, we call:

The second law of event horizons: The area of any connected two-surface in a predictable event horizon cannot decrease with time. The area may be infinite if the two-dimensional cross section is not compact. However, in the examples in Sec. II, the natural two-sections are compact and have constant area.

In the case of gravitational collapse in asymptotically flat spacetimes one expects the spacetime eventually to settle down to a quasistationary state because all the available energy will either fall through the event horizon of the black hole (thereby increasing its area) or be radiated away to infinity. In a similar way one would expect that where the intersection of $I^-(\lambda)$ with a spacelike surface $\mathscr{S}$ had compact closure (which we shall assume henceforth), there would only be a finite amount of energy available to be radiated through the cosmological event horizon of the observer and that therefore this spacetime would eventually approach a stationary state. One is thus lead to consider solutions in which there is a Killing vector K which is timelike in at least some region of $I^-(\lambda) \cap J^+(\mathscr{S})$. Such solutions would represent the asymptotic future limit of general spacetimes with predictable event horizons.

Several results about stationary empty asymptotically flat black-hole solutions can be generalized to stationary solutions of the Einstein equations, with cosmological constant, which contain predictable event horizons. The first such theorem is that the null geodesic generators of each connected component of the event horizon must coincide with orbits of some Killing vector.[24,21]

These Killing vectors may not coincide with the original Killing vector K and may be different for different components of the horizon. In either of these two cases there are at least two Killing vectors. One can chose a linear combination $\bar{K}$ whose orbits are spacelike closed curves in $I^-(\lambda) \cap J^+(\mathcal{S})$. One could interpret this as implying that the solution is axisymmetric as well as being stationary, though we have not been able to prove that there is necessarily any axis on which $\bar{K}$ vanishes.

Let $\hat{K}$ be the Killing vector which coincides with the generators of one component of the event horizon. If $\hat{K}$ is not hypersurface orthogonal and if then space is empty or contains only an electromagentic field, one can apply a generalized Lichnerowicz theorem[12,21] to show that $\hat{K}$ must be spacelike in some "ergoregion" of $I^-(\lambda)$. One can then apply energy extraction arguments[24,12] or the results of Hajicek[25] to show that this ergoregion contains another component of the event horizon whose generators do not coincide with the orbits of $\hat{K}$. It therefore follows that either $\hat{K}$ is hypersurface orthogonal (in which case the solution is static) or that there are at least two Killing vectors (in which case the solution is axisymmetric as well as stationary). If there is only a cosmological horizon and no black-hole horizon, then the solution is necessarily static.

One would expect that in the static vacuum case one could generalize Israel's theorem[1,2] to prove that the space was spherically symmetric. One could then generalize Birkhoff's theorem to include a cosmological constant and show that the space was necessarily the Schwarzschild–de Sitter space described in Sec. II. In the case that there was only a cosmological event horizon, it would be de Sitter space. In the stationary axisymmetric case one would expect that one could generalize and extend the results of Carter and Robinson[3,5] to show that vacuum solutions were members of the Kerr–de Sitter family described in Sec. II. If there is matter present it will distort the spacetime from the Schwarzschild–de Sitter or Kerr–de Sitter solution just as matter around a black hole in asymptotically flat space will distort the spacetime away from the Schwarzschild or Kerr solution.

The proof given in Ref. 13 of the zeroth law of black holes can be generalized immediately to the case of nonzero cosmological constant. One thus has:

The zeroth law of event horizons: The surface gravity of a connected component of the event horizon $I^-(\lambda)$ is constant over that component. This is analogous to the zeroth law of nonrelativistic thermodynamics which states that the temperature is constant over a body in thermal equilibrium. We shall show in Secs. IV and V that quantum effects cause each component of the event horizon to radiate thermally with a temperature proportional to its surface gravity.

One can also generalize the first law of black holes. We shall do this for stationary axisymmetric solutions with no electromagnetic field and where $I^-(\lambda) \cap J^+(\mathcal{S})$ consists of two components, a black-hole event horizon and a cosmological event horizon. Let K be the Killing vector which is null on the cosmological event horizon. The orbits of K will constitute the stationary frame which appears to be nonrotating with respect to distant objects near the cosmological event horizon. In the general case the normalization of K is somewhat arbitrary but we shall assume that some particular normalization has been chosen. The Killing vector $\hat{K}$ which coincides with the generators of the black-hole horizon can be expressed in the form

$$\hat{K} = K + \Omega_H \bar{K} , \tag{3.1}$$

where Ω_H is the angular velocity of the black-hole horizon relative to the cosmological horizon in the units of time defined by the normalization of K and $\bar{K}$ is the uniquely defined axial Killing vector whose orbits are closed curves with parameter length 2π.

For any Killing vector field ξ^a one has

$$\xi^{a;b}_{\;\;;b} = R^a_{\;b} \xi^b . \tag{3.2}$$

Choose a three-surface $\mathcal{S}$ which is tangent to $\bar{K}$, and integrate (3.2) over it with $\xi = \bar{K}$. On using Einstein's equations this gives

$$(8\pi)^{-1} \int_H \bar{K}^{a;b} d\Sigma_{ab} + (8\pi)^{-1} \int_C \bar{K}^{a;b} d\Sigma_{ab} = \int T^{ab} \bar{K}_a d\Sigma_b , \tag{3.3}$$

where the three-surface integral on the right-hand side is taken over the portions of $\mathcal{S}$ between the black-hole and cosmological horizons and the two-surface integrals marked H and C are taken over the intersections of $\mathcal{S}$ with the respective horizons, the orientation being given by the direction out of $I^-(\lambda)$. One can interpret the right-hand side of (3.3) as the angular momentum of the matter between the two horizons. One can therefore regard the second term on the left-hand side of (3.3) as being the total angular momentum, J_C, contained in the cosmological horizon, and the first on the left-hand side term as the negative of the angular momentum of the black hole, J_H.

One can also apply Eq. (3.2) to the Killing vector K to obtain

$$(4\pi)^{-1} \int K^{a;b} d\Sigma_{ab} + (4\pi)^{-1} \int K^{a;b} d\Sigma_{ab}$$

$$= \int 2(T_{ab} - \tfrac{1}{2} T^c_c g_{ab}) K^a d\Sigma_b + \int \Lambda (4\pi)^{-1} K_a d\Sigma^a . \tag{3.4}$$

One can regard the terms on the right-hand side of Eq. (3.4) as representing respectively the (positive) contribution of the matter and the (negative) contribution of the Λ term to the mass within the cosmological horizon. One can therefore regard the second term on the left-hand side as the (negative) mass M_C within the cosmological horizon and the first term on the left-hand side as the negative of the (positive) mass M_H of the black hole. As in Ref. 13, one can express M_H and M_C as

$$M_H = \kappa_H A_H (4\pi)^{-1} + 2\Omega_H J_H , \tag{3.5}$$

$$M_C = -\kappa_C A_C (4\pi)^{-1} . \tag{3.6}$$

One therefore has the Smarr-type[26] formulas

$$M_C = -\kappa_C A_C (4\pi)^{-1}$$

$$= \kappa_H A_H (4\pi)^{-1} + 2\Omega_H J_H + \int 2(T_{ab} - T_c^c g_{ab}) K^b d\Sigma^a$$

$$+ (4\pi)^{-1} \int \Lambda K_a d\Sigma^a . \tag{3.7}$$

One can take the differential of the mass formula in a manner similar to that in Ref. 13. One obtains:

The first law of event horizons.

$$\int \delta T_{ab} K^a d\Sigma^b = -\kappa_c \delta A_C (8\pi)^{-1} - \kappa_H \delta A_H (8\pi)^{-1} - \Omega_H \delta J_H , \tag{3.8}$$

where δT_{ab} is the variation in the matter energy-momentum tensor between the horizons in a gauge in which $\delta K^a = \delta \bar{K}^a = 0$.

From this law one sees that if one regards the area of a horizon as being proportional to the entropy beyond that horizon, then the corresponding surface gravity is proportional to the effective temperature of that horizon, that is, the temperature at which that horizon would be in thermal equilibrium and therefore the temperature at which that horizon radiates. In the next section we shall show that the factor of proportionality between temperature and surface gravity is $(2\pi)^{-1}$. This means that the entropy is $\frac{1}{4}$ the area. In the case of the cosmological horizon in de Sitter space the entropy is $3\pi\Lambda^{-1} \geq 10^{120}$ because $\Lambda < 10^{-120}$.

IV. PARTICLE CREATION IN DE SITTER SPACE

In this section we shall calculate particle creation in solutions of the Einstein equations with positive cosmological constant. The simplest example is de Sitter space and particle production in this situation has been studied by Nachtmann,[27] Tagirov,[28] Candelas and Raine,[29] and Dowker and Critchley,[30] among others. They all used definit-

ions of particles that were observer-independent and invariant under the de Sitter group. Under these conditions only two answers are possible for the rate of particle creation per unit volume, zero or infinity, because if there is nonzero production of particles with a certain energy, then by de Sitter group invariance there must be the same rate of creation of particles with all other energies. It is therefore not surprising that the authors mentioned above chose their definitions of particles to get the zero answer.

An observer-independent definition of particles is, however, not relevant to what a given observer would measure with a particle detector. This depends not only on the spacetime and the quantum state of the system, but also on the observer's world line. For example, Unruh[14] has shown that in Minkowski space in the normal vacuum state accelerated observers can detect and absorb particles. To a nonaccelerating observer such an absorption will appear to be emission from the accelerated observer's detector. In a similar manner, an observer at a constant distance from a black hole will detect a steady flux of particles coming out from the hole with a thermal spectrum while an observer who falls into the hole will not see many particles.

A feature common to the examples of a uniformly accelerated observer in Minkowski space and an observer at constant distance from the black hole is that both observers have event horizons which prevent them from seeing the whole of the spacetime and from measuring the complete quantum state of the system. It is this loss of information about the quantum state which is responsible for the thermal radiation that the observers see. Because any observer in de Sitter space also has an event horizon, one would expect that such an observer would also detect thermal radiation. We shall show that this is indeed the case. This can be done either by the frequency-mixing method in which the thermal radiation from black holes was first derived,[31,32] or by the path-integral method of Hartle and Hawking.[10] We shall adopt the latter approach because it is more elegant and gives a clearer intuitive picture of what is happening. The same results can, however, be obtained by the former method.

As in the method of Hartle and Hawking,[10] we construct the propagator for a scalar field of mass m by the path integral

$$G(x, x') = \lim_{\epsilon \to 0} \int_0^\infty dW F(W, x, x') \exp[-(im^2 W + \epsilon W^{-1})] , \tag{4.1}$$

where

$$F(W, x, x') = \int \delta x[w] \exp\left[\frac{i}{4}\int_0^W g(\dot{x}, \dot{x})dw\right] \quad (4.2)$$

and the integral is taken over all paths $x(w)$ from x to x'.

As in the Hartle and Hawking paper,[10] this path integral can be given a well-defined meaning by analtyically continuing the parameter W to negative imaginary values and analytically continuing the coordinates to a region where the metric is positive-definite. A convenient way of doing this is to embed de Sitter space as the hyperboloid

$$-T^2 + S^2 + X^2 + Y^2 + Z^2 = 3\Lambda^{-1} \quad (4.3)$$

in the five-dimensional space with a Lorentz metric:

$$ds^2 = -dT^2 + dS^2 + dX^2 + dY^2 + dZ^2. \quad (4.4)$$

Taking T to be $i\tau$ (τ real), we obtain a sphere in five-dimensional Euclidean space. On this sphere the function F satisfies the diffusion equation

$$\frac{\partial F}{\partial\Omega} = \tilde{\Box}^2 F, \quad (4.5)$$

where $\Omega = iW$ and $\tilde{\Box}^2$ is the Laplacian on the four-sphere. Because the four-sphere is compact there is a unique solution of (4.5) for the initial condition

$$F(0, x, x') = \delta(x, x'), \quad (4.6)$$

where $\delta(x, x')$ is the Dirac δ function on the four-sphere. One can then define the propagator $G(x, x')$ from (4.1) by analytically continuing the solution for F back to real values of the parameter W and real coordinates x and x'. Because the function F is analytic for finite points x and x', any singularities which occur in $G(x, x')$ must come from the end points of the integration in (4.1). As shown in Ref. 10, there will be singularities in $G(x, x')$ when, and only when, x and x' can be joined by a null geodesic. This will be the case if and only if

$$(T - T')^2 = (S - S')^2 + (X - X')^2 + (Y - Y')^2 + (Z - Z')^2. \quad (4.7)$$

The coordinates, T, S, X, Y, Z can be related to the static coordinates t, r, θ, ϕ used in Sec. II by

$$T = (\Lambda 3^{-1} - r^2)^{1/2} \sinh\Lambda^{1/2}3^{-1/2}t, \quad (4.8)$$

$$S = (\Lambda 3^{-1} - r^2)^{1/2} \cosh\Lambda^{1/2}3^{-1/2}t, \quad (4.9)$$

$$X = r\sin\theta\cos\phi, \quad (4.10)$$

$$Y = r\sin\theta\sin\phi, \quad (4.11)$$

$$Z = r\cos\theta. \quad (4.12)$$

The horizons $\Lambda r^2 = 3$ are the intersection of the hyperplanes $T = \pm S$ with the hyperboloid. As in

Ref. 10 we define the complexified horizon by $\Lambda r^2 = 3$, θ, ϕ real. On the complexified horizon X, Y, and Z are real and either $T = S = \Lambda^{-1/2}3^{1/2}V$, $U = 0$ or $T = -S = \Lambda^{-1/2}3^{1/2}U$, $V = 0$. By Eq. (4.7) a complex null geodesic from a real point (T', S', X', Y', Z') on the hyperboloid can intersect the complex horizon only on the real sections $T = \pm S$ real. If the point (T', S', X', Y', Z') is in region I ($S > |T|$) the propagator $G(x', x)$ will have a singularity on the past horizon at the point where the past-directed null geodesic from x' intersects the horizon. As shown in Ref. 10, the ϵ convergence factor in (4.1) will displace the pole slightly below the real axis in the complex plane on the complexified past horizon. The propagator $G(x', x)$ is therefore analytic in the upper half U plane on the past horizon. Similarly, it will be analytic in the lower half V plane on the future horizon.

The propagator $G(x', x)$ satisfies the wave equation

$$(\Box_x^2 - m^2)G(x', x) = -\delta(x, x') \quad (4.13)$$

Thus if x' is a fixed point in region I, the value $G(x', x)$ for a point in region II will be determined by the values of $G(x', x)$ on a characteristic Cauchy surface for region II consisting of the section of the $U = 0$ horizon for real $V \geq 0$ and the section of the $V = 0$ horizon for real $U \geq 0$. The coordinates r and t of the point x are related to U and V by

$$e^{2\kappa_c t} = VU^{-1} \quad (4.14)$$

$$r = (1 + UV)(1 - UV)^{-1}\kappa_c^{-1} \quad (4.15)$$

If one holds r fixed at a real value but lets $t = \tau + i\sigma$, then

$$U = |U| \exp(-i\sigma\kappa_c), \quad (4.16)$$

$$V = |V| \exp(+i\sigma\kappa_c). \quad (4.17)$$

For a fixed value of σ the metric (2.3) of de Sitter space remains real and unchanged. Thus the value of $G(x', x)$ at a complex coordinate t of the point x but real r, θ, ϕ can be obtained by solving the Klein-Gordon equation with real coefficients and with initial data on the Cauchy surface $V = 0$, $U = |U| \exp(-i\kappa_c\sigma)$ and $U = 0$, $V = |V| \exp(+i\kappa_c\sigma)$. Because $G(x', x)$ is analytic in the upper half U plane on $V = 0$ and the lower half V plane on $U = 0$, the data and hence the solution will be regular provided that

$$-\pi\kappa_c^{-1} \leq \sigma \leq 0. \quad (4.18)$$

The operator

$$\left(\frac{\partial}{\partial t}\right)_r = \kappa_c\left(V\frac{\partial}{\partial\overline{V}} - U\frac{\partial}{\partial\overline{U}}\right) \quad (4.19)$$

commutes with the Klein-Gordon operator $\Box_x^2 - m^2$ and is zero when acting on the initial data for σ

satisfying (4.18). Thus the solution $G(x', x)$ determined by the initial data will be analytic in the coordinates t of the point x for σ satisfying Eq. (4.18).

This is the basic result which enables us to show that an observer moving on a timelike geodesic in de Sitter space will detect thermal radiation.

The propagator we have defined appears to be similar to that constructed by other authors.[28-30] However, our use of the propagator will be different: Instead of trying to obtain some observer-independent measure of particle creation, we shall be concerned with what an observer moving on a timelike geodesic in de Sitter space would measure with a particle detector which is confined to a small tube around his world line. Without loss of generality we can take the observer's world line to be at the origin of polar coordinates in region I. Within the world tube of the particle detector the spacetime can be taken as flat.

The results we shall obtain are independent of the detailed nature of the particle detector. However, for explicitness we shall consider a particle model of a detector similar to that discussed by Unruh[14] for uniformly accelerated observers in flat space. This will consist of some system such as an atom which can be described by a nonrelativistic Schrödinger equation

$$i \frac{\partial \Psi}{\partial t'} = H_0 \Psi + g \phi \Psi ,$$

where t' is the proper time along the observer's world line, H_0 is the Hamiltonian of the undisturbed particle detector and $g \phi \Psi$ is a coupling term to the scalar field ϕ. The undisturbed particle detector will have energy levels E_i and wave functions $\Psi_i(\vec{R}')e^{-iE_i t}$, where $\vec{R}'$ represents the spatial position of a point in the detector.

By first-order perturbation theory the amplitude to excite the detector from energy level E_i to a higher-energy level E_j is proportional to

$$\int dt' \int d^3\vec{R}' \overline{\Psi}_j g \phi \Psi_i \exp[-i(E_i - E_j)t'] .$$

In other words, the detector responds to components of field ϕ which are positive frequency along the observer's world line with respect to his proper time. By superimposing detector levels with different energies one can obtain a detector response function of a form

$$f(t')h(\vec{R}) ,$$

where $f(t')$ is a purely positive-frequency function of the observer's proper time t' and h is zero outside some value of r' corresponding to the radius of the particle detector. Let $\mathcal{P}$ be a three-

surface which completely surrounds the observer's world line. If the observer detects a particle, it must have crossed $\mathcal{P}$ in some mode k_j which is a solution of the Klein-Gordon equation with unit Klein-Gordon norm over the hypersurface $\mathcal{P}$. The amplitude for the observer to detect such a particle will be

$$\int \int f h(x') G(x', x) \overline{\delta}_a \overline{k}_j(x) dV' d\Sigma^a , \qquad (4.20)$$

where the volume integral in x' is taken over the volume of the particle detector and the surface integral in x is taken over $\mathcal{P}$.

The hypersurface $\mathcal{P}$ can be taken to be a spacelike surface of large constant r in the past in region III and a spacelike surface of large constant r in the future in region II. In the limit that r tends to infinity these surfaces tend to past infinity $\mathcal{I}^-$ and future infinity $\mathcal{I}^+$, respectively. We shall assume that there were no particles present on the surface in the distant past. Thus the only contribution to the amplitude (4.20) comes from the surface in the future. One can interpret this as the spontaneous creation of a pair of particles, one with positive and one with negative energy with respect to the Killing vector $K = \partial/\partial t$. The particle with positive energy propagates to the observer and is detected. The particle with negative energy crosses the event horizon into region II where K is spacelike. It can exist there as a real particle with timelike four-momentum. Equivalently, one can regard the world lines of the two particles as being the world line of a single particle which tunnels through the event horizon out of region II and is detected by the observer.

Suppose the detector is sensitive to particles of a certain energy E. In this case the positive-frequency-response function $f(t)$ will be proportional to e^{-iEt}. By the stationarity of the metric, the propagator $G(x', x)$ can depend on the coordinates t' and t only through their difference. This means that the amplitude (4.20) will be zero except for modes k_j of the form $\chi(r, \theta, \varphi) e^{-iEt}$. If one takes out a δ function which arises from the integral over $t - t'$, the amplitude for detection is proportional to

$$\mathcal{E}_E(\vec{R}', \vec{R}) = \int_{-\infty}^{+\infty} dt\, e^{-iEt} G(0, \vec{R}'; t, \vec{R}) , \qquad (4.21)$$

where $\vec{R}'$ and $\vec{R}$ denote respectively (r', θ', ϕ') and (r, θ, φ) and the radial and angular integrals over the functions h and χ have been factored out. Using the result derived above that $G(x', x)$ is analytic in a strip of width $\pi \kappa_c^{-1}$ below the real t axis, one can displace the contour in (4.21) down $\pi \kappa_c^{-1}$ to obtain

$$\mathcal{S}_E(\vec{R}', \vec{R}) = \exp(-\pi E_C \kappa_C^{-1}) \int_{-\infty}^{+\infty} dt \, e^{-iEt} G(0, \vec{R}', t - i\pi\kappa_C^{-1}, \vec{R}).$$ (4.22)

By Eqs. (4.16) and (4.17) the point $(t - i\pi\kappa_C^{-1}, r, \theta, \varphi)$ is the point in region III obtained by reflecting in the origin of the U, V plane. Thus

$$\begin{pmatrix} \text{amplitude for particle of energy } E \text{ to propagate} \\ \text{from region II and be absorbed by observer} \end{pmatrix} = \exp(-\pi E \kappa_C^{-1}) \begin{pmatrix} \text{amplitude for particle with energy} \\ E \text{ to propagate from region III and} \\ \text{be absorbed by observer} \end{pmatrix}.$$

(4.23)

By time-reversal invariance the latter amplitude is equal to the amplitude for the observer's detector in an excited state to emit a particle with energy E which travels to region II. Therefore

$$\begin{pmatrix} \text{probability for detector to absorb} \\ \text{a particle from region II} \end{pmatrix} = \exp(-2\pi E \kappa_C^{-1}) \begin{pmatrix} \text{probability for detector to emit} \\ \text{a particle to region II} \end{pmatrix}.$$ (4.24)

This is just the condition for the detector to be in thermal equilibrium at a temperature

$$T = (2\pi)^{-1} \kappa_C = (12)^{-1/2} \pi^{-1} \Lambda^{1/2}.$$ (4.25)

The observer will therefore measure an isotropic background of thermal radiation with the above temperature. Because all timelike geodesics are equivalent under the de Sitter group, any other observer will also see an isotropic background with the same temperature even though he is moving relative to the first observer. This is yet another illustration of the fact that different observers have different definitions of particles. It would seem that one cannot, as some authors have attempted, construct a unique observer-independent renormalized energy-momentum tensor which can be put on the right-hand side of the classical Einstein equations. This subject will be dealt with in another paper.[16]

Another way in which one can derive the result that a freely moving observer in de Sitter space will see thermal radiation is to note that the propagator $G(x, x')$ is an analytic function of the

coordinates T, S, T', S', or alternatively U, V, U', V' except when x and x' can be joined by null geodesics. On the other hand, the static-time coordinate t is a multivalued function of T and S or U and V, being defined only up to an integral multiple of $2\pi i \kappa_C^{-1}$. Thus the propagator $G(x', x)$ is a periodic function of t with period $2\pi i \kappa_C^{-1}$. This behavior is characteristic of what are known as "thermal Green's functions."[33] These may be defined (for interacting fields as well as the noninteracting case considered here) as the expectation value of the time-ordered product of the field operators, where the expectation value is taken not in the vacuum state but over a grand canonical ensemble at some temperature $T = \beta^{-1}$. Thus

$$G_T(x', x) = i \, \text{Tr}[e^{-\beta H} \mathcal{J} \varphi(\dot{x}) \varphi(x)] / \text{Tr} e^{-\beta H},$$

(4.26)

where $\mathcal{J}$ denotes Wick time-ordering and H is the Hamiltonian in the observer's static frame. ϕ is the quantum field operator and Tr denotes the trace taken over a complete set of states of the system. Therefore

$$\begin{aligned} -iG_T(\vec{R}', t', \vec{R}, t) &= \text{Tr}[e^{-\beta H} \mathcal{J} \varphi(\vec{R}, t) \varphi(\vec{R}, t')] / \text{Tr} e^{-\beta H} \\ &= \text{Tr}[e^{-\beta H} \mathcal{J} \varphi(\vec{R}', t') e^{\beta H} e^{-\beta H} \phi(\vec{R}, t)] / \text{Tr} e^{-\beta H} \\ &= \text{Tr}[e^{-\beta H} \mathcal{J} \phi(\vec{R}', t' + i\beta) \phi(\vec{R}', t)] / \text{Tr} e^{-\beta H} \\ &= -iG_T(\vec{R}', t' + i\beta; \vec{R}, t). \end{aligned}$$

(4.27)

Since

$$\phi(\vec{R}, t) = e^{-\beta H} \phi(\vec{R}, t - i\beta) e^{\beta H}.$$ (4.28)

Thus the thermal propagator is periodic in $t - t'$ with period iT^{-1}. One would expect $G_T(x', x)$ to have singularities when x and x' can be connected by a null geodesic and these singularities would be repeated periodically in the complex $t' - t$ plane. It therefore seems that the propagator

$G(x', x)$ that we have defined by a path integral is the same as the thermal propagator $G_T(x', x)$ for a grand canonical ensemble at temperature T $T = (2\pi)^{-1} \kappa_C$ in the observer's static frame. Thus to the observer it will seem as if he is in a bath of blackbody radiation at the above temperature. It is interesting to note that a similar result was found for two-dimensional de Sitter space by Figari, Hoegh-Krohn, and Nappi[34] although they did not appreciate its significance in terms of particle creation.

The correspondence between $G(x', x)$ and the thermal Green's function is the same as that which has been pointed out in the black-hole case by Gibbons and Perry.[35] As in their paper, one can argue that because the free-field propagator $G(x', x)$ is identical with the free-field thermal propagator $G_T(x', x)$, any n-point interacting Green's function $\hat{G}$ which can be constructed by perturbation theory from G in a renormalizable field theory will be identical to the n-point inter- acting thermal Green's function constructed from G_T in a similar manner. This means that the re- sult that an observer will think himself to be immersed in blackbody radiation at temperature $T = \kappa_C (2\pi)^{-1}$ will be true not only in the free-field case that we have treated but also for fields with mutual interactions and self-interactions. In particular, one would expect it to be true for the gravitational field, though this is, of course, not renormalizable, at least in the ordinary sense.

It is more difficult to formulate the propagator for higher-spin fields in terms of a path integral. However, it seems reasonable to define the prop- agators for such fields as solutions of the relevant inhomogeneous wave equation with the boundary conditions that the propagator from a point x' in region I is an analytic function of x in the upper half U plane and lower half V plane on the com- plexified horizon. With this definition one ob- tains thermal radiation just as in the scalar case.

V. PARTICLE CREATION IN BLACK-HOLE DE SITTER SPACES

For the reasons given in Sec. III one would ex- pect that a solution of Einstein's equations with positive cosmological constant which contained a black hole would settle down eventually to one of the Kerr-Newman-de Sitter solutions described in Sec. II. We shall therefore consider what would be seen by an observer in such a solution. Con- sider first the Schwarzschild-de Sitter solution. Suppose the observer moves along a world line λ of constant r, θ, and ϕ in region I of Fig. 4. The world line λ coincides with an orbit of the static Killing vector $K = \partial/\partial t$. Let $\varphi^2 = g(K, K)$ on λ. One would expect that the observer would see thermal radiation with a temperature $T_C = (2\pi\psi)^{-1}\kappa_C$ coming from all directions except that of the black hole and thermal radiation of temperature $T_H = (2\pi\varphi)^{-1}\kappa_H$ coming from the black hole. The factor ψ appears in order to normalize the static Killing vector to have unit magnitude at the observer. The varia- tion of ψ with r can be interpreted as the normal red-shifting of temperature.

There are, however, certain problems in show- ing that this is the case. These difficulties arise from the fact that when one has two or more sets of horizons with different surface gravities one has to introduce separate Kruskal-type coordi- nate patches to cover each set of horizons. The coordinates of one patch will be real analytic func- tions of the coordinates of the next patch in some overlap region between the horizons in the real manifold. However, branch cuts arise if one continues the coordinates to complex values. To see this, let U_1, V_1 be Kruskal coordinates in a patch covering a pair of intersecting horizons with a surface gravity κ_1 and let U_2, V_2 be a neigh- boring coordinate patch covering horizons with surface gravity κ_2. In the overlap region one has

$$V_1 U_1^{-1} = -e^{2\kappa_2 t}, \tag{5.1}$$

$$V_2 U_2^{-1} = -e^{2\kappa_2 t}. \tag{5.2}$$

Thus

$$-V_2 U_2^{-1} = (-V_1)^P U_1^{-P}, \tag{5.3}$$

where $P = \kappa_2 \kappa_1^{-1}$. There is thus a branch cut in the relation between the two coordinate patches if $\kappa_2 \neq \kappa_1$.

One way of dealing with this problem would be to imagine perfectly reflecting walls between each black-hole horizon and each cosmological horizon. These walls would divide the manifold up into a number of separate regions each of which could be covered by a single Kruskal-coordinate patch. In each region one could construct a propagator as before but with perfectly reflecting boundary conditions at the walls. By arguments similar to those given in the previous section, these prop- agators will have the appropriate periodic and analytic properties to be thermal Green's functions with temperatures given by the surface gravities of the horizons contained within each region. Thus an observer on the black-hole side of a wall will see thermal radiation with the black-hole tempera- ture, while an observer on the cosmological side of the wall will see radiation with the cosmological temperature. One would expect that, if the walls were removed, an observer would see a mixture of radiation as described above.

Another way of dealing with the problem would be to define the paopagator $G(x', x)$ to be a solution of the inhomogeneous wave equation on the real manifold which was such that if the point were extended to complex values of a Kruskal-type- coordinate patch covering one set of intersecting horizons, it would be analytic on the complexified horizon in the upper half or lower half U or V plane depending on whether the point x was respectively to the future or the past of $V = 0$ or $U = 0$. Then, using a similar argument to that in the previous section about the dependence of the propagator on initial data on the complexified horizon, one can show that the propagator $G(x', x)$ between a point x' in region I and a point x in re-

gion II$_C$ is analytic in a strip of width $\pi \kappa_C^{-1}$ below the real axis of the complex t plane. Similarly, the propagator $G(x', x)$ between a point x' in region I and a point x in region II$_H$ will be analytic in a strip of width $\pi \kappa_H^{-1}$. Using these results one can show that

$$\begin{pmatrix} \text{probability of a particle of energy } E, \\ \text{relative to the observer, propagating} \\ \text{from } \mathscr{I}^+ \text{ to observer} \end{pmatrix} = \exp[-(E 2\pi \psi \kappa_C^{-1})] \begin{pmatrix} \text{probability of a particle of energy } E, \\ \text{relative to the observer, propagating} \\ \text{from observer to } \mathscr{I}^+ \end{pmatrix},$$

(5.4)

and similarly the probability of propagating from the future singularity of the black hole will be related by the appropriate factor to the probability for a similar particle to propagate from the observer into the black hole. These results establish the picture described at the beginning of this section.

One can derive similar results for the Kerr–de Sitter spaces. There is an additional complication in this case because there is a relative angular velocity between the black hole and the cosmological horizon. An observer in region I who is at a constant distance r from the black hole and who is nonrotating with respect to distant stars will move on an orbit of the Killing vector K which is null on the cosmological horizon. For such an observer the probability of a particle of energy E, relative to the observer, propagating to him from beyond the future cosmological horizon will be $\exp[-(2\pi \psi E \kappa_C^{-1})]$ times the probability for a similar particle to propagate from the observer to beyond the cosmological horizon. The probabilities for emission and absorption by the black hole will be similarly related except that in this case the energy E will be replaced by $E - n\Omega_H$, where n is the azimuthal quantum number or angular momentum of the particle about the axis of rotation of the black hole and Ω_H is the angular velocity of the black-hole horizon relative to the cosmological horizon. As in the ordinary black-hole case, the black hole will exhibit superradiance for modes for which $E < n\Omega_H$. In the case that the observer is moving on the orbit of a Killing vector K which is rotating with respect to the cosmological horizon, one again gets similar results for the radiation from the cosmological and black-hole horizons with E replaced by $E - n\Omega_C$ and $E - n\Omega_H$, respectively. Where Ω_C and Ω_H are the angular velocities of the cosmological and black-hole horizons relative to the observers frame and are defined by the requirement that $K + \Omega_C \tilde{K}$ and $K + \Omega_H \tilde{K}$ should be null on the cosmological and black-hole horizons.

VI. IMPLICATIONS AND CONCLUSIONS

We have shown that the close connection between event horizons and thermodynamics has a wider validity than the ordinary black-hole situations in which it was first discovered. As observer in a cosmological model with a positive cosmological constant will have an event horizon whose area can be interpreted as the entropy or lack of information that the observer has about the regions of the universe that he cannot see. When the solution has settled down to a stationary state, the event horizon will have associated with it a surface gravity κ which plays a role similar to temperature in the classical first law of event horizons derived in Sec. III. As was shown in Sec. IV., this similarity is more than an analogy: The observer will detect an isotropic background of thermal radiation with temperature $(2\pi)^{-1}\kappa$ coming, apparently, from the event horizon. This result was obtained by considering what an observer with a particle detector would actually measure rather than by trying to define particles in an observer-independent manner. An illustration of the observer dependence of the concept of particle is the result that the thermal radiation in de Sitter space appears isotropic and at the same temperature to every geodesic observer. If particles had an observer-independent existence and if the radiation appeared isotropic to one geodesic observer, it would not appear isotropic to any other geodesic observer. Indeed, as an observer approached the first observer's future event horizon the radiation would diverge. It seems clear that this observer dependence of particle creation holds in the case of black holes as well: An observer at constant distance from a black hole will observe a steady emission of thermal radiation but an observer falling into a black hole will not observe any divergence in the radiation as he approaches the first-observer's event horizon.

A consequence of the observer dependence of particle creation would seem to be that the back reaction must be observer-dependent also, if one assumes, as seems reasonable, that the mass of the detector increases when it absorbs a particle and therefore the gravitational field changes. This will be discussed further in another paper,[16] but we remark here that it involves the abandoning of the concept of an observer-independent metric for spacetime and the adoption of something like the Everett-Wheeler interpretation of quantum mechanics.[36] The latter viewpoint seems to be required anyway when dealing with the quantum mechanics of the whole universe rather than an isolated system.

If a geodesic observer in de Sitter space chooses not to absorb any of the thermal radiation, his energy and entropy do not change and so one would not expect any change in the solution. However, if he does absorb some of the radiation, his energy and hence his gravitational mass will increase. If the solution now settles down again to a new stationary state, it follows from the first law of event horizons that the area of the cosmological event horizon will be less than it appeared to be before. One can interpret this as a reduction in the entropy of the universe beyond the event horizon caused by the propagation of some radiation from this region to the observer. Unlike the black-hole case, the surface gravity of the cosmological horizon decreases as the horizon shrinks. There is thus no danger of the observer's cosmological event horizon shrinking catastrophically around him because of his absorbing too much thermal radiation. He has, however, to be careful that he does not absorb so much radiation that his particle detector undergoes gravitational collapse to produce a black hole. If this were to happen, the black hole would always have a higher temperature than the surrounding universe and so would radiate energy faster than it absorbs it. It would therefore evaporate, leaving the universe as it was before the observer began to absorb radiation.

---

*Present address: Max-Planck-Institute für Physik and Astrophysik, 8 München 40, Postfach 401212, West Germany. Telephone: 327001.

[1] W. Israel, Phys. Rev. 164, 1776 (1967).

[2] H. Muller zum Hagen et al., Gen. Relativ. Gravit. 4, 53 (1973).

[3] B. Carter, Phys. Rev. Lett. 26, 331 (1970).

[4] S. W. Hawking, Commun. Math. Phys. 25, 152 (1972).

[5] D. C. Robinson, Phys. Rev. Lett. 34, 905 (1975).

[6] D. C. Robinson, Phys. Rev. D 10, 458 (1974).

[7] J. Bekenstein, Phys. Rev. D 7, 2333 (1973).

[8] J. Bekenstein, Phys. Rev. D 9, 3292 (1974).

[9] S. W. Hawking, Phys. Rev. D 13, 191 (1976).

[10] J. Hartle and S. W. Hawking, Phys. Rev. D 13, 2188 (1976).

[11] R. Penrose, in Relativity, Groups and Topology, edited by C. DeWitt and B. DeWitt (Gordon and Breach, New York, 1964).

[12] S. W. Hawking and G. F. R. Ellis, Large Scale Structure of Spacetime (Cambridge Univ. Press, New York, 1973).

[13] J. Bardeen, B. Carter, and S. W. Hawking, Commun. Math Phys. 31, 162 (1973).

[14] W. Unruh, Phys. Rev. D 14, 870 (1976).

[15] A. Ashtekar and A. Magnon, Proc. R. Soc. London A346, 375 (1975).

[16] S. W. Hawking, in preparation.

[17] J. D. North, The Measure of the Universe (Oxford Univ. Press, New York, 1965).

[18] C. Kahn and F. Kahn, Nature 257, 451 (1975).

[19] E. Schrödinger, Expanding Universes (Cambridge Univ. Press, New York, 1956).

[20] B. Carter, Commun. Math. Phys. 17, 233 (1970).

[21] B. Carter, in Les Astre Occlus (Gordon and Breach, New York, 1973).

[22] R. Geroch, E. H. Kronheimer, and R. Penrose, Proc. R. Soc. London A327, 545 (1972).

[23] S. W. Hawking, in preparation.

[24] S. W. Hawking, Commun. Math. Phys. 25, 152 (1972).

[25] P. Hajicek, Phys. Rev. D 7, 2311 (1973).

[26] L. Smarr, Phys. Rev. Lett. 30, 71 (1973); 30, 521(E) (1973).

[27] O. Nachtmann, Commun. Math. Phys. 6, 1 (1967).

[28] E. A. Tagirov, Ann. Phys. (N.Y.) 76, 561 (1973).

[29] P. Candelas and D. Raine, Phys. Rev. D 12, 965 (1975).

[30] J. S. Dowker and R. Critchley, Phys. Rev. D 13, 224 (1976).

[31] S. W. Hawking, Nature 248, 30 (1974).

[32] S. W. Hawking, Commun. Math. Phys. 43, 199 (1975).

[33] A. L. Fetter and J. P. Walecka, Quantum Theory of Many Particle Systems (McGraw-Hill, New York, 1971).

[34] R. Figari, R. Hoegh-Krohn, and C. Nappi, Commun. Math. Phys. 44, 265 (1975).

[35] G. W. Gibbons and M. J. Perry, Phys. Rev. Lett. 36, 985 (1976).

[36] The Many Worlds Interpretation of Quantum Mechanics edited by B. S. DeWitt and N. Graham (Princeton Univ. Press, Princeton, N. J., 1973).

Volume 115B, number 4   PHYSICS LETTERS   9 September 1982

THE DEVELOPMENT OF IRREGULARITIES
IN A SINGLE BUBBLE INFLATIONARY UNIVERSE

S.W. HAWKING

University of Cambridge, DAMTP, Silver Street, Cambridge, UK

Received 25 June 1982

The horizon, flatness and monopole problems can be solved if the universe underwent an exponentially expanding stage which ended with a Higgs scalar field running slowly down an effective potential. In the downhill phase irregularities would develop in the scalar field. These would lead to fluctuations in the rate of expansion which would have the right spectrum to account for the existence of galaxies. However the amplitude would be too high to be consistent with observations of the isotropy of the microwave background unless the effective coupling constant of the Higgs scalar was very small.

Observations of the microwave background and of the abundances of helium and deuterium indicate that the standard hot big-bang model is probably a good description of the universe, at least back to the time when the temperature was 10^{10} K. However this model leaves unanswered a number of questions including the following.

(1) Why does the ratio of the number of baryons to the number of photons in the universe have the observed value of about 10^{-8}–10^{-10}?

(2) Why is the universe so homogeneous and isotropic on a large scale if different regions were out of contact with each other at early times as they are in the standard model?

(3) Why is the present density of the universe so near the critical value that divides recollapse from indefinite expansion?

(4) Why are there not many more superheavy monopoles formed when the grand unified symmetry was broken?

(5) Why, despite the large scale homogeneity and isotropy, were there sufficient irregularities to give rise to stars and galaxies?

A possible answer to the first question has been provided by grand unified theories which predict the creation of a non-zero baryon number if there are *CP* violating interactions. In attempts to answer the second, third and fourth questions, various authors have suggested that the universe underwent a period of exponential expansion [1–3]. In order not to conflict with the explanation of the baryon number, this exponential expansion would have had to have taken place before the last time the universe was at a temperature of the order of the grand unification energy, $M \sim 10^{14}$–10^{15} GeV. A detailed scenario for such an inflationary or exponential expansion phase has been provided by Guth [3]. At very early times the temperature of the universe is supposed to have been above the grand unification energy M and the symmetry of the grand unified theory would have been unbroken. As the universe expanded, the temperature T would have fallen below M but the Higgs scalar fields may have been prevented from acquiring a symmetry breaking expectation value by the existence of a barrier in the effective potential. In this situation the universe would supercool in the metastable unbroken symmetry phase. The vacuum energy of the unbroken phase would act as an effective Λ term and would lead to an exponential expansion of the universe with a Hubble constant H of the order of $(\frac{8}{3}\pi)^{1/2}M^2/m_p$ where m_p is the Planck mass, 10^{19} GeV.

The universe is obviously not now expanding at this rate, so something must have happened to end the exponentially expanding stage. Guth's original suggestion was that there would be a first order phase transition in which bubbles of the broken symmetry phase would form. Most of the vacuum energy of the unbroken phase would be converted into the kinetic energy of walls of the bubble which would expand at nearly the speed of light. The region inside the bubble would be at a low temperature and would be nearly empty. The energy in the bubble walls would be released and the universe reheated to the GUT temperature only if the bubble collided with other bubbles. However, because of the exponential expansion of the universe, the probability of more than a few bubbles colliding would be small. This would lead to a very inhomogeneous universe which would not be compatible with the observations of the microwave background.

In order to avoid this difficulty several authors [4–8] have suggested that the barrier in the effective potential becomes very small or disappears altogether:

$$V(\phi) = \tfrac{1}{2}\mu^2\phi^2 + \alpha^2\phi^4\left[\log(\phi^2/\phi_0^2) - \tfrac{1}{2}\right] + V_0,$$

where μ^2 includes rest mass, thermal and curvature contributions and $|\mu|$ is small compared to the expansion rate H. Quantum field theory in the exponentially expanding stage can be defined on S^4, the euclidean version of de Sitter space [6,9]. Thus the fluctuations of the scalar field around the unbroken symmetry value can be decomposed into four-sphere harmonics. The homogeneous $l = 0$ mode will be unstable if $\mu^2 < 0$ and neutrally stable if $\mu^2 = 0$. If $0 < \mu^2 \ll H^2$, the universe will make a quantum transition to a state of constant ϕ at the maximum of the potential [6]. In all three cases the $l = 0$ mode will then start to run down the hill to the global minimum at $\phi = \phi_0$ in a timescale $t_1 = cH_1^{-1}$. Provided that $|\mu^2| \lesssim \tfrac{1}{20}H^2$, the constant c will be greater than about 60. In this case any initial spatial curvature of the hypersurfaces of constant time on which ϕ is constant will be reduced to a sufficiently small value by the time the scalar field reaches ϕ_0 that the universe will expand until the present time as a nearly $k = 0$ model.

The higher l modes on the four-sphere will be stable because of the gradient terms in the action. However they will have quantum fluctuations which will be superimposed on the downhill career of the $l = 0$ mode. They are described by a two-point correlation function $G(x,y) = \langle\phi(x)\phi(y)\rangle'$ where the prime indicates that the $l = 0$ modes have been projected out. This obeys the equation

$$(-\Box + \mu^2)G(x,y) = \delta(x,y) - (3/8\pi^2)H^4$$

on the four-sphere of radius H^{-1}. When analytically continued to de Sitter space

$$G(x,y) \approx -\pi^{-2}H^2\log(H|x - y|),$$

where the points x, y lie on a surface of constant time in a $k = 0$ coordinate system at a separation greater than the horizon H^{-1}. The Fourier transform of $G(x,y)$ in a surface of constant time is

$$\widetilde{G}(k) \approx -H^2k^{-3}, \quad k \ll H.$$

Thus a Fourier component of the ϕ field with a wave number k has an amplitude of the order of $Hk^{-3/2}$.

These inhomogeneous fluctuations mean that on a surface of constant time there will be some regions in which the ϕ field has run further down the hill of the effective potential than in other regions. However, when one is dealing with fluctuations with wavelengths much longer than the horizon H^{-1} i.e. with $k \ll H$, such variations can be removed by choosing a new time coordinate such that the surfaces of constant time are surfaces of constant ϕ. The amount by which the time coordinate has to be shifted is of the order of

$$\widetilde{\delta t} = Hk^{-3/2}[(d/dt)\langle\phi\rangle]^{-1}.$$

The $l = 0$ mode represented by $\langle\phi\rangle$ will obey the equation

$$(\partial^2/\partial t^2)\langle\phi\rangle + 3H(\partial/\partial t)\langle\phi\rangle = -\partial V/\partial\langle\phi\rangle.$$

Thus

$$\langle\phi\rangle \approx [3H/8\alpha^2(t_0 - t)]^{1/2},$$

for $t < t_0 - H$ where t_0 is the time at which the field reaches the global minimum $\langle\phi\rangle = \phi_0$. A comoving region with a present length k^{-1} will cross the event horizon of the de Sitter space at a time $t = t_0 - H^{-1}\log(Hk^{-1})$. Thus

$$\widetilde{\delta t} \approx \alpha H^{-1}[k^{-1}\log(Hk^{-1})]^{3/2}.$$

The surfaces of constant time will now be surfaces of nearly constant energy–momentum tensor. However, the change of time coordinate will have introduced inhomogeneous fluctuations in the rate of expansion H.

$$\delta\widetilde{H} \approx k^2\delta t \approx \alpha H^{-1}k^{1/2}[\log(Hk^{-1})]^{3/2}.$$

The two-point correlation function of δH^2 therefore has a Fourier transform of the order of

$$\alpha^2 [\log(Hk^{-1})]^3 .$$

This is just the scale-independent spectrum of fluctuations that Harrison and Zeldovich [10,11] have suggested could account for galaxy formation. However observations of the microwave background place an upper limit of about 10^{-8} on the dimensionless amplitude of these fluctuations on scales of the order of the present Hubble radius. For such scales $\log(Hk^{-1}) \approx 50$ so the fluctuations would be too large to be compatible with observations unless the coupling constant α were very small. What is needed is a potential of a different form with a region of nearly constant slope $-\partial V/\partial\phi \gg H^3$. Such a potential might arise in a supersymmetric theory.

References

[1] A.A. Starobinskii, Phys. Lett. 91B (1980) 99.

[2] J.R. Gott, Nature 295 (1982) 304.

[3] A.H. Guth, Phys. Rev. D23 (1981) 347.

[4] W.H. Press, Galaxies may be single particle fluctuations from an early, false-vacuum era, Harvard preprint (1981) 1491.

[5] A.D. Linde, Phys. Lett. 108B (1982) 389; 114B (1982) 431.

[6] S.W. Hawking and I.G. Moss, Phys. Lett. 110B (1982) 35.

[7] A. Albrecht and P.J. Steinhardt, Phys. Rev. Lett. 48 (1982) 1220.

[8] A. Albrecht, P.J. Steinhardt, M.S. Turner and F. Wilczek, Phys. Rev. Lett. 48 (1982) 1437.

[9] G.W. Gibbons and S.W. Hawking, Phys. Rev. D15 (1977) 273.

[10] E.R. Harrison, Phys. Rev. D1 (1970) 2726.

[11] Ya.B. Zeldovich, Mon. Not. R. Astr. Soc. 160 (1972) 1P.

Commun. math. Phys. 55, 133—148 (1977)

Communications in
Mathematical Physics
© by Springer-Verlag 1977

Zeta Function Regularization of Path Integrals in Curved Spacetime

S. W. Hawking

Department of Applied Mathematics and Theoretical Physics, University of Cambridge,
Cambridge CB3 9EW, England

Abstract. This paper describes a technique for regularizing quadratic path integrals on a curved background spacetime. One forms a generalized zeta function from the eigenvalues of the differential operator that appears in the action integral. The zeta function is a meromorphic function and its gradient at the origin is defined to be the determinant of the operator. This technique agrees with dimensional regularization where one generalises to n dimensions by adding extra flat dimensions. The generalized zeta function can be expressed as a Mellin transform of the kernel of the heat equation which describes diffusion over the four dimensional spacetime manifold in a fith dimension of parameter time. Using the asymptotic expansion for the heat kernel, one can deduce the behaviour of the path integral under scale transformations of the background metric. This suggests that there may be a natural cut off in the integral over all black hole background metrics. By functionally differentiating the path integral one obtains an energy momentum tensor which is finite even on the horizon of a black hole. This energy momentum tensor has an anomalous trace.

1. Introduction

The purpose of this paper is to describe a technique for obtaining finite values to path integrals for fields (including the gravitational field) on a curved spacetime background or, equivalently, for evaluating the determinants of differential operators such as the four-dimensional Laplacian or D'Alembertian. One forms a gemeralised zeta function from the eigenvalues λ_n of the operator

$$\zeta(s) = \sum_n \lambda_n^{-s} . \tag{1.1}$$

In four dimensions this converges for $\mathrm{Re}(s) > 2$ and can be analytically extended to a meromorphic function with poles only at $s = 2$ and $s = 1$. It is regular at $s = 0$. The derivative at $s = 0$ is formally equal to $-\sum_n \log \lambda_n$. Thus one can define the determinant of the operator to be $\exp(-d\zeta/ds)|_{s=0}$.

In situations in which one knows the eigenvalues explicitly one can calculate the zeta function directly. This will be done in Section 3, for the examples of thermal radiation or the Casimir effect in flat spacetime. In more complicated situations one can use the fact that the zeta function is related by an inverse Mellin transform to the trace of the kernel of the heat equation, the equation that describes the diffusion of heat (or ink) over the four dimensional spacetime manifold in a fifth dimension of parameter time t. Asymptotic expansions for the heat kernel in terms of invariants of the metric have been given by a number of authors [1–4].

In the language of perturbation theory the determinant of an operator is expressed as a single closed loop graph. The most commonly used method for obtaining a finite value for such a graph in flat spacetime is dimensional regularization in which one evaluates the graph in n spacetime dimensions, treats n as a complex variable and subtracts out the pole that occurs when n tends to four. However it is not clear how one should apply this procedure to closed loops in a curved spacetime. For instance, if one was dealing with the four sphere, the Euclidean version of de Sitter space, it would be natural to generalize that S^4 to S^n [5, 6]. On the other hand if one was dealing with the Schwarzschild solution, which has topology $R^2 \times S^2$, one might generalize to $R^2 \times S^{n-2}$. Alternatively one might add on extra dimensions to the R^2. These additional dimensions might be either flat or curved. The value that one would obtain for a closed loop graph, would be different in these different extensions to n dimensions so that dimensional regularization is ambiguous in curved spacetime. In fact it will be shown in Section 5 that the answer given by the zeta function technique agrees up to a multiple of the undetermined renormalization parameter with that given by dimensional regularization where the generalization to n dimensions is given by adding on extra flat dimensions.

The zeta function technique can be applied to calculate the partition functions for thermal gravitons and matter quanta on black hole and de Sitter backgrounds. It gives finite values for these despite the infinite blueshift of the local temperature on the event horizons. Using the asymptotic expansion for the heat kernel, one can relate the behaviour of the partition function under changes of scale of the background spacetime to an integral of a quadratic expression in the curvature tensor. In the case of de Sitter space this completely determines the partition function up to a multiple of the renormalization parameter while in the Schwarzschild solution it determines the partition function up to a function of r_0/M where r_0 is the radius of the box containing a black hole of mass M in equilibrium with thermal radiation. The scaling behaviour of the partition function suggests that there may be a natural cut off at small masses when one integrates over all masses of the black hole background.

By functional differentiating the partition function with respect to the background metric one obtains the energy momentum tensor of the thermal radiation. This can be expressed in terms of derivatives of the heat kernel and is finite even on the event horizon of a black hole background. The trace of the energy momentum is related to the behaviour of the partition function under scale transformations. It is given by a quadratic expression in the curvature and is non zero even for conformally invariant fields [7–12].

The effect of the higher order terms in the path integrals is discussed in Section 9. They are shewn to make an insignificant contribution to the partition function for thermal radiation in a black hole background that is significantly bigger than the Planck mass. Generalised zeta functions have also been used by Dowker and Critchley [11] to regularize one-loop graphs. Their approach is rather different from that which will be given here.

2. Path Integrals

In the Feynmann sum over histories approach to quantum theory one considers expressions of the form

$$Z = \int d[g]d[\phi]\exp\{iI[g,\phi]\} , \tag{2.1}$$

where $d[g]$ is a measure on the space of metrics g, $d[\phi]$ is a measure on the space of matter fields ϕ and $I[g,\phi]$ is the action. The integral is taken over all fields g and ϕ that satisfy certain boundary or periodicity conditions. A situation which is of particular interest is that in which the fields are periodic in imaginary time on some boundary at large distance with period β [13]. In this case Z is the partition function

for a canonical ensemble at the temperature $T = \dfrac{1}{\beta}$.

The dominant contribution to the path integral (2.1) will come from fields that are near background fields g_0 and ϕ_0 which satisfy the boundary or periodicity conditions and which extremise the action i.e. they satisfy the classical field equations. One can expand the action in a Taylor series about the background fields:

$$I[g,\phi] = I[g_0,\phi_0] + I_2[\tilde{g}] + I_2[\tilde{\phi}] + \text{higher order terms}, \tag{2.2}$$

where

$$g = g_0 + \tilde{g}, \quad \phi = \phi_0 + \tilde{\phi}$$

and $I_2[\tilde{g}]$ and $I_2[\tilde{\phi}]$ are quadratic in the fluctuations $\tilde{g}$ and $\tilde{\phi}$. Substituting (2.2) into (2.1) and neglecting the higher order terms one has

$$\begin{aligned} \log Z = iI[g_0,\phi_0] + \log \int d[\tilde{g}]\exp iI_2[\tilde{g}] \\ + \log \int d[\tilde{\phi}]\exp iI_2[\tilde{\phi}] . \end{aligned} \tag{2.3}$$

The background metric g_0 will depend on the situation under consideration but in general it will not be a real Lorentz metric. For example in de Sitter space one complexifies the spacetime and goes to a section (the Euclidean section) on which the metric is the real positive definite metric on a four sphere. Because the imaginary time coordinate is periodic on this four sphere, Z will be the partition function for a canonical ensemble. The action $I[g_0,\phi_0]$ of the background de Sitter metric gives the contribution of the background metric to the partition function while the second and third terms in Equation (2.3) give the contributions of thermal gravitons and matter quanta respectively on this background. In the case of the canonical ensemble for a spherical box with perfectly reflecting walls the background metric can either be that of a Euclidean space or it can be that of a section (the Euclidean section) of the complexified Schwarzschild solution on which the metric is real positive definite. Again the action of the background metric gives the contribution of the background metric to the partition function. This corresponds to an entropy equal to one quarter of the area of the event horizon in units in which $G = c = \hbar = k = 1$. The second and third terms in Equation (2.3) give the contributions of thermal gravitons and matter quanta on a Schwarzschild background. In the case of the grand canonical ensemble for a box with temperature $T = \beta^{-1}$ and angular velocity Ω one considers fields which, on the walls of the box, have the same value at the point (t, r, θ, ϕ) and at the point $(t + i\beta, r, \theta, \phi + i\beta\Omega)$. This boundary cannot be filled in with any real metric but it can be filled in with a complex flat metric or with a complex section (the quasi Euclidean section [13]) of the Kerr solution. In both cases the metric is strongly elliptic (I am grateful to Dr. Y. Manor for this point) [14] if the rotational velocity of the boundary is less than that of light. A metric g is said to be strongly elliptic if there is a function f such that $\mathrm{Re}(fg)$ is positive definite. It seems necessary to use such strongly elliptic background metrics to make the path integrals well defined. One could take this to be one of the basic postulates of quantum gravity.

The quadratic term $I_2[\phi]$ will have the form

$$I_2[\tilde{\phi}] = - \int \tfrac{1}{2}\tilde{\phi} A \tilde{\phi} (-g_0)^{1/2} d^4 x \, , \tag{2.4}$$

where A is a second order differential operator constructed out of the background fields g_0, ϕ_0. (In the case of the fermion fields the operator A is first order. For simplicity I shall deal only with boson fields but the results can easily be extended to fermions.) The quadratic term $I_2[\tilde{g}]$ in the metric fluctuations can be expressed similarly. Here however, the second order differential operator is degenerate i.e. it does not have an inverse. This is because of the gauge freedom to make coordinate transformations. One deals with this by taking the path integral only over metrics that satisfy some gauge condition which picks out one metric from each equivalence class under coordinate transformations. The Jacobian from the space of all metrics to the space of those satisfying the gauge condition can be regarded in perturbation theory as introducing fictitious particles known as Feynmann-de Witt [15, 16] or Fadeev-Popov ghosts [17]. The path integral over the gravitational fluctuations will be treated in another paper by methods similar to those used here for matter fields without gauge degrees of freedom.

In the case when the background metric g_0 is Euclidean i.e. real and positive definite the operator A in the quadratic term $I_2[\phi]$ will be real, elliptic and self-adjoint. This means that it will have a complete spectrum of eigenvectors ϕ_n with real eigenvalues λ_n:

$$A\phi_n = \lambda_n \phi_n \, . \tag{2.5}$$

The eigenvectors can be normalized so that

$$\int \phi_n \phi_m (g_0)^{1/2} d^4 x = \delta_{nm} \, . \tag{2.6}$$

Note that the volume element which appears in the (2.6) is $(g_0)^{1/2}$ because g_0 is positive definite. On the other hand the volume element that appears in the action I is $(-g)^{1/2} = -i(g)^{1\ 2}$ where the minus sign corresponds to a choice of the direction of Wick rotation of the time axis into the complex plane.

If the background metric g_0 is not Euclidean, the operator A will not be self-adjoint. However I shall assume that the eigen functions ϕ_n are still complete. If this is so, one can express the fluctuation $\tilde{\phi}$ in terms of the eigen functions.

$$\tilde{\phi} = \sum a_n \phi_n \ . \tag{2.7}$$

The measure $d[\phi]$ on the space of all fields $\tilde{\phi}$ can then be expressed in terms of the coefficients a_n:

$$d[\phi] = \prod_n \mu da_n \ , \tag{2.8}$$

where μ is some normalization constant with dimensions of mass or inverse length. From (2.5)–(2.8) it follows that

$$
\begin{aligned}
Z[\tilde{\phi}] &= \int d[\phi] \exp i I_2[\tilde{\phi}] \\
&= \prod_n \int \tfrac{1}{2}\mu da_n \exp(-\lambda_n a_n^2) \\
&= \prod_n \tfrac{1}{2}\mu\pi^{1/2}\lambda_n^{-1/2} \\
&= (\det(4\mu^{-2}\pi^{-1}A))^{-1/2} \ .
\end{aligned} \tag{2.9}
$$

3. The Zeta Function

The determinant of the operator A clearly diverges because the eigenvalues λ_n increase without bound. One therefore has to adopt some regularization procedure. The technique that will be used in this paper will be called the zeta function method. One forms a generalized zeta function from the eigenvalues of the operator A:

$$\zeta(s) = \sum_n \lambda_n^{-s} \ . \tag{3.1}$$

In four dimensions this will converge for $\mathrm{Re}(s) > 2$. It can be analytically extended to a merophorphic function of s with poles only at $s = 2$ and $s = 1$ [18]. In particular it is regular at $s = 0$. The gradient of zeta at $s = 0$ is formally equal to $-\sum_n \log\lambda_n$. One can therefore *define* $\det A$ to be $\exp(-d\zeta/ds|_{s=0})$ [19]. Thus the partition function

$$\log Z[\tilde{\phi}] = \tfrac{1}{2}\zeta'(0) + \tfrac{1}{2}\log(\tfrac{1}{4}\pi\mu^2)\zeta(0) \ . \tag{3.2}$$

In situations in which the eigenvalues are known, the zeta function can be computed explicitly. To illustrate the method, I shall treat the case of a zero rest mass scalar field ϕ contained in a box of volume V in flat spacetime at the temperature $T = \beta^{-1}$. The partition function will be defined by a path integral over all fields ϕ on the Euclidean space obtained by putting $\tau = it$ which are zero on the walls of the box and which are periodic in τ with period β. The operator A in the action is the negative of the four dimensional Laplacian on the Euclidean space. If the dimensions of the box are large compared to the characteristic wavelength β, one can approximate the spatial dependence of the eigenfunctions by plane waves with periodic boundary conditions. The eigenvalues are then

$$\lambda_n = (2\pi\beta^{-1}n)^2 + k^2 \tag{3.3}$$

and the density of eigenvalues in the continuum limit is

$$\frac{2V}{(2\pi)^3}\int d^3k \tag{3.4}$$

when $n>0$ and half that when $n=0$. The zeta function is therefore

$$\zeta(s) = \frac{4\pi V}{(2\pi)^3}\left\{\int dk\, k^{2-2s} + 2\sum_{n=1}^{\infty}\int dk\, k^2(4\pi^2\beta^{-2}n^2 + k^2)^{-s}\right\}. \tag{3.5}$$

The second term can be integrated by parts to give

$$-\frac{8\pi V}{(2\pi)^3}\sum_{n=1}^{\infty}\int dk(4\pi^2\beta^{-2}n^2 + k^2)^{-s+1}(2-2s)^{-1}. \tag{3.6}$$

Put $k = 2\pi n\beta^{-1}\sinh y$. This gives

$$-\frac{8\pi V}{(2\pi)^3}\sum_{n=1}^{\infty}\int dy(2\pi\beta^{-1}n)^{-2s+3}(2-2s)^{-1}(\cosh y)^{-2s+3}$$

$$= -\frac{8\pi V}{(2\pi)^3}(2\pi\beta^{-1})^{3-2s}\times\zeta_R(2s-3)$$

$$\cdot(2-2s)^{-1}\times\frac{1}{2}\frac{\Gamma(1/2)\Gamma(s-3/2)}{\Gamma(s-1)}, \tag{3.7}$$

where ζ_R is the usual Riemann zeta function $\sum_n n^{-s}$. The first term in (3.5) seems to diverge at $k=0$ when s is large and positive. This infra red divergence can be removed if one assumes that the box containing the radiation is large but finite. In this case the k integration has a lower cut off at some small value ε. If s is large, the k integration then gives a term proportional to ε^{3-2s}. When analytically continued to $s=0$, this can be neglected in the limit $\varepsilon\to0$, corresponding to a large box.

The gamma function $\Gamma(s-1)$ has a pole at $s=0$ with residue -1. Thus the generalised zeta function is zero at $s=0$ and

$$\zeta'(0) = 2\pi V\beta^{-3}\zeta_R(-3)\Gamma(1/2)\Gamma(-3/2) \tag{3.8}$$

$$= \frac{\pi^2}{45}VT^3$$

thus the partition function for scalar thermal radiation at temperature T in a box of volume V is given by

$$\log Z = \frac{\pi^2 VT^3}{90}. \tag{3.9}$$

Note that because $\zeta(0)=0$, the partition function does not depend on the undetermined normalization parameter μ. However, this will not in general be the case in a curved space background.

From the partition function one can calculate the energy, entropy and pressure of the radiation.

$$E = -\frac{d}{d\beta}\log Z = \frac{\pi^2}{30}VT^4 \, , \tag{3.10}$$

$$S = \beta E + \log Z = \frac{2\pi^2}{45}VT^3 \, , \tag{3.11}$$

$$P = \beta^{-1}\frac{d}{dV}\log Z = \frac{\pi^2}{90}T^4 \, . \tag{3.12}$$

One can calculate the partition functions for other fields in flat space in a similar manner. For a charged scalar field there are twice the number of eigenfunctions so that $\log Z$ is twice the value given by Equation (3.9). In the case of the electromagnetic field the operator A in the action integral is degenerate because of the freedom to make electromagnetic gauge transformations. One therefore has, as in the gravitational case, to take the path integral only over fields which satisfy some gauge condition and to take into account the Jacobian from the space of all fields satisfying the gauge condition. When this is done one again obtains a value $\log Z$ which is twice that of Equation (3.9). This corresponds to the fact that the electromagnetic field has two polarization states.

One can also use the zeta function technique to calculate the Casimir effect between two parallel reflecting planes. In this case instead of summing over all field configurations which are periodic in imaginary time, one sums over fields which are zero on the plates. Defining Z to be the path integral over all such fields over an interval of imaginary time τ one has

$$\log Z = \frac{\pi^2 A\tau b^{-3}}{720} \, , \tag{3.13}$$

where b is the separation and A the area of the plates. Thus the force between the plates is

$$F = \tau^{-1}\frac{d}{db}\log Z = -\frac{\pi^2 Ab^{-4}}{240} \, . \tag{3.14}$$

4. The Heat Equation

In situations in which one does not know the eigenvalues of the operator A, one can obtain some information about the generalized zeta function by studying the heat equation.

$$\frac{d}{dt}F(x,y,t) + AF(x,y,t) = 0 \tag{4.1}$$

here x and y represent points in the four dimensional spacetime manifold, t is a fifth dimension of parameter time and the operator A is taken to act on the first argument of F. With the initial conditions

$$F(x, y, 0) = \delta(x, y) \tag{4.2}$$

the heat kernel F represents the diffusion over the spacetime manifold in parameter time t of a unit quantity of heat (or ink) placed at the point y at $t = 0$. One can express F in terms of the eigenvalues and eigenfunctions of A:

$$F(x, y, t) = \sum_n \exp(-\lambda_n t) \phi_n(x) \phi_n(y) . \tag{4.3}$$

In the case of a field ϕ with tensor or spinor indices, the eigenfunctions will carry a set of indices at the point x and a set at the point y. If one puts $x = y$, contracts over the indices at x and y and integrates over all the manifold one obtains

$$Y(t) \equiv \int \mathrm{Tr} F(x, x, t)(g_0)^{1/2} d^4x = \sum_n \exp(-\lambda_n t) . \tag{4.4}$$

The generalized zeta function is related to $Y(t)$ by a Mellin transform:

$$\zeta(s) = \sum_n' \lambda_n^{-s} = \frac{1}{\Gamma(s)} \int_0^\infty t^{s-1} Y(t) dt . \tag{4.5}$$

A number of authors e.g. [1–4] have obtained asymptotic expansions for F and Y valid as $t \to 0^+$. In the case that the operator A is a second order Laplacian type operator on a four dimensional compact manifold.

$$Y(t) \sim \sum_n B_n t^{n-2} , \tag{4.6}$$

where the coefficients B_n are integrals over the manifold of scalar polynomials in the metric, the curvature tensor and its covariant derivatives, which are of order $2n$ in the derivatives of the metric

i.e. $$B_n = \int b_n (g_0)^{1/2} d^4x . \tag{4.7}$$

DeWitt [1, 2] has calculated the b_n for the operator $-\Box + \xi R$ acting on scalars,

$$b_0 = (4\pi)^{-2}$$
$$b_1 = (4\pi)^{-2}(\tfrac{1}{6} - \xi)R$$
$$b_2 = (2880\pi^2)^{-1}$$
$$\cdot [R^{abcd} R_{abcd} - R^{ab} R_{ab} + 30(1 - 6\xi)^2 R^2 + (6 - 30\xi)\Box R] . \tag{4.8}$$

Note that b_1 is zero when $\xi = \tfrac{1}{6}$ which corresponds to a conformally invariant scalar field.

In the case of a non-compact spacetime manifold one has to impose boundary conditions on the heat equation and on the eigenfunctions of the operator A. This can be done by adding a boundary to the manifold and requiring the field or its normal derivative to be zero on the boundary. An example is the case of a black hole metric such as the Euclidean section of the Schwarzschild solution in which one adds a boundary at some radius $r = r_0$. This boundary represents the walls of a perfectly reflecting box enclosing the black hole. For a manifold with boundary the asymptotic expansion for Y takes the form [20].

$$Y(t) = \sum_n (B_n + C_n) t^{n-2} ,\tag{4.9}$$

where, as before, B_n has the form (4.7) and

$$C_n = \int c_n(h)^{1/2} d^3 x ,$$

where c_n is a scalar polynomial in the metric, the normal to the boundary and the curvature and their covariant derivatives of order $2n-1$ in the derivatives of the metric and h is the induced metric on the boundary. The first coefficient c_0 is zero because their is no polynomial of order -1. McKean and Singer [3] showed that $c_1 = \dfrac{-1}{48\pi^2} K$ when $\xi = 0$ where K is the trace of the second fundamental form of the boundary. In the case of a Schwarzschild black hole in a spherical box of radius r_0, c_2 must be zero in the limit of large r_0 because all polynomials of degree 3 in the derivatives of the metric go down faster than r_0^{-2}.

In a compact manifold with or without boundary with a strongly elliptical metric g_0 the eigenvalues of a Laplacian type operator A will be discrete. If there are any zero eigenvalues they have to be omitted from the definition of the generalized zeta function and dealt with separately. This can be done by defining a new operator $\tilde{A} = A - P$ where P denotes projection on the zero eigenfunctions. Zero eigenvalues have important physical effects such as the anomaly in the axial vector current conservation [21, 22]. Let $\varepsilon > 0$ be the lowest eigenvalue of $\tilde{A}$ (from now on I shall simply use A and assume that any zero eigenfunctions have been projected out). Then

$$\zeta(s) = \frac{1}{\Gamma(s)} \left[\int_0^1 t^{s-1} Y(t) dt + \int_1^\infty t^{s-1} Y(t) dt \right] .\tag{4.10}$$

As $t \to \infty$, $Y \to e^{-\varepsilon t}$. Thus the second integral in Equation (4.10) converges for all s. In the first integral one can use the asymptotic expression (4.9). This gives

$$\sum_n \frac{B_n + C_n}{n+s-2} .\tag{4.11}$$

Thus ζ has a pole at $s=2$ with residue B_0 and a pole at $s=1$ with residue $B_1 + C_1$. There would be a pole at $s=0$ but it is cancelled out by the pole in $\Gamma(s)$. Thus $\zeta(0) = B_2 + C_2$. Similarly the values of ζ at negative integer values of s are given by (4.11) and (4.10).

5. Other Methods of Regularization

A commonly used method to evaluate the determinant of the operator A is to start with the integrated heat kernel

$$Y(t) = \sum_n \exp(-\lambda_n t) .\tag{5.1}$$

Multiply by $\exp(-m^2 t)$ and integrate from $t=0$ to $t=\infty$

$$\int_0^\infty Y(t)\exp(-m^2 t)dt = \sum_n' (\lambda_n + m^2)^{-1} \tag{5.2}$$

then integrate over m^2 from $m^2 = 0$ to $m^2 = \infty$ and interchange the orders of integration to obtain

$$\int_0^\infty t^{-1}Y(t)dt = \left[\sum_n' \log(\lambda_n + m^2)\right]_0^\infty . \tag{5.3}$$

One then throws away the value of the righthand side of (5.3) at the upper limit and claims that

$$\log \det A = \sum \log \lambda_n$$
$$= -\int_0^\infty t^{-1}Y(t)dt . \tag{5.4}$$

This is obviously a very dubious procedure. One can obtain the same result from the zeta function method in the following way. One has

$$\log \det A = -\zeta'(0)$$
$$= -\frac{d}{ds}\left[\frac{1}{\Gamma(s)}\int_0^\infty t^{s-1}Y(t)dt\right]. \tag{5.5}$$

Near $s=0$

$$\frac{1}{\Gamma(s)} = s + \gamma s^2 + O(s^3) , \tag{5.6}$$

where γ is Euler's constant.

Thus

$$\log \det A = -\underset{s\to 0}{\text{Lim}}\left[(1+2\gamma s)\int_0^\infty t^{s-1}Y(t)dt \right.$$
$$\left. +(s+\gamma s^2)\int_0^\infty t^{s-1}\log t\, Y(t)dt\right]. \tag{5.7}$$

If one ignores the fact that the two integrals in Equation (5.7) diverged when $s=0$, one would obtain Equation (5.4). Using the asymptotic expansion for Y, one sees that the integral in Equation (5.4) has a t^{-2}, t^{-1}, and a $\log t$ divergence at the lower limit with coefficients $\frac{1}{2}B_0$, B_1, and B_2 respectively. The first of these is often subtraced out by adding an infinite cosmological constant to the action while the second is cancelled by adding an infinite multiple of the scalar curvature which is interpreted as a renormalization of the gravitational constant. The logarithmic term requires an infinite counter term of a new type which is quadratic in the curvature.

To obtain a finite answer from Equation (5.4) dimensional regularization is often used. One generalizes the heat equation from $4+1$ dimensions to $2\omega+1$ dimensions and then subtracts out the pole that occurs in (5.4) at $2\omega=4$. As mentioned in the introduction, this is ambiguous because there are many ways that one could generalize a curved spacetime to 2ω dimensions. The simplest generalization would be to take the product of the four dimensional spacetime manifold with $2\omega - 4$ flat dimensions. In this case the integrated heat kernel Y would be multiplied by $(4\pi t)^{2-\omega}$. Then (5.4) would become

$$\log \det A = - \int_0^\infty t^{1-\omega}(4\pi)^{2-\omega}Y(t)dt \ . \tag{5.8}$$

This has a pole at $2\omega = 4$ with residue $\zeta(0)$ and finite part $-\zeta'(0) + (2\gamma + \log 4\pi) \times \zeta(0)$. Thus, the value of the $\log Z$ derived by the dimensional regularization using flat dimensions agrees with the value obtained by the zeta function method up to a multiple of $\zeta(0)$ which can be absorbed in the normalization constant. However, if one extended to $2\omega + 1$ dimensions in some more general way than merely adding flat dimensions, the integrated heat kernel would have the form

$$Y(t_0\omega) = \sum_n B_n(\omega)t^{n-\omega} \ , \tag{5.9}$$

where the coefficients $B_n(\omega)$ depend on the dimensions 2ω. The finite part at $\omega = 2$ would then acquire an extra term $B'_2(2)$. This could not be absorbed in the normalization constant μ. One therefore sees that the zeta function method has the conceptual advantages that it avoids the dubious procedures used to obtain Equation (5.4), it does not require the subtraction of any pole term or the addition of infinite counter terms, and it is unambiguous unlike dimensional regularization which depends on how one generalizes to 2ω dimensions.

6. Scaling

In this Section I shall consider the behaviour of the partition function Z under a constant scale transformation of the metric

$$\tilde{g}_{ab} = kg_{ab} \ . \tag{6.1}$$

If A is a Laplacian type operator for a zero rest mass field, the eigenvalues transform as

$$\lambda_n = k^{-1}\lambda_n \ . \tag{6.2}$$

Thus the new generalized zeta function is

$$\tilde{\zeta}(s) = k^s\zeta(s) \tag{6.3}$$

and

$$\log \det \tilde{A} = \log \det A - \log k\zeta(0) \ . \tag{6.4}$$

Thus

$$\log \tilde{Z} = \log Z + \tfrac{1}{2}\log k\zeta(0)$$
$$+ (\log \tilde{\mu} - \log \mu)\zeta(0) \ . \tag{6.5}$$

If one assumed that the normalization constant μ remained unchanged under a scale transformation, the last term would vanish. This assumption is equivalent to assuming that the measure in the path integral over all configurations of the field ϕ is defined not on a scalar field but on a scalar density of weight $\frac{1}{2}$. This is because the eigenfunctions of the operator A would have to transform according to

$$\tilde{\phi}_n = k^{-1}\phi_n \tag{6.6}$$

in order to maintain the normalization condition (2.6). The coefficients a_n of the expansion of a given scalar field ϕ would therefore transform according to

$$\tilde{a}_n = ka_n \tag{6.7}$$

and the normalization constant μ would transform according to

$$\tilde{\mu} = k^{-1}\mu \tag{6.8}$$

if the measure is defined on the scalar field itself, i.e. if

$$d[\phi] = \Pi_x d\phi(x) . \tag{6.9}$$

However if the measure is defined on densities of weight $\frac{1}{2}$, i.e.

$$d[\phi] = \Pi_x (g(x))^{1/4} d\phi(x) \tag{6.10}$$

then the normalization parameter μ is unchanged.

The weight of the measure can be deduced from considerations of unitarity. In the case of a scalar field one can use the manifestly unitary formalism of summing over all particle paths. This gives the conformally invariant scalar wave equation if the fields are taken to be densities of weight $\frac{1}{2}$ [23]. By contrast, the "minimally coupled" wave equation $\Box\phi = 0$ will be obtained if the weight is 1. In the case of a gravitational field itself one can use the unitary Hamiltonian formalism. From this Fadeev and Popov [17] deduce that the measure is defined on densities of weight $\frac{1}{2}$ and is scale invariant. Similar procedures could be used to find the weight of the measure for other fields. One would expect it to be $\frac{1}{2}$ for massless fields.

These scaling arguments give one certain amounts of information about the partition function. In DeSitter space they determine it up to the arbitrariness of the normalization parameter μ because DeSitter space is completely determined by the scale. Thus

$$\log Z = B_2 \log r/r_0 , \tag{6.11}$$

where r is the radius of the space and r_0 is related to μ. In the case of a Schwarzschild black hole of mass M in a large spherical box of radius r_0,

$$\log Z = B_2 \log M/M_0 + f(r_0 M^{-1}) , \tag{6.12}$$

where again M_0 is related to μ. If the radius of the box is large compared to M, one would expect that the partition function should approach that for thermal radiation at temperature $T = (8\pi M)^{-1}$ in flat space. Thus one would expect

$$f = \frac{r_0^3}{34560 M^3} + O\left(\frac{r_0^2}{M^2}\right) . \tag{6.13}$$

It should be possible to verify this and to calculate the lower order terms by developing suitable approximations to the eigenvalues of the radial equation in the Schwarzschild solution. In particular f and $\log Z$ will be finite. This contrasts with the result that one would obtain if one naively assumed that the thermal radiation could be described as a fluid with a density of $\log Z$ equal to $\pi^2/_{90}\bar{T}^3$ where $\bar{T} = T(1 - 2Mr^{-1})^{-1/2}$ is the local temperature. Near the horizon $\bar{T}$ would get very large because of a blueshift effect and so $\log Z$ would diverge.

For a conformally invariant scalar field $B_2 = -\frac{2}{45}$ for DeSitter space and $\frac{1}{45}$ for the Schwarzschild solution. The fact that B_2 is positive in the latter case may provide a natural cut off in the path integral when one integrates over background metrics will all masses M. If the measure on the space of gravitational fields is scale invariant then the action of the background fields will give an integral of the form

$$\int_0^\infty \exp(-4\pi M^2) M^{-1} dM \ . \tag{6.14}$$

This converges nicely at large M but has a logarithmic divergence at $M = 0$. However if one includes a contribution of the thermal radiation the integral is modified to

$$\int_0^\infty \exp(-4\pi M^2) M^{-1+B_2} dM \ . \tag{6.15}$$

This converges if B_2 is positive. Such a cut off can however be regarded as suggestive only because it ignores the contributions of high order terms which will be important near $M = 0$. One might hope that these terms might in turn be represented by further black hole background metrics.

7. Energy-Momentum Tensor

By functionally differentiating the partition function one obtains the energy momentum tensor of the thermal radiation

$$T_{ab} = -2(g_0)^{-1/2} \frac{\delta \log Z}{\delta g_0^{ab}} \ . \tag{7.1}$$

The energy momentum tensor will be finite even on the event horizon of a black hole background metric despite the fact that the blueshifted temperature $\bar{T}$ diverges there. This shows that the energy momentum tensor cannot be that of a perfect fluid with pressure equal to one third the energy density.

One can express the energy momentum tensor in terms of derivatives of the heat kernel F:

$$\delta \log Z = \tfrac{1}{2}\delta\zeta'(0) - \mu^{-1}\delta\mu\zeta(0) - \tfrac{1}{2}\log(\tfrac{1}{4}\pi\mu^2)\delta\zeta(0) \ . \tag{7.2}$$

The second term on the right of (7.2) will vanish if one assumes that μ does not change under variations of the metric. This will be the case if the measure is defined on densities of weight $\frac{1}{2}$. The third term can be expressed as the variation of an integral quadratic in the curvature tensor and can be evaluated directly. To calculate the first term one writes

$$\zeta'(0) = \frac{d}{ds}\left[\frac{1}{\Gamma(s)}\int\int_0^\infty t^{s-1}F(x,x,t)(g_0)^{1/2}d^4xdt\right]\Bigg|_{s=0}.$$ (7.3)

Therefore

$$\delta\zeta'(0) = \frac{d}{ds}\left[\frac{1}{\Gamma(s)}\int\int_0^\infty t^{s-1}\delta[F(x,x,t)(g_0)^{1,2}d^4xdt]\right]\Bigg|_{s=0}.$$ (7.4)

To calculate δF one uses the varied heat equation

$$\left(A+\frac{\partial}{\partial t}\right)\delta F(x,y,t)+\delta AF(x,y,t)=0$$ (7.5)

with $\delta[(g_0(y))^{1/2}F(x,y,0)]=0$. The solution is

$$\delta[(g_0(y)^{1/2}F(x,y,t)] = -\int\int_0^t F(x,z,t-t')\delta AF(z,y,t')g_0(y)g_0(z)^{1/2}d^4zdt'.$$ (7.6)

Therefore

$$\delta\int F(x,x,t)(g_0)^{1/2}d^4x = -t\int\delta AF(z,z,t)(g_0)^{1/2}d^4z.$$ (7.7)

Where the operator δA acts on the first argument of F.

The operator δA involves δg^{ab} and its covariant derivatives in the background metric. Integrating by parts, one obtains an expression for T^{ab} in terms of F and its covariant derivatives. For a conformally invariant scalar field.

$$T_{ab} = \frac{d}{ds}\left[\frac{1}{\Gamma(s)}\int_0^\infty t^s(\tfrac{2}{3}{}_aF_b - \tfrac{1}{6}g_{ab\,c}F^c - \tfrac{1}{3}F_{ab}\right.$$
$$\left. + \tfrac{1}{3}g_{ab}F_c^c + \tfrac{1}{6}R_{ab}F - \tfrac{1}{12}g_{ab}RF)dt\right]$$
$$- \log(\tfrac{1}{4}\pi\mu^2)\frac{\delta B_2}{\delta g_{ab}}(g_0)^{-1/2}.$$ (7.8)

Where indices placed before or after F indicates differentiation with respect to the first or second arguments respectively and the two arguments are taken at the point x at which the energy momentum tensor is to be evaluated. In an empty spacetime the quantity B_2 is the integral of a pure divergence so B_2 vanishes.

8. The Trace Anomaly

Naively one would expect T_a^a, the trace of the energy momentum tensor, would be zero for a zero rest mass field. However this is not the case as can be seen either directly from (7.8) or by the following simple argument. Consider a scale transformation in which the metric is multiplied by a factor $k = 1 + \varepsilon$. Then $\delta g_{ab} = \varepsilon g_{ab}$ and

$$\int T_a^a (g_0)^{1/2} d^4 x = 2 \frac{d \log Z}{dk}$$
$$= B_2 (1 + \tfrac{1}{2} \mu^{-1} d\mu/dk)$$
$$= B_2 \qquad\qquad\qquad (8.1)$$

if the measure is defined on densities of weight $\frac{1}{2}$. Thus for the case of a conformally invariant scalar field

$$T_a^a = \frac{1}{2880\pi^2} \left[R_{abcd} R^{abcd} - R_{ab} R^{ab} + \Box R \right] . \qquad\qquad (8.2)$$

The trace anomalies for other zero rest mass fields can be calculated in a similar manner.

These results for the trace anomaly agree with those of a number of other authors [7–12]. However, they disagree with some calculations by the point separation method [24] which do not obtain any anomaly. The trace anomaly for DeSitter completely determines the energy momentum because it must be a multiple of the metric by the symmetry. In a two dimensional black hole in a box the trace anomaly also determines the energy momentum tensor and in the four dimensional case it determines it up to one function of position [25].

9. Higher Order Terms

The path integral over the terms in the action which are quadratic in the fluctuations about the background fields are usually represented in perturbation theory by a single closed loop without any vertices. Functionally differentiating with respect to the background metric to obtain the energy momentum tensor corresponds to introducing a vertex coupling the field to the gravitational field. If one then feeds this energy momentum tensor as a perturbation back into the Einstein equations for the background field, the change in the $\log Z$ would be described by a diagram containing two closed loops each with a gravitational vertex and with the two vertices joined by a gravitational propagator. Under a scale transformation in which the metric was multiplied by a constant factor k, such a diagram would be multiplied by k^{-2}. Another diagram which would have the same scaling behaviour could be obtained by functionally differentiating $\log Z$ with respect to the background metric at two different points and then connecting these points by a gravitational propagator. In fact all the higher order terms have scaling behaviour k^{-n} where $n \geq 2$. Thus one would expect to make a negligible contribution to the partition function for black holes of significantly more than the Planck mass. The higher order terms will however be important near the Planck mass and will cause the scaling argument in Section 6 to break down. One might nevertheless hope that just as a black hole background metric corresponds to an infinite sequence of higher order terms in a perturbation expansion around flat space, so the higher order terms in expansion about a black hole background might in turn be represented by more black holes.

Acknowledgement. I am grateful for discussions with a number of colleagues including G. W. Gibbons, A. S. Lapedes, Y. Manor, R. Penrose, M. J. Perry, and I. M. Singer.

References

1. DeWitt,B.S.: Dynamical theory of groups and fields in relativity, groups and topology (eds. C. M. and B. S. DeWitt). New York: Gordon and Breach 1964
2. DeWitt,B.S.: Phys. Rep. **19**C, 295 (1975)
3. McKean,H.P., Singer,J.M.: J. Diff. Geo. **5**, 233—249 (1971)
4. Gilkey,P.B.: The index theorem and the heat equation. Boston: Publish or Perish 1974
5. Candelas,P., Raine,D.J.: Phys. Rev. D **12**, 965—974 (1975)
6. Drummond,I.T.: Nucl. Phys. **94**B, 115—144 (1975)
7. Capper,D., Duff,M.: Nuovo Cimento **23**A, 173 (1974)
8. Duff,M., Deser,S., Isham,C.J.: Nucl. Phys. **111**B, 45 (1976)
9. Brown,L.S.: Stress tensor trace anomaly in a gravitational metric: scalar field. University of Washington, Preprint (1976)
10. Brown,L.S., Cassidy,J.P.: Stress tensor trace anomaly in a gravitational metric: General theory, Maxwell field. University of Washington, Preprint (1976)
11. Dowker,J.S., Critchley,R.: Phys. Rev. D **13**, 3224 (1976)
12. Dowker,J.S., Critchley,R.: The stress tensor conformal anomaly for scalar and spinor fields. University of Manchester, Preprint (1976)
13. Gibbons,G.W., Hawking,S.W.: Action integrals and partition functions in quantum gravity. Phys. Rev. D (to be published)
14. Manor,Y.: Complex Riemannian sections. University of Cambridge, Preprint (1977)
15. Feynman,R.P.: Magic without magic, (eds. J. A. Wheeler and J. Klaunder). San Francisco: W. H. Freeman 1972.
16. DeWitt,B.S.: Phys. Rev. **162**, 1195—1239 (1967)
17. Fadeev,L.D., Popov,V.N.: Usp. Fiz. Nauk **111**, 427—450 (1973) [English translation in Sov. Phys. Usp. **16**, 777—788 (1974)]
18. Seeley,R.T.: Amer. Math. Soc. Proc. Symp. Pure Math. **10**, 288—307 (1967)
19. Ray,D.B., Singer,I.M.: Advances in Math. **7**, 145—210 (1971)
20. Gilkey,P.B.: Advanc. Math. **15**, 334—360 (1975)
21. 't Hooft,G.: Phys. Rev. Letters **37**, 8—11 (1976)
22. 't Hooft,G.: Computation of the quantum effects due to a four dimensional pseudoparticle. Harvard University, Preprint
23. Hartle,J.B., Hawking,S.W.: Phys. Rev. D **13**, 2188—2203 (1976)
24. Adler,S., Lieverman,J., Ng,N.J.: Regularization of the stress-energy tensor for vector and scalar particles. Propagating in a general background metric. IAS Preprint (1976)
25. Fulling,S.A., Christensen,S.: Trace anomalies and the Hawking effect. Kings College London, Preprint (1976)

Communicated by R. Geroch

Received February 10, 1977

15. The path-integral approach to quantum gravity

S. W. HAWKING

15.1 Introduction

Classical general relativity is a very complete theory. It prescribes not only the equations which govern the gravitational field but also the motion of bodies under the influence of this field. However it fails in two respects to give a fully satisfactory description of the observed universe. Firstly, it treats the gravitational field in a purely classical manner whereas all other observed fields seem to be quantized. Second, a number of theorems (see Hawking and Ellis, 1973) have shown that it leads inevitably to singularities of spacetime. The singularities are predicted to occur at the beginning of the present expansion of the universe (the big bang) and in the collapse of stars to form black holes. At these singularities, classical general relativity would break down completely, or rather it would be incomplete because it would not prescribe what came out of a singularity (in other words, it would not provide boundary conditions for the field equations at the singular points). For both the above reasons one would like to develop a quantum theory of gravity. There is no well defined prescription for deriving such a theory from classical general relativity. One has to use intuition and general considerations to try to construct a theory which is complete, consistent and which agrees with classical general relativity for macroscopic bodies and low curvatures of spacetime. It has to be admitted that we do not yet have a theory which satisfies the above three criteria, especially the first and second. However, some partial results have been obtained which are so compelling that it is difficult to believe that they will not be part of the final complete picture. These results relate to the conection between black holes and thermodynamics which has already been described in chapters 6 and 13 by Carter and Gibbons. In the present article it will be shown how this relationship between gravitation and thermodynamics appears also when one quantizes the gravitational field itself.

There are three main approaches to quantizing gravity:

1   The operator approach

In this one replaces the metric in the classical Einstein equations by a distribution-valued operator on some Hilbert space. However this would not seem to be a very suitable procedure to follow with a theory like gravity, for which the field equations are non-polynomial. It is difficult enough to make sense of the product of the field operators at the same spacetime point let alone a non-polynomial function such as the inverse metric or the square root of the determinant.

2   The canonical approach

In this one introduces a family of spacelike surfaces and uses them to construct a Hamiltonian and canonical equal-time commutation relations. This approach is favoured by a number of authors because it seems to be applicable to strong gravitational fields and it is supposed to ensure unitarity. However the split into three spatial dimensions and one time dimension seems to be contrary to the whole spirit of relativity. Moreover, it restricts the topology of spacetime to be the product of the real line with some three-dimensional manifold, whereas one would expect that quantum gravity would allow all possible topologies of spacetime including those which are not products. It is precisely these other topologies that seem to give the most interesting effects. There is also the problem of the meaning of equal-time commutation relations. These are well defined for matter fields on a fixed spacetime geometry but what sense does it make to say that two points are spacelike-separated if the geometry is quantized and obeying the Uncertainty Principle?

For these reasons I prefer:

3   The path-integral approach

This too has a number of difficulties and unsolved problems but it seems to offer the best hope. The starting point for this approach is Feynman's idea that one can represent the amplitude

$$\langle g_2, \phi_2, S_2 | g_1, \phi_1, S_1 \rangle,$$

to go from a state with a metric g_1 and matter fields ϕ_1 on a surface S_1 to a state with a metric g_2 and matter fields ϕ_2 on a surface S_2, as a sum over all field configurations g and ϕ which take the given values on the surfaces S_1

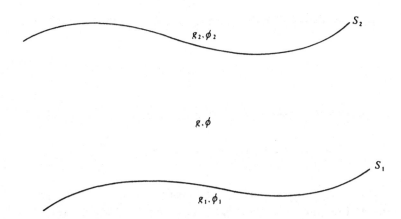

Figure 15.1. The amplitude $\langle g_2, \phi_2, S_2 | g_1, \phi_1, S_1 \rangle$ to go from a metric g_1 and matter fields ϕ_1, on a surface S_1 to a metric g_2 and matter fields ϕ_2 on a surface S_2 is given by a path integral over all fields g, ϕ which have the given values on S_1 and S_2.

and S_2 (figure 15.1). More precisely

$$\langle g_2, \phi_2, S_2 | g_1, \phi_1, S_1 \rangle = \int D[g, \phi] \exp (iI[g, \phi]),$$

where $D[g, \phi]$ is a measure on the space of all field configurations g and ϕ, $I[g, \phi]$ is the action of the fields, and the integral is taken over all fields which have the given values on S_1 and S_2.

In the above it has been implicitly assumed either that the surfaces S_1 and S_2 and the region between them are compact (a 'closed' universe) or that the gravitational and matter fields die off in some suitable way at spatial infinity (the asymptotically flat space). To make the latter more precise one should join the surfaces S_1 and S_2 by a timelike tube at large radius so that the boundary and the region contained within it are compact, as in the case of a closed universe. It will be seen in the next section that the surface at infinity plays an essential role because of the presence of a surface term in the gravitational action.

Not all the components of the metrics g_1 and g_2 on the boundary are physically significant, because one can give the components $g^{ab}n_b$ arbitrary values by diffeomorphisms or gauge transformations which move points in the interior, M, but which leave the boundary, ∂M, fixed. Thus one need specify only the three-dimensional induced metric h on ∂M and that only up to diffeomorphisms which map the boundary into itself.

In the following sections it will be shown how the path integral approach can be applied to the quantization of gravity and how it leads to the concepts of black hole temperature and intrinsic quantum mechanical entropy.

15.2 The action

The action in general relativity is usually taken to be

$$I = \frac{1}{16\pi G} \int (R - 2\Lambda)(-g)^{1/2} \, d^4x + \int L_m (-g)^{1/2} \, d^4x, \quad (15.1)$$

where R is the curvature scalar, Λ is the cosmological constant, g is the determinant of the metric and L_m is the Lagrangian of the matter fields. Units are such that $c = \hbar = k = 1$. G is Newton's constant and I shall sometimes use units in which this also has a value of one. Under variations of the metric which vanish and whose normal derivatives also vanish on ∂M, the boundary of a compact region M, this action is stationary if and only if the metric satisfies the Einstein equations:

$$R_{ab} - \tfrac{1}{2} g_{ab} R + \Lambda g_{ab} = 8\pi G T_{ab}, \quad (15.2)$$

where $T^{ab} = \tfrac{1}{2}(-g)^{-1/2}(\delta L_m / \delta g_{ab})$ is the energy-momentum tensor of the matter fields. However this action is not an extremum if one allows variations of the metric which vanish on the boundary but whose normal derivatives do not vanish there. The reason is that the curvature scalar R contains terms which are linear in the second derivatives of the metric. By integration by parts, the variation in these terms can be converted into an integral over the boundary which involves the normal derivatives of the variation on the boundary. In order to cancel out this surface integral, and so obtain an action which is stationary for solutions of the Einstein equations under all variations of the metric that vanish on the boundary, one has to add to the action a term of the form (Gibbons and Hawking, 1977a):

$$\frac{1}{8\pi G} \int K (\pm h)^{1/2} \, d^3x + C, \quad (15.3)$$

where K is the trace of the second fundamental form of the boundary, h is the induced metric on the boundary, the plus or minus signs are chosen according to whether the boundary is spacelike or timelike, and C is a term which depends only on the boundary metric h and not on the values of g at the interior points. The necessity for adding the surface term (15.3) to the action in the path-integral approach can be seen by considering the

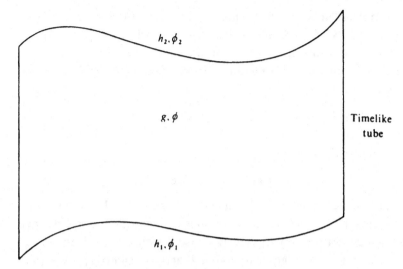

Figure 15.2. Only the induced metric h need be given on the boundary surface. In the asymptotically flat case the initial and final surfaces should be joined by a timelike tube at large radius to obtain a compact region over which to perform the path integral.

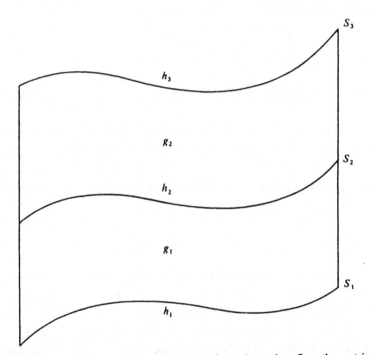

Figure 15.3. The amplitude to go from the metric h_1 on the surface S_1 to the metric h_3 on the surface S_3 should be the sum of the amplitude to go by all metrics h_2 on the intermediate surface S_2. This will be true only if the action contains a surface term.

situation depicted in figure 15.3, where one considers the transition from a metric h_1, on a surface S_1, to a metric h_2 on a surface S_2 and then to a metric h_3 on a later surface S_3. One would want the amplitude to go from the initial to the final state to be obtained by summing over all states on the intermediate surface S_2, i.e.

$$\langle h_3, S_3 | h_1, S_1 \rangle = \sum_{h_2} \langle h_2, S_2 | h_1, S_1 \rangle \langle h_3, S_3 | h_2, S_2 \rangle. \tag{15.4}$$

This will be true if and only if

$$I[g_1 + g_2] = I[g_1] + I[g_2], \tag{15.5}$$

where g_1 is the metric between S_1 and S_2, g_2 is the metric between S_2 and S_3, and $[g_1 + g_2]$ is the metric on the regions between S_1 and S_3 obtained by joining together the two regions. Because the normal derivative of g_1 at S_2 will not in general be equal to that of g_2 at S_2, the metric $[g_1 + g_2]$ will have a δ-function in the Ricci tensor of strength $2(K_{ab}^1 - K_{ab}^2)$, where K_{ab}^1 and K_{ab}^2 are the second fundamental forms of the surface S_2 in the metrics g_1 and g_2 respectively, defined with respect to the future-directed normal. This means that the relation (15.5) will hold if and only if the action is the sum of (15.1) and (15.3), i.e.

$$I = \frac{1}{16\pi G} \int (R - 2\Lambda)(-g)^{1/2} \, \mathrm{d}^4 x + \int L_\mathrm{m}(-g)^{1/2} \, \mathrm{d}^4 x$$

$$+ \frac{1}{8\pi G} \int K(\pm h)^{1/2} \, \mathrm{d}^3 x + C. \tag{15.6}$$

The appearance of the term C in the action is somewhat awkward. One could simply absorb it into the renormalization of the measure $D[g, \phi]$. However, in the case of asymptotically flat metrics it is natural to treat it so that the contribution from the timelike tube at large radius is zero when g is the flat-space metric, η. Then

$$C = -\frac{1}{8\pi G} \int K^0 (\pm h)^{1/2} \, \mathrm{d}^3 x, \tag{15.7}$$

where K^0 is the second fundamental form of the boundary imbedded in flat space. This is not a completely satisfactory prescription because a general boundary metric h cannot be imbedded in flat space. However in an asymptotically flat situation one can suppose that the boundary will become asymptotically imbeddable as one goes to larger and larger radii. Ultimately I suspect that one should do away with all boundary surfaces and should deal only with closed spacetime manifolds. However, at the present state of development it is very convenient to use non-compact, asymptotically flat metrics and to evaluate the action using a boundary at large radius.

A metric which is asymptotically flat in the three spatial directions but not in time can be written in the form

$$ds^2 = -(1 - 2M_t r^{-1}) dt^2 + (1 + 2M_s r^{-1}) dr^2$$
$$+ r^2 (d\theta^2 + \sin^2 \theta \, d\phi^2) + O(r^{-2}). \tag{15.8}$$

If the metric satisfies the vacuum Einstein equations ($\Lambda = 0$) near infinity then $M_t = M_s$, but in the path integral one considers all asymptotically flat metrics, whether or not they satisfy the Einstein equation. In such a metric it is convenient to choose the boundary ∂M to be the t-axis times a sphere of radius r_0. The area of ∂M is

$$\int (-h)^{1/2} d^3x = 4\pi r_0^2 \int (1 - M_t r_0^{-1} + O(r_0^{-2})) \, dt. \tag{15.9}$$

The integral of the trace of the second fundamental form of ∂M is given by

$$\int K(-h)^{1/2} d^3x = \frac{\partial}{\partial n} \int (-h)^{1/2} d^3x, \tag{15.10}$$

where $\partial/\partial n$ indicates the derivative when each point of ∂M is moved out along the unit normal. Thus

$$\int K(-h)^{1/2} d^3x = \int (8\pi r_0 - 4\pi M_t - 8\pi M_s + O(r_0^{-2})) \, dt. \tag{15.11}$$

For the flat space metric, η, $K^0 = 2r_0^{-1}$. Thus

$$\frac{1}{8\pi G} \int (K - K^0)(-h)^{1/2} d^3x = \frac{1}{2G} \int (M_t - 2M_s) \, dt. \tag{15.12}$$

In particular for a solution of the Einstein equation with mass M as measured from infinity, $M_s = M_t = M$ and the surface term is

$$-\frac{M}{2G} \int dt + O(r_0^{-1}). \tag{15.13}$$

15.3 Complex spacetime

For real Lorentzian metrics g (i.e. metrics with signature $- + + +$) and real matter fields ϕ, the action $I[g, \phi]$ will be real and so the path integral will oscillate and will not converge. A related difficulty is that to find a field configuration which extremizes the action between given initial and final surfaces, one has to solve a hyperbolic equation with initial and final boundary values. This is not a well-posed problem: there may not be any solution or there may be an infinite number, and if there is a solution it will not depend smoothly on the boundary values.

In ordinary quantum field theory in flat spacetime one deals with this difficulty by rotating the time axis 90° clockwise in the complex plane, i.e. one replaces t by $-i\tau$. This introduces a factor of $-i$ into the volume integral for the action I. For example, a scalar field of mass m has a Lagrangian

$$L = -\tfrac{1}{2}\phi_{,a}\phi_{,b}g^{ab} - \tfrac{1}{2}m^2\phi^2. \tag{15.14}$$

Thus the path integral

$$Z = \int D[\phi] \exp\left(iI[\phi]\right) \tag{15.15}$$

becomes

$$Z = \int D[\phi] \exp\left(-\hat{I}[\phi]\right), \tag{15.16}$$

where $\hat{I} = -iI$ is called the 'Euclidean' action and is greater than or equal to zero for fields ϕ which are real on the Euclidean space defined by real τ, x, y, z. Thus the integral over all such configurations of the field ϕ will be exponentially damped and should therefore converge. Moreover the replacement of t by an imaginary coordinate τ has changed the metric η^{ab} from Lorentzian (signature $-+++$) to Euclidean (signature $++++$). Thus the problem of finding an extremum of the action becomes the well-posed problem of solving an elliptic equation with given boundary values.

The idea, then, is to perform all path integrals on the Euclidean section (τ, x, y, z real) and then analytically continue the results anticlockwise in the complex t-plane back to Lorentzian or Minkowski section (t, x, y, z real). As an example consider the quantity

$$Z[J] = \int D[\phi] \exp -\left(\tfrac{1}{2}\phi A\phi + J\phi\right) dx\, dy\, dz\, d\tau, \tag{15.17}$$

where A is the second-order differential operator $-\Box + m^2$, $\Box$ is the four-dimensional Laplacian and $J(x)$ is a prescribed source field which dies away at large Euclidean distances. The path integral is taken over all fields ϕ that die away at large Euclidean distances. One can write $Z[J]$

symbolically as

$$Z[J] = \exp\left(\tfrac{1}{2}JA^{-1}J\right)\int D[\phi]\exp\left(-\tfrac{1}{2}(\phi - A^{-1}J)A(\phi - A^{-1}J)\right), \quad (15.18)$$

where $A^{-1}(x_1, x_2)$ is the unique inverse or Green's function for A that dies away at large Euclidean distances,

$$A^{-1}J(x) = \int A^{-1}(x, x')J(x')\,\mathrm{d}^4x' \qquad (15.19)$$

$$JA^{-1}J = \int\int J(x)A^{-1}(x, x')J(x')\,\mathrm{d}^4x\,\mathrm{d}^4x'. \qquad (15.20)$$

The measure $D[\phi]$ is invariant under the translation $\phi \to \phi - A^{-1}J$. Thus

$$Z[J] = \exp\left(\tfrac{1}{2}JA^{-1}J\right)Z[0]. \qquad (15.21)$$

Then one can define the Euclidean propagator or two-point correlation function

$$\langle 0|\phi(x_2)\phi(x_1)|0\rangle = \left.\frac{\delta^2 \log Z}{\delta J(x_1)\,\delta J(x_2)}\right|_{J=0}$$

$$= A^{-1}(x_2, x_1). \qquad (15.22)$$

One obtains the Feynman propagator by analytically continuing $A^{-1}(x_2, x_1)$ anticlockwise in the complex $t_2 - t_1$-plane.

It should be pointed out that this use of the Euclidean section has enabled one to define the vacuum state by the property that the fields ϕ die off at large positive and negative imaginary times τ. The time-ordering operation usually used in the definition of the Feynman propagator has been automatically achieved by the direction of the analytic continuation from Euclidean space, because if $\mathrm{Re}\,(t_2 - t_1) > 0$, $\langle 0|\phi(x_2), \phi(x_1)|0\rangle$ is holomorphic in the lower half $t_2 - t_1$-plane, i.e. it is positive-frequency (a positive-frequency function is one which is holomorphic in the lower half t-plane and which dies off at large negative imaginary t).

Another use of the Euclidean section that will be important in what follows is to construct the canonical ensemble for a field ϕ. The amplitude to propagate from a configuration ϕ_1 on a surface at time t_1 to a configuration ϕ_2 on a surface at time t_2 is given by the path integral

$$\langle \phi_2, t_2|\phi_1, t_1\rangle = \int D[\phi]\exp\left(iI[\phi]\right). \qquad (15.23)$$

Using the Schrödinger picture, one can also write this amplitude as

$$\langle \phi_2|\exp\left(-iH(t_2 - t_1)\right)|\phi_1\rangle.$$

Put $t_2 - t_1 = -i\beta$, $\phi_2 = \phi_1$ and sum over a complete orthonormal basis of configurations ϕ_n. One obtains the partition function

$$Z = \sum \exp{(-\beta E_n)} \tag{15.24}$$

of the field ϕ at a temperature $T = \beta^{-1}$, where E_n is the energy of the state ϕ_n. However from (15.23) one can also represent Z as a Euclidean path integral

$$Z = \int D[\phi] \exp{(-i\hat{I}[\phi])}, \tag{15.25}$$

where the integral is taken over all fields ϕ that are real on the Euclidean section and are periodic in the imaginary time coordinate τ with period β. As before one can introduce a source J and obtain a Green's function by functionally differentiating $Z[J]$ with respect to J at two different points. This will represent the two-point correlation function or propagator for the field ϕ, not this time in the vacuum state but in the canonical ensemble at temperature $T = \beta^{-1}$. In the limit that the period β tends to infinity, this thermal propagator tends to the normal vacuum Feynman propagator.

It seems reasonable to apply similar complexification ideas to the gravitational field, i.e. the metric. For example, supposing one was considering the amplitude to go from a metric h_1 on a surface S_1 to a metric h_2 on a surface S_2, where the surfaces S_1 and S_2 are asymptotically flat, and are separated by a time interval t at infinity. As explained in section 15.1, one would join S_1 and S_2 by a timelike tube of length t at large radius. One could then rotate this time interval into the complex plane by introducing an imaginary time coordinate $\tau = it$. The induced metric on the timelike tube would now be positive-definite so that one would be dealing with a path integral over a region M on whose boundary the induced metric h was positive-definite everywhere. One could therefore take the path integral to be over all positive-definite metrics g which induced the given positive-definite metric h on ∂M. With the same choice of the direction of rotation into the complex plane as in flat-space Euclidean theory, the factor $(-g)^{1/2}$ which appears in the volume element becomes $-i(g)^{1/2}$, so that the Euclidean action, $\hat{I} = -iI$, becomes

$$\hat{I} = -\frac{1}{16\pi G}\int (R - 2\Lambda)(g)^{1/2}\,\mathrm{d}^4 x - \frac{1}{8\pi G}\int (K - K^0)(h)^{1/2}\,\mathrm{d}^3 x$$

$$- \int L_\mathrm{m}(g)^{1/2}\,\mathrm{d}^4 x. \tag{15.26}$$

172

The problem arising from the fact that the gravitational part of this Euclidean action is not positive-definite will be discussed in section 15.4.

The state of the system is determined by the choice of boundary conditions of the metrics that one integrates over. For example, it would seem reasonable to expect that the vacuum state would correspond to integrating over all metrics which were asymptotically Euclidean, i.e. outside some compact set as they approached the flat Euclidean metric on R^4. Inside the compact set the curvature might be large and the topology might be different from that of R^4.

As an example, one can consider the canonical ensemble for the gravitational fields contained in a spherical box of radius r_0 at a temperature T, by performing a path integral over all metrics which would fit inside a boundary consisting of a timelike tube of radius r_0 which was periodically identified in the imaginary time direction with period $\beta = T^{-1}$.

In complexifying the spacetime manifold one has to treat quantities which are complex on the real Lorentzian section as independent of their complex conjugates. For example, a charged scalar field in real Lorentzian spacetime may be represented by a complex field ϕ and its complex conjugate $\bar{\phi}$. When going to complex spacetime one has to analytically continue $\bar{\phi}$ as a new field $\tilde{\phi}$ which is independent of ϕ. The same applies to spinors. In real Lorentzian spacetime one has unprimed spinors λ_A which transform under SL(2, C) and primed spinors $\mu_{A'}$ which transform under the complex conjugate group $\overline{\text{SL}(2, \text{C})}$. The complex conjugate of an unprimed spinor is a primed spinor and vice versa. When one goes to complex spacetime, the primed and unprimed spinors become independent of each other and transform under independent groups SL(2, C) and $\widetilde{\text{SL}}(2, \text{C})$ respectively. If one analytically continues to a section on which the metric is positive-definite and restricts the spinors to lie in that section, the primed and unprimed spinors are still independent but these groups become SU(2) and $\widetilde{\text{SU}}(2)$ respectively. For example, in a Lorentzian metric the Weyl tensor can be represented as

$$C_{AA'BB'CC'DD'} = \psi_{ABCD}\varepsilon_{A'B'}\varepsilon_{C'D'} + \bar{\psi}_{A'B'C'D'}\varepsilon_{AB}\varepsilon_{CD}. \quad (15.27)$$

When one complexifies, $\bar{\psi}_{A'B'C'D'}$ is replaced by an independent field $\tilde{\psi}_{A'B'C'D'}$. In particular one can have a metric in which $\psi_{ABCD} \neq 0$, but $\tilde{\psi}_{A'B'C'D'} = 0$. Such a metric is said to be conformally self-dual and satisfies

$$C_{abcd} = {}^*C_{abcd} = \tfrac{1}{2}\varepsilon_{abef}C^{ef}{}_{cd}. \quad (15.28)$$

The metric is said to be self-dual if

$$R_{abcd} = {}^*R_{abcd}$$

which implies

$$R_{ab} = 0, \quad C_{abcd} = {}^*C_{abcd}. \tag{15.29}$$

A complexified spacetime manifold M with a complex self-dual or conformally self-dual metric g_{ab} may admit a section on which the metric is real and positive definite (a 'Euclidean' section) but it will not admit a Lorentzian section, i.e. a section on which the metric is real and has a signature $-+++$.

15.4 The indefiniteness of the gravitational action

The Euclidean action for scalar or Yang–Mills fields is positive-definite. This means that the path integral over all configurations of such fields that are real on the Euclidean section converges, and that only those configurations contribute that die away at large Euclidean distances, since otherwise the action would be infinite. The action for fermion fields is not positive-definite. However, one treats them as anticommuting quantities (Berezin, 1966) so that the path integral over them converges. On the other hand, the Euclidean gravitational action is not positive-definite even for real positive-definite metrics. The reason is that although gravitational waves carry positive energy, gravitational potential energy is negative because gravity is attractive. Despite this, in classical general relativity it seems that the total energy or mass, as measured from infinity, of any asymptotically flat gravitational field is always non-negative. This is known as the *positive energy conjecture* (Brill and Deser, 1968; Geroch, 1973). What seems to happen is that whenever the gravitational potential energy becomes too large, an event horizon is formed and the region of high gravitational binding undergoes gravitational collapse, leaving behind a black hole of positive mass. Thus one might expect that the black holes would play a role in controlling the indefiniteness of the gravitational action in quantum theory and there are indications that this is indeed the case.

To see that the action can be made arbitrarily negative, consider a conformal transformation $\tilde{g}_{ab} = \Omega^2 g_{ab}$, where Ω is a positive function which is equal to one on the boundary ∂M.

$$\tilde{R} = \Omega^{-2}R - 6\Omega^{-3}\Box\Omega \tag{15.30}$$

$$\tilde{K} = \Omega^{-1}K + 3\Omega^{-2}\Omega_{;a}n^a, \tag{15.31}$$

where n^a is the unit outward normal to the boundary ∂M. Thus

$$\hat{I}[\tilde{g}] = -\frac{1}{16\pi G} \int_M (\Omega^2 R + 6\Omega_{;a}\Omega_{;b}g^{ab} - 2\Lambda\Omega^4)(g)^{1/2}\, d^4x$$
$$-\frac{1}{8\pi G} \int_{\partial M} \Omega^2(K - K^0)(h)^{1/2}\, d^3x. \qquad (15.32)$$

One sees that $\hat{I}$ may be made arbitrarily negative by choosing a rapidly varying conformal factor Ω.

To deal with this problem it seems desirable to split the integration over all metrics into an integration over conformal factors, followed by an integration over conformal equivalence classes of metrics. I shall deal separately with the case in which the cosmological constant Λ is zero but the spacetime region has a boundary ∂M, and the case in which Λ is nonzero but the region is compact without boundary.

In the former case, the path integral over the conformal factor Ω is governed by the conformally invariant scalar wave operator, $A = -\Box + \frac{1}{6}R$. Let $\{\lambda_n, \phi_n\}$ be the eigenvalues and eigenfunctions of A with Dirichlet boundary conditions, i.e.

$$A\phi_n = \lambda_n\phi_n, \ \phi_n = 0 \text{ on } \partial M.$$

If $\lambda_1 = 0$, then $\Omega^{-1}\phi_1$ is an eigenfunction with zero eigenvalue for the metric $\tilde{g}_{ab} = \Omega^2 g_{ab}$. The nonzero eigenvalues and corresponding eigenfunctions do not have any simple behaviour under conformal transformation. However they will change continuously under a smooth variation of the conformal factor which remains positive everywhere. Because the zero eigenvalues are conformally invariant, this shows that the number of negative eigenvalues (which will be finite) remains unchanged under a conformal transformation Ω which is positive everywhere.

Let $\Omega = 1 + y$, where $y = 0$ on ∂M. Then

$$\hat{I}[\tilde{g}] = -\frac{6}{16\pi g} \int (yAy + 2Ry)(g)^{1/2}\, d^4x + \hat{I}[g]$$

$$= -\frac{6}{16\pi G} \int \{(y - A^{-1}R)A(y - A^{-1}R)\}(g)^{1/2}\, d^4x$$

$$+ \frac{6}{16\pi G} RA^{-1}R + \hat{I}[g]$$

$$= \frac{6}{16\pi G} RA^{-1}R + \hat{I}[g] - \frac{6}{16\pi G} \int \gamma A\gamma(g)^{1/2}\, d^4x, \quad (15.33)$$

where $\gamma = (y - A^{-1}R)$.

Thus one can write

$$\hat{I}[\tilde{g}] = I^1 + I^2,$$

where I^1 is the first and second term on the right of (15.33) and I^2 is the third term.

I^1 depends only on the conformal equivalence class of the metric g, while I^2 depends on the conformal factor. One can thus define a quantity X to be the path integral of $\exp{(-I^2)}$ over all conformal factors in one conformal equivalence class of metrics.

If the operator A has no negative or zero eigenvalues, in particular if g is a solution of the Einstein equations, the inverse, A^{-1}, will be well defined and the metric $g'_{ab} = (1 + A^{-1}R)^2 g_{ab}$ will be a regular metric with $R' = 0$ everywhere. In this case I^1 will equal $\hat{I}[g']$, which in turn will be given by a surface integral of K' on the boundary. It seems plausible to make the *positive action conjecture*: any asymptotically Euclidean, positive-definite metric with $R = 0$ has positive or zero action (Gibbons, Hawking and Perry, 1978). There is a close connection between this and the positive energy conjecture in classical Lorentzian general relativity. This claims that the mass or energy as measured from infinity of any Lorentzian, asymptotically flat solution of the Einstein equations is positive or zero if the solution develops from a non-singular initial surface, the mass being zero if and only if the metric is identically flat. Although no complete proof exists, the positive energy conjecture has been proved in a number of restricted cases or under certain assumptions (Brill, 1959; Brill and Deser, 1968; Geroch, 1973; Jang and Wald, 1977) and is generally believed. If it held also for classical general relativity in five dimensions (signature $-++++$), it would imply the positive action conjecture, because a four-dimensional asymptotically Euclidean metric with $R = 0$ could be taken as time-symmetric initial data for a five-dimensional solution and the mass of such a solution would be equal to the action of the four-dimensional metric. Page (1978) has obtained some results which support the positive action conjecture. However he has also shown that it does not hold for metrics like the Schwarzschild solution which are asymptotically flat in the spatial directions, but are not in the Euclidean time direction. The significance of this will be seen later.

Let g_0 be a solution of the field equations. If I^1 increases under all perturbations away from g_0 that are not purely conformal transformations, the integral over conformal classes will tend to converge. If there is some non-conformal perturbation, δg, of g_0 which reduces I^1, then in order to make the path integral converge one will have to integrate over the metrics of the form $g_0 + i\delta g$. This will introduce a factor i into Z for each mode of non-conformal perturbations which reduces I^1. This will be discussed in the next section. For metrics which are far from a solution of the field equation, the operator A may develop zero or negative eigenvalues. When an eigenvalue passes through zero, the inverse, A^{-1}, will become undefined and I^1 will become infinite. When there are negative eigenvalues but not zero eigenvalues, A^{-1} and I^1 will be well defined, but the conformal factor $\Omega = 1 + A^{-1}R$, which transforms g to the metric g' with $R' = 0$, will pass through zero and so g' will be singular. This is very similar to what happens with three-dimensional metrics on time-symmetric initial surfaces (Brill, 1959). If h is a three-dimensional positive-definite metric on the initial surface, one can make a conformal transformation $\tilde{h} = \Omega^4 h$ to obtain a metric with $\tilde{R} = 0$ which will satisfy the constraint equations. If the three-dimensional conformally invariant operator $B = -\Delta + R/8$ has no zero or negative eigenvalues (which will be the case for metrics h sufficiently near flat space) the conformal factor Ω needed will be finite and positive everywhere. If, however, one considers a sequence of metrics h for which one of the eigenvalues of B passes through zero and becomes negative, the corresponding Ω will first diverge and then will become finite again but will pass through zero so that the metric $\tilde{h}$ will be singular. The interpretation of this is that the metric h contained a region with so much negative gravitational binding energy that it cut itself off from the rest of the universe by forming an event horizon. To describe such a situation one has to use initial surfaces with different topologies.

It seems that something analogous may be happening in the four-dimensional case. In some sense one could think that metrics g for which the operator A had negative eigenvalues contained regions which cut themselves off from the rest of the spacetime because they contained too much curvature. One could then represent their effect by going to manifolds with different topologies. Anyway, metrics for which A has negative eigenvalues are in some sense far from solutions of the field equations, and we shall see in the next section that one can in fact evaluate path integrals only over metrics near solutions of the field equations.

The operator A appears in I^2 with a minus sign. This means that in order to make the path integral over the conformal factors converge at a solution of the field equations, and in particular at flat space, one has to take γ to be purely imaginary. The prescription, therefore, for making the path integral converge is to divide the space of all metrics into conformal equivalence classes. In each equivalence class pick the metric g' for which $R' = 0$. Integrate over all metrics $\tilde{g} = \Omega^2 g'$, where Ω is of the form $1 + i\xi$. Then integrate over conformal equivalence classes near solutions of the field equations, with the non-conformal perturbation being purely imaginary for modes which reduce I^1.

The situation is rather similar for compact manifolds with a Λ-term. In this case there is no surface term in the action and no requirement that $\Omega = 1$ on the boundary. If $\tilde{g} = \Omega^2 g$,

$$\hat{I}[\tilde{g}] = -\frac{6}{16\pi G} \int (\Omega^2 R + 6\Omega_{;a}\Omega_{;b}g^{ab} - 2\Lambda\Omega^4)(g)^{1/2}\, \mathrm{d}^4 x. \quad (15.34)$$

Thus quantum gravity with a Λ-term on a compact manifold is a sort of average of $\lambda\phi^4$ theory over all background metrics. However unlike ordinary $\lambda\phi^4$ theory, the kinetic term $(\nabla\Omega)^2$, appears in the action with a minus sign. This means that the integration over the conformal factors has to be taken in a complex direction just as in the previous case.

One can again divide the space of all the positive-definite metrics g on the manifold M into conformal equivalence classes. In each equivalence class the action will have one extremum at the vanishing metric for which $\Omega = 0$. In general there will be another extremum at a metric g' for which $R' = 4\Lambda$, though in some cases the conformal transformation $g' = \Omega^2 g$, where g is a positive-definite metric, may require a complex Ω. Putting $\tilde{g} = (1 + y)^2 g'$, one obtains

$$\hat{I}[\tilde{g}] = -\frac{\Lambda V}{8\pi G} - \frac{6}{16\pi G} \int (6y_{;a}y_{;b}g^{ab} - 8y^2\Lambda - 8y^3\Lambda - 2y^4\Lambda)(g')^{1/2}\, \mathrm{d}^4 x,$$

$$(15.35)$$

where $V = \int (g')^{1/2}\, \mathrm{d}^4 x$.

If Λ is negative and one neglects the cubic and quartic terms in y, one obtains convergence in the path integral by integrating over purely imaginary y in a similar manner to what was done in the previous case. It therefore seems reasonable to adopt the prescription for evaluating path integrals with Λ-terms that one picks the metric g' in each conformal equivalence class for which $R' = 4\Lambda$, and one then integrates over conformal factors of the form $\Omega = 1 + i\xi$ about g'.

If Λ is positive, the operator $-6\Box - 8\Lambda$, which acts on the quadratic terms in ξ, has at least one negative eigenvalue, $\xi = $ constant. In fact it seems that this is the only negative eigenvalue. Its significance will be discussed in section 15.10.

15.5 The stationary-phase approximation

One expects that the dominant contribution to the path integral will come from metrics and fields which are near a metric g_0, and fields ϕ_0 which are an extremum of the action, i.e. a solution of the classical field equations. Indeed this must be the case if one is to recover classical general relativity in the limit of macroscopic systems. Neglecting for the moment, questions of convergence, one can expand the action in a Taylor series about the background fields g_0, ϕ_0,

$$\hat{I}[g, \phi] = \hat{I}[g_0, \phi_0] + I_2[\bar{g}, \bar{\phi}] + \text{higher-order terms}, \qquad (15.36)$$

where

$$g_{ab} = g_{0ab} + \bar{g}_{ab}, \quad \phi = \phi_0 + \bar{\phi},$$

and $I_2[\bar{g}, \bar{\phi}]$ is quadratic in the perturbations $\bar{g}$ and $\bar{\phi}$. If one ignores the higher-order terms, the path integral becomes

$$\log Z = -\hat{I}[g_0, \phi_0] + \log \int D[\bar{g}, \bar{\phi}] \exp(-I_2[\bar{g}, \bar{\phi}]). \qquad (15.37)$$

This is known variously as the stationary-phase, WKB or one-loop approximation. One can regard the first term on the right of (15.37) as the contribution of the background fields to $\log Z$. This will be discussed in sections 15.7 and 15.8. The second term on the right of (15.37) is called the one-loop term and represents the effect of quantum fluctuations around the background fields. The remainder of this section will be devoted to describing how one evaluates it. For simplicity I shall consider only the case in which the background matter fields, ϕ_0, are zero. The quadratic term $I_2[\bar{g}, \bar{\phi}]$ can then be expressed as $I_2[\bar{g}] + I_2[\phi]$ and

$$\log Z = -\hat{I}[g_0] + \log \int D[\phi] \exp(-I_2[\phi]) + \log \int D[\bar{g}] \exp(-I_2[\bar{g}]).$$

$$(15.38)$$

I shall consider first the one-loop term for the matter fields, the second term on the right of (15.38). One can express $I_2[\phi]$ as

$$I_2[\phi] = \tfrac{1}{2} \int \phi A \phi (g_0)^{1/2} \, d^4 x, \qquad (15.39)$$

where A is a differential operator depending on the background metric g_0. In the case of boson fields, which I shall consider first, A is a second-order differential operator. Let $\{\lambda_n, \phi_n\}$ be the eigenvalues and the corresponding eigenfunctions of A, with $\phi_n = 0$ on ∂M in the case where there is a boundary surface. The eigenfunctions, ϕ_n, can be normalized so that

$$\int \phi_n \phi_m \cdot (g_0)^{1/2} \, d^4x = \delta_{mn}. \tag{15.40}$$

One can express an arbitrary field ϕ which vanishes on ∂M as a linear combination of these eigenfunctions:

$$\phi = \sum_n y_n \phi_n. \tag{15.41}$$

Similarly one can express the measure on the space of all fields ϕ as

$$D[\phi] = \prod_n \mu \, dy_n. \tag{15.42}$$

Where μ is a normalization factor with dimensions of mass or $(\text{length})^{-1}$. One can then express the one-loop matter term as

$$
\begin{aligned}
Z_\phi &= \int D[\phi] \exp\left(-I_2[\phi]\right) \\
&= \prod_n \int \mu \, dy_n \exp\left(-\tfrac{1}{2}\lambda_n y_n^2\right) \\
&= \prod_n (2\pi\mu^2 \lambda_n^{-1})^{1/2} \\
&= (\det\left(\tfrac{1}{2}\pi^{-1}\mu^{-2}A\right))^{-1/2}.
\end{aligned}
\tag{15.43}
$$

In the case of a complex field ϕ like a charged scalar field, one has to treat ϕ and the analytic continuation $\tilde{\phi}$ of its complex conjugate as independent fields. The quadratic term then has the form

$$I_2[\phi, \tilde{\phi}] = \tfrac{1}{2} \int \tilde{\phi} A \phi (g_0)^{1/2} \, d^4x. \tag{15.44}$$

The operator A will not be self-adjoint if there is a background electromagnetic field. One can write $\tilde{\phi}$ in terms of eigenfunctions of the adjoint operator $A^\dagger$:

$$\tilde{\phi} = \sum_n \tilde{y}_n \tilde{\phi}_n. \tag{15.45}$$

The measure will then have the form

$$D[\phi, \tilde{\phi}] = \prod_n \mu^2 \, dy_n \, d\tilde{y}_n. \tag{15.46}$$

Because one integrates over y_n and $\tilde{y}_n$ independently, one obtains

$$Z_\phi = (\det (\tfrac{1}{2}\pi^{-1}\mu^{-2}A))^{-1}. \tag{15.47}$$

To treat fermions in the path integrals one has to regard the spinor ψ and its independent adjoint $\tilde{\psi}$ as anticommuting Grassman variables (Berezin, 1966). For a Grassman variable x one has the following (formal) rules of integration

$$\int dx = 0, \quad \int x\, dx = 1, \tag{15.48}$$

These suffice to determine all integrals, since x^2 and higher powers of x are zero by the anticommuting property. Notice that (15.48) implies that if $y = ax$, where a is a real constant, then $dy = a^{-1}\, dx$.

One can use these rules to evaluate path integrals over the fermion fields ψ and $\tilde{\psi}$. The operator A in this case is just the ordinary first-order Dirac operator. If one expands $\exp(-I_2)$ in a power series, only the term linear in A will survive because of the anticommuting property. Integration of this respect to $d\psi$ and $d\tilde{\psi}$ gives

$$Z_\psi = \det (\tfrac{1}{2}\mu^{-2}A). \tag{15.49}$$

Thus the one-loop terms for fermion fields are proportional to the determinant of their operator while those for bosons are inversely proportional to determinants.

One can obtain an asymptotic expansion for the number of eigenvalues $N(\lambda)$ of an operator A with values less than λ:

$$N(\lambda) \sim \tfrac{1}{2}B_0\lambda^2 + B_1\lambda + B_2 + O(\lambda^{-1}), \tag{15.50}$$

where B_0, B_1 and B_2 are the 'Hamidew' coefficients referred to by Gibbons in chapter 13. They can be expressed as $B_n = \int b_n (g_0)^{1/2}\, d^4x$, where the b_n are scalar polynomials in the metric, the curvature and its covariant derivatives (Gilkey, 1975). In the case of the scalar wave operator, $A = -\Box + \xi R + m^2$, they are

$$b_0 = \frac{1}{16\pi^2} \tag{15.51}$$

$$b_1 = \frac{1}{16\pi^2}((1/6-\xi)R - m^2) \tag{15.52}$$

$$b_2 = \frac{1}{2880\pi^2}(R^{abcd}R_{abcd} - R_{ab}R^{ab} + (6-30\xi)\Box R + \tfrac{5}{2}(6\xi-1)^2R^2$$

$$+ 30m^2(1-6\xi)R + 90m^4). \tag{15.53}$$

When there is a boundary surface ∂M, this introduces extra contributions into (15.50) including a $\lambda^{1/2}$-term. This would seem an additional reason for trying to do away with boundary surfaces and working simply with closed manifolds.

From (15.50) one can see that the determinant of A, the product of its eigenvalues, is going to diverge badly. In order to obtain a finite answer one has to regularize the determinant by dividing out by the product of the eigenvalues corresponding to the first two terms on the right of (15.50) (and those corresponding to a $\lambda^{1/2}$-term if it is present). There are various ways of doing this – dimensional regularization (t'Hooft and Veltman, 1972), point splitting (DeWitt, 1975), Pauli–Villars (Zeldovich and Starobinsky, 1972) and the zeta function technique (Dowker and Critchley, 1976; Hawking, 1977). The last method seems the most suitable for regularizing determinants of operators on a curved space background. It will be discussed further in the next section.

For both fermion and baryon operators the term B_0 is $(nV/16\pi^2)$, where V is the volume of the manifold in the background metric, g_0, and n is the number of spin states of the field. If, therefore, there are an equal number of fermion and boson spin states, the leading divergences in Z produced by the B_0-terms will cancel between the fermion and boson determinants without having to regularize. If in addition the B_1-terms either cancel or are zero (which will be the case for zero-rest-mass, conformally invariant fields), the other main divergence in Z will cancel between fermions and bosons. Such a situation occurs in theories with supersymmetry, such as supergravity (Deser and Zumino, 1976; Freedman, van Nieuwenhuizen and Ferrara, 1976) or extended supergravity (Ferrara and van Nieuwenhuizen, 1976). This may be a good reason for taking these theories seriously, in particular for the coupling of matter fields to gravity.

Whether or not the divergences arising from B_0 and B_1 cancel or are removed by regularization, the net B_2 will in general be nonzero, even in supergravity, if the topology of the spacetime manifold is non-trivial (Perry, 1978). This means that the expression for Z will contain a finite number (not necessarily an integer) of uncancelled eigenvalues. Because the eigenvalues have dimensions (length)$^{-2}$, in order to obtain a dimensionless result for Z each eigenvalue has to be divided by μ^2, where μ is the normalization constant or regulator mass. Thus Z will depend on μ. For renormalizable theories such as $\lambda\phi^4$, quantum electrodynamics or Yang–Mills in flat spacetime, B_2 is proportional to the action of the field. This means that one can absorb the μ-dependence into an effective coupling constant $g(\mu)$ which depends on the scale at which it is measured. If $g(\mu) \to 0$ as $\mu \to \infty$, i.e. for very short length scales or high energies, the theory is said to be asymptotically free.

In curved spacetime however, B_2 involves terms which are quadratic in the curvature tensor of the background space. Thus unless one supposes that the gravitational action contains terms quadratic in the curvature (and this seems to lead to a lot of problems including negative energy, fourth-order equations and no Newtonian limit (Stelle, 1977, 1978)) one cannot remove the μ-dependence. For this reason gravity is said to be unrenormalizable because new parameters occur when one regularizes the theory.

If one tried to regularize the higher-order terms in the Taylor series about a background metric, one would have to introduce an infinite sequence of regularization parameters whose values could not be fixed by the theory. However it will be argued in section 15.9 that the higher-order terms have no physical meaning and that one ought to consider only the one-loop quadratic terms. Unlike $\lambda\phi^4$ or Yang–Mills theory, gravity has a natural length scale, the Planck mass. It might therefore seem reasonable to take some multiple of this for the one-loop normalization factor μ.

15.6 Zeta function regularization

In order to regularize the determinant of an operator A with eigenvalues and eigenfunctions $\{\lambda_n, \phi_n\}$, one forms a generalized zeta function from the eigenvalues

$$\zeta_A(s) = \sum \lambda_n^{-s}. \tag{15.54}$$

From (15.50) it can be seen that ζ will converge for $\mathrm{Re}\, s > 2$. It can be analytically extended to a meromorphic function of s with poles only at $s = 2$ and $s = 1$. In particular it is regular at $s = 0$. Formally one has

$$\zeta_A'(0) = -\sum \log \lambda_n. \tag{15.55}$$

Thus one can *define* the regularized value of the determinant of A to be

$$\det A = \exp\left(-\zeta_A'(0)\right). \tag{15.56}$$

The zeta function can be related to the kernel $F(x, x', t)$ of the heat or diffusion equation

$$\frac{\partial F}{\partial t} + A_x F = 0, \tag{15.57}$$

where A_x indicates that the operator acts on the first argument of F. With the initial condition

$$F(x, x', 0) = \delta(x, x'), \tag{15.58}$$

F represents the diffusion over the manifold M, in a fifth dimension of parameter time t, of a point source of heat placed at x' at $t = 0$. The heat equation has been much studied by a number of authors including DeWitt (1963), McKean and Singer (1967) and Gilkey (1975). A good exposition can be found in Gilkey (1974).

It can be shown that if A is an elliptic operator, the heat kernel $F(x, x', t)$ is a smooth function of x, x', and t, for $t > 0$. As $t \to 0$, there is an asymptotic expression for $F(x, x, t)$:

$$F(x, x, t) \sim \sum_{n=0}^{\infty} b_n t^{n-2}, \tag{15.59}$$

where again the b_n are the 'Hamidew' coefficients and are scalar polynomials in the metric, the curvature and its covariant derivatives of order $2n$ in derivatives of the metrics.

One can represent F in terms of the eigenfunctions and eigenvalues of A

$$F(x, x', t) = \sum \phi_n(x)\phi_n(x') \exp(-\lambda_n t). \tag{15.60}$$

Integrating this over the manifold, one obtains

$$Y(t) = \int F(x, x, t)(g_0)^{1/2} \, d^4x = \sum \exp(-\lambda_n t). \tag{15.61}$$

The zeta function can be obtained from $Y(t)$ by an inverse Mellin transform

$$\zeta(s) = \frac{1}{\Gamma(s)} \int_0^{\infty} Y(t) t^{s-1} \, dt. \tag{15.62}$$

Using the asymptotic expansion for F, one sees that $\zeta(s)$ has a pole at $s = 2$ with residue B_0 and a pole at $s = 1$ with residue B_1. There would be a pole at $s = 0$ but it is cancelled by the pole in the gamma function. Thus $\zeta(0) = B_2$. In a sense the poles at $s = 2$ and $s = 1$ correspond to removing the divergences caused by the first two terms in (15.50).

If one knows the eigenvalue explicitly, one can calculate the zeta function and evaluate its derivative at $s = 0$. In other cases one can obtain some information from the asymptotic expansion for the heat kernel. For example, suppose the background metric is changed by a constant scale factor $\tilde{g}_0 = k^2 g_0$, then the eigenvalues, λ_n, of a zero-rest-mass operator A will become $\tilde{\lambda}_n = k^{-2}\lambda_n$. Thus

$$\zeta_{\tilde{A}}(s) = k^{2s}\zeta_A(s)$$

and

$$\zeta'_{\tilde{A}}(0) = 2 \log k \zeta_A(0) + \zeta'_A(0), \tag{15.63}$$

therefore

$$\log (\det \tilde{A}) = -2\zeta(0) \log k + \log (\det A). \tag{15.64}$$

Because B_2, and hence $\zeta(0)$, are not in general zero, one sees that the path integral is not invariant under conformal transformations of the background metric, even for conformally invariant operators A. This is known as a conformal anomaly and arises because in regularizing the determinant one has to introduce a normalization quantity, μ, with dimensions of mass or inverse length. Alternatively, one could say that the measure $D[\phi] = \prod \mu \, dy_n$ is not conformally invariant.

Further details of zeta function regularization of matter field determinants will be found in Hawking (1977), Gibbons (1977c), and Lapedes (1978).

The zeta function regularization of the one-loop gravitational term about a vacuum background has been considered by Gibbons, Hawking and Perry (1978). I shall briefly describe this work and generalize it to include a Λ-term.

The quadratic term in the fluctuations $\bar{g}$ about a background metric, g_0, is

$$I_2[\bar{g}] = \tfrac{1}{2} \int \bar{g}^{ab} A_{abcd} \bar{g}^{cd} (g_0)^{1/2} \, d^4x, \tag{15.65}$$

where

$$g^{ab} = g_0^{ab} + \bar{g}^{ab} \tag{15.66}$$

and

$$16\pi A_{abcd} = \tfrac{1}{4}g_{cd}\nabla_a\nabla_b - \tfrac{1}{4}g_{ac}\nabla_d\nabla_b + \tfrac{1}{8}(g_{ac}g_{bd} + g_{ab}g_{cd})\nabla_e\nabla^e + \tfrac{1}{2}R_{ad}g_{bc}$$

$$-\tfrac{1}{4}R_{ab}g_{cd} + \tfrac{1}{16}Rg_{ab}g_{cd} - \tfrac{1}{8}Rg_{ac}g_{bd} - \tfrac{1}{8}\Lambda g_{ab}g_{cd} + \tfrac{1}{4}\Lambda g_{ac}g_{bd}$$

$$+(a\leftrightarrow b)+(c\leftrightarrow d)+(a\leftrightarrow b, c\leftrightarrow d). \tag{15.67}$$

One cannot simply take the one-loop term to be $(\det (\tfrac{1}{2}\pi^{-1}\mu^{-1}A))^{1/2}$, because A has a large number of zero eigenvalues corresponding to the fact that the action is unchanged under an infinitesimal diffeomorphism

(gauge transformation)

$$x^a \rightarrow x^a + \varepsilon \xi^a$$

$$g_{ab} \rightarrow g_{ab} + 2\varepsilon \xi_{(a;b)}.$$

(15.68)

One would like to factor out the gauge freedom by integrating only over gauge-inequivalent perturbations $\tilde{g}$. One would then obtain an answer which depended on the determinant of A on the quotient of all fields $\tilde{g}$ modulo infinitesimal gauge transformations. The way to do this has been indicated by Feynman (1972), DeWitt (1967) and Fade'ev and Popov (1967). One adds a gauge-fixing term to the action

$$I_f = \tfrac{1}{2} \int \tilde{g}^{ab} B_{abcd} \tilde{g}^{cd} (g_0)^{1/2} \, \mathrm{d}^4 x.$$

(15.69)

The operator B is chosen so that for any sufficiently small perturbation $\tilde{g}$ which satisfies the appropriate boundary condition there is a unique transformation, ξ^a, which vanishes on the boundary such that

$$B_{abcd}(\tilde{g}^{cd} + 2\xi^{(c;d)}) = 0.$$

(15.70)

I shall use the harmonic gauge in the background metric

$$16\pi B_{abcd} = \tfrac{1}{4} g_{bd} \nabla_a \nabla_c - \tfrac{1}{8} g_{cd} \nabla_a \nabla_b - \tfrac{1}{8} g_{ab} \nabla_c \nabla_d$$

$$+ \tfrac{1}{16} g_{ab} g_{cd} \Box + (a \leftrightarrow b) + (c \leftrightarrow d) + (a \leftrightarrow b, c \leftrightarrow d).$$

(15.71)

The operator $(A + B)$ will in general have no zero eigenvalues. However, $\det (A + B)$ contains the eigenvalues of the arbitrarily chosen operator B. To cancel them out one has to divide by the determinant of B on the subspace of all $\tilde{g}$ which are pure gauge transformations, i.e. of the form $\tilde{g}^{ab} = 2\xi^{(a;b)}$ for some ξ which vanishes on the boundary. The determinant of B on this subspace is equal to the square of the determinant of the operator C on the space of all vector fields which vanish on the boundary, where

$$16\pi C_{ab} = -g_{ab} \Box - R_{ab}.$$

(15.72)

Thus one obtains

$$\log Z = -\hat{I}[g_0] - \tfrac{1}{2} \log \det (\tfrac{1}{2}\pi^{-1}\mu^{-2}(A + B)) + \log \det (\tfrac{1}{2}\pi^{-1}\mu^{-2}C).$$

(15.73)

The last term is the so-called ghost contribution.

In order to use the zeta function technique it is necessary to express $A + B$ as $K - L$ where K and L each have only a finite number of negative eigenvalues. To do this, let

$$A + B = -F + G, \tag{15.74}$$

where

$$F = -\tfrac{1}{16}(\nabla_a \nabla^a + 2\Lambda), \tag{15.75}$$

which operates on the trace, ϕ, of $\bar{g}$, $\phi = \bar{g}^{ab} g_{0ab}$

$$G_{abcd} = -\tfrac{1}{8}(g_{ac}g_{bd} + g_{ad}g_{bc})\nabla^e \nabla_e - \tfrac{1}{4}(C_{dcab} + C_{dbac}) + \tfrac{1}{6}\Lambda g_{ab}g_{cd}, \tag{15.76}$$

which operates on the trace-free part, $\tilde{\phi}$, of $\bar{g}$, $\tilde{\phi}^{ab} = \bar{g}^{ab} - \tfrac{1}{4}g_0{}^{ab}\phi$.

If $\Lambda \leq 0$, the operator F will have only positive eigenvalues. Therefore in order to make the one-loop term converge, one has to integrate over purely imaginary ϕ. This corresponds to integrating over conformal factors of the form $\Omega = 1 + i\xi$. if $\Lambda > 0$, F will have some finite number, p, of negative eigenvalues. Because a constant function will be an eigenfunction of F with negative eigenvalue (in the case where there is no boundary), p will be at least one. In order to make the one-loop term converge, one will have to rotate the contour of integration of the coefficient of each eigenfunction, with a negative eigenvalue to lie along the real axis. This will introduce a factor of i^p into Z.

If the background metric g_0 is flat, the operator G will be positive-definite. Thus one will integrate the trace-free perturbations $\tilde{\phi}$ along the real axis. This corresponds to integrating over real conformal equivalence classes. However for non-flat background metrics, G may have some finite number, q, of negative eigenvalues because of the Λ and Weyl tensor terms. Again one will have to rotate the contour of integration for these modes (this time from real to imaginary) and this will introduce a factor of i^{-q} into Z.

The ghost operator is

$$C_{ab} = -g_{ab}(\nabla^e \nabla_e + \Lambda). \tag{15.77}$$

If $\Lambda > 0$, C will have some finite number, r, of negative eigenvalues. Because it is the determinant of C that appears in Z rather than its square root, the negative eigenvalues will contribute a factor $(-1)^r$.

One has

$$\log Z = -\hat{I}[g_0] + \tfrac{1}{2}\zeta'_F(0) + \tfrac{1}{2}\zeta'_G(0) - \zeta'_C(0)$$

$$+ \tfrac{1}{2}\log(2\pi\mu^2)(\zeta_F(0) + \zeta_G(0) - 2\zeta_C(0)). \tag{15.78}$$

From the asymptotic expansion for the heat kernel one has to evaluate the zeta functions at $s = 0$. From the results of Gibbons and Perry (1979) one has

$$\zeta_F(0) + \zeta_G(0) - 2\zeta_C(0) = \int \left(\frac{53}{720\pi^2} C_{abcd} C^{abcd} + \frac{763}{540\pi^2} \Lambda^2 \right) (g_0)^{1/2} \, \mathrm{d}^4 x.$$

(15.79)

From this one can deduce the behaviour of the one-loop term under scale transformations of the background metric. Let $\tilde{g}_{0ab} = k^2 g_{0ab}$, then

$$\log \check{Z} = \log Z + (1 - k^2)\hat{I}[g_0] + \tfrac{1}{2}\gamma \log k,$$

(15.80)

where γ is the right-hand side of (15.79). Providing $\hat{I}[g_0]$ is positive, $\check{Z}$ will be very small for large scales, k. The fact that γ is positive will mean that it is also small for very small scales. Thus quantum gravity may have a cut-off at short length scales. This will be discussed further in section 15.10.

15.7 The background fields

In this section I shall describe some positive-definite metrics which are solutions of the Einstein equations in vacuum or with a Λ-term. In some cases these are analytic continuations of well-known Lorentzian solutions, though their global structure may be different. In particular the section through the complexified manifold on which the metric is positive-definite may not contain the singularities present on the Lorentzian section. In other cases the positive-definite metrics may occur on manifolds which do not have any section on which the metric is real and Lorentzian. They may nevertheless be of interest as stationary-phase points in certain path integrals.

The simplest non-trivial example of a vacuum metric is the Schwarzschild solution (Hartle and Hawking, 1976; Gibbons and Hawking, 1977a). This is normally given in the form

$$\mathrm{d}s^2 = -\left(1 - \frac{2M}{r}\right) \mathrm{d}t^2 + \left(1 - \frac{2M}{r}\right)^{-1} \mathrm{d}r^2 + r^2 \, \mathrm{d}\Omega^2. \quad (15.81)$$

Putting $t = -i\tau$ converts this into a positive-definite metric for $r > 2M$. There is an apparent singularity at $r = 2M$ but this is like the apparent singularity at the origin of polar coordinates, as can be seen by defining a new radial coordinate $x = 4M(1 - 2Mr^{-1})^{1/2}$. Then the metric becomes

$$\mathrm{d}s^2 = \left(\frac{x}{4M}\right)^2 \mathrm{d}\tau^2 + \left(\frac{r^2}{4M^2}\right)^2 \mathrm{d}x^2 + r^2 \, \mathrm{d}\Omega^2.$$

This will be regular at $x = 0$, $r = 2M$, if τ is regarded as an angular variable and is identified with period $8\pi M$ (I am using units in which the gravitational constant $G = 1$). The manifold defined by $x \geq 0$, $0 \leq \tau \leq 8\pi M$ is called the Euclidean section of the Schwarzschild solution. On it the metric is positive-definite, asymptotically flat and non-singular (the curvature singularity at $r = 0$ does not lie on the Euclidean section).

Because the Schwarzschild solution is periodic in imaginary time with period $\beta = 8\pi M$, the boundary surface ∂M at radius r_0 will have topology $S^1 \times S^2$ and the metric will be a stationary-phase point in the path integral for the partition function of a canonical ensemble at temperature $T = \beta^{-1} = (8\pi M)^{-1}$. As shown in section 15.2, the action will come entirely from the surface term, which gives

$$\hat{I} = \tfrac{1}{2}\beta M = 4\pi M^2. \tag{15.82}$$

One can find a similar Euclidean section for the Reissner–Nordström solution with $Q^2 + P^2 < M^2$, where Q is the electric charge and P is the magnetic monopole charge. In this case the radial coordinate has the range $r_+ \leq r < \infty$. Again the outer horizon, $r = r_+$, is an axis of symmetry in the $r - \tau$-plane and the imaginary time coordinate, τ, is identified with period $\beta = 2\pi\kappa^{-1}$, where κ is the surface gravity of the outer horizon. The electromagnetic field, F_{ab}, will be real on the Euclidean section if Q is imaginary and P is real. In particular if $Q = iP$, the field will be self-dual or anti-self-dual,

$$F_{ab} = \pm {}^*F_{ab} = \tfrac{1}{2}\varepsilon_{abcd}F^{cd}, \tag{15.83}$$

where ε_{abcd} is the alternating tensor. If F_{ab} is real on the Euclidean section, the operators governing the behaviour of charged fields will be elliptic and so one can evaluate the one-loop terms by the zeta function method. One can then analytically continue the result back to real Q just as one analytically continues back from positive-definite metrics to Lorentzian ones.

Because $R = 0$, the gravitational part of the action is unchanged. However there is also a contribution from the electromagnetic Lagrangian, $-(1/8\pi)F_{ab}F^{ab}$. Thus

$$\hat{I} = \tfrac{1}{2}\beta(M - \Phi Q + \psi P), \tag{15.84}$$

where $\Phi = Q/r_+$ is the electrostatic potential of the horizon and $\psi = P/r_+$ is the magnetostatic potential.

In a similar manner one can find a Euclidean section for the Kerr metric provided that the mass M is real and the angular momentum J is imaginary. In this case the metric will be periodic in the frame that co-rotates with the horizon, i.e. the point (τ, r, θ, ϕ) is identified with $(\tau + \beta, r, \theta, \phi + i\beta\Omega)$ where Ω is the angular velocity of the horizon (Ω will be imaginary if J is imaginary). As in the electromagnetic case, it seems best to evaluate the one-loop terms with J imaginary and then analytically continue to real J. The presence of angular momentum does not affect the asymptotic metric to leading order to that the action is

$$\hat{I} = \tfrac{1}{2}\beta M \quad \text{with } \beta = 2\pi\kappa^{-1},$$

where κ is the surface gravity of the horizon.

Another interesting class of vacuum solutions are the Taub–NUT metrics (Newman, Unti and Tamburino, 1963; Hawking and Ellis, 1973). These can be regarded as gravitational dyons with an ordinary 'electric' type mass M and a gravitational 'magnetic' type mass N. The metric can be written in the form

$$ds^2 = -V\left(dt + 4N \sin^2\frac{\theta}{2} d\phi\right)^2 + V^{-1} dr^2 + (r^2 + N^2)(d\theta^2 + \sin^2\theta \, d\phi^2),$$

$$(15.85)$$

where $V = 1 - (2Mr + N^2)/(r^2 + N^2)$. This metric is regular on half-axis $\theta = 0$ but it has a singularity at $\theta = \pi$ because the $\sin^2(\theta/2)$ term in the metric means that a small loop around the axis does not shrink to zero length as $\theta = \pi$. This singularity can be regarded as the analogue of a Dirac string in electrodynamics, caused by the presence of a magnetic monopole charge. One can remove this singularity by introducing a new coordinate

$$t' = t + 4N\phi. \qquad (15.86)$$

The metric then becomes

$$ds^2 = -V\left(dt' - 4N \cos^2\frac{\theta}{2} d\phi\right)^2 + V^{-1} dr^2 + (r^2 + N^2)(d\theta^2 + \sin^2\theta \, d\phi^2).$$

$$(15.87)$$

This is regular at $\theta = \pi$ but not at $\theta = 0$. One can therefore use the (t, r, θ, ϕ) coordinates to cover the north pole ($\theta = 0$) and the (t', r, θ, ϕ) co-ordinates to cover the south pole ($\theta = \pi$). Because ϕ is identified with period 2π, (15.86) implies that t and t' have to be identified with period $8\pi N$. Thus if ψ is a regular field with t-dependence of the form $\exp(-i\omega t)$, then ω must satisfy

$$4N\omega = \text{an integer}. \qquad (15.88)$$

This is the analogue of the Dirac quantization condition and relates the 'magnetic' charge, N, of the Taub–NUT solution to the 'electric' charge or energy, ω, of the field ψ. The process of removing the Dirac string singularity by introducing coordinates t and t' and periodically identifying, changes the topology of the surfaces of constant r from $S^2 \times R^1$ to S^3 on which $(t/2N)$, θ and ϕ are Euler angle coordinates.

The metric (15.85) also has singularities where $V = 0$ or ∞. As in the Schwarzschild case $V = \infty$ corresponds to an irremovable curvature singularity but $\dot{V} = 0$ corresponds to a horizon and can be removed by periodically identifying the imaginary time coordinate. This identification is compatible with the one to remove the Dirac string if the two periods are equal, which occurs if $N = \pm iM$. If this is the case, and if M is real, the metric is real and is positive-definite in the region $r > M$ and the curvature is self-dual or anti-self-dual

$$R_{abcd} = \pm {}^*R_{abcd} = \pm \tfrac{1}{2}\varepsilon_{abef}R^{ef}{}_{cd}. \tag{15.89}$$

The apparent singularity at $r = M$ becomes a single point, the origin of hyperspherical coordinates, as can be seen by introducing new radial and time variables

$$x = 2(2M(r-M))^{1/2},$$
$$\psi = -\frac{it}{2M}. \tag{15.90}$$

The metric then becomes

$$ds^2 = \frac{Mx^2}{2(r+M)}(d\psi + \cos\theta\, d\phi)^2$$
$$+ \frac{r+M}{2M}dx^2 + \frac{x^2(r+M)}{8M}(d\theta^2 + \sin^2\theta\, d\phi^2). \tag{15.91}$$

Thus the manifold defined by $x \geq 0$, $0 \leq \psi \leq 4\pi$, $0 \leq \theta \leq \pi$, $0 \leq \phi \leq 2\pi$, with ψ, θ, ϕ interpreted as hyperspherical Euler angles, is topologically R^4 and has a non-singular, positive-definite metric. The metric is asymptotically flat in the sense that the Riemann tensor decreases as r^{-3} as $r \to \infty$ but it is not asymptotically Euclidean, which would require curvature proportional to r^{-4}. The surfaces of the constant r are topologically S^3 but their metric is that of a deformed sphere. The orbits of the $\partial/\partial\psi$ Killing vector define a Hopf fibration $\pi; S^3 \to S^2$, where the S^2 is parametrized by the coordinates θ and ϕ. The induced metric on the S^2 is that of a

2-sphere of radius $(r^2 - M^2)^{1/2}$, while the fibres are circles of circumference $8\pi M V^{1/2}$. Thus, in a sense the boundary at large radius is $S^1 \times S^2$ but is a twisted product.

It is also possible to combine self-dual Taub–NUT solutions (Hawking, 1977). The reason is that the attraction between the electric type masses M is balanced by the repulsion between the imaginary magnetic type masses N. The metric is

$$ds^2 = U^{-1}(d\tau + \boldsymbol{\omega} \cdot d\boldsymbol{x})^2 + U \, d\boldsymbol{x} \cdot d\boldsymbol{x}, \qquad (15.92)$$

where

$$U = 1 + \sum \frac{2M_i}{r_i}$$

and

$$\mathrm{curl}\,\boldsymbol{\omega} = \mathrm{grad}\, U. \qquad (15.93)$$

Here r_i denotes the distance from the ith 'NUT' in the flat, three-dimensional metric $d\boldsymbol{x} \cdot d\boldsymbol{x}$. The curl and grad operations refer to this 3-metric, as does the vector $\boldsymbol{v}$. Each NUT has $N_i = iM_i$.

The vector fields $\boldsymbol{\omega}$ will have Dirac string singularities running from each NUT. If the masses M_i are all equal, these string singularities and the horizon-type singularities at $r_i = 0$ can all be removed by identifying τ with period $8\pi M$. The boundary surface at large radius is then a lens space (Steenrod, 1951). This is topologically an S^3 with n points identified in the fibre S^1 of the Hopf fibration $S^3 \to S^2$, where n is the number of NUTs.

The boundary surface cannot be even locally imbedded in flat space so that one cannot work out the correction term K^0 in the action. If one tries to imbed it as nearly as one can, one obtains the value of $4\pi n M^2$ for the action, the same as Schwarzschild for $n = 1$ (Davies, 1978). In fact the presence of a gravitational magnetic mass alters the topology of the space and prevents it from being asymptotically flat in the usual way. One can, however, obtain an asymptotically flat space containing an equal number, n, of NUTs ($N = iM$) and anti-NUTs ($N = -iM$). Because the NUTs and the anti-NUTs attract each other, they have to be held apart by an electromagnetic field. This solution is in fact one of the Israel–Wilson metrics (Israel and Wilson, 1972; Hartle and Hawking, 1972). The gravitational part of the action is $8\pi n M^2$, so that each NUT and anti-NUT contributes $4\pi M^2$.

I now come on to positive-definite metrics which are solutions of the Einstein equations with a Λ-term on manifolds which are compact without boundary. The simplest example is an S^4 with the metric induced by imbedding it as a sphere of radius $(3\Lambda^{-1})^{1/2}$ in five-dimensional Euclidean space. This is the analytic continuation of de Sitter space (Gibbons and Hawking, 1977b). The metric can be written in terms of a Killing vector $\partial/\partial\tau$:

$$ds^2 = (1 - \tfrac{1}{3}\Lambda r^2)\,d\tau^2 + (1 - \tfrac{1}{3}\Lambda r^2)^{-1}\,dr^2 + r^2\,d\Omega^2. \qquad (15.94)$$

There is a horizon-type singularity at $r = (3\Lambda^{-1})^{1/2}$. This is in fact a 2-sphere of area $12\pi\Lambda^{-1}$ which is the locus of zeros of the Killing vector $\partial/\partial\tau$. The action is $-3\pi\Lambda^{-1}$.

One can also obtain black hole solutions which are asymptotically de Sitter instead of asymptotically flat. The simplest of these is the Schwarzschild–de Sitter (Gibbons and Hawking, 1977b). The metric is

$$ds^2 = V\,d\tau^2 + V^{-1}\,dr^2 + r^2\,d\Omega^2, \qquad (15.95)$$

where

$$V = 1 - 2Mr^{-1} - \tfrac{1}{3}\Lambda r^2.$$

If $\Lambda < (9M^2)^{-1}$, there are two positive values of r for which $V = 0$. The smaller of these corresponds to the black hole horizon, while the larger is similar to the 'cosmological horizon' in de Sitter space. One can remove the apparent singularities at each horizon by identifying τ periodically. However, the periodicities required at the two horizons are different, except in the limiting case $\Lambda = (9M^2)^{-1}$. In this case, the manifold is $S^2 \times S^2$ with the product metric and the action is $-2\pi\Lambda^{-1}$.

One can also obtain a Kerr–de Sitter solution (Gibbons and Hawking, 1977b). This will be a positive-definite metric for values of r lying between the cosmological horizon and the outer black hole horizon, if the angular momentum is imaginary. Again, one can remove the horizon singularities by periodic identifications and the periodicities will be compatible for a particular choice of the parameters (Page, 1978). In this case one obtains a singularity-free metric on an S^2 bundle over S^2. The action is $-0.9553\,(2\pi\Lambda^{-1})$.

One can also obtain Taub–de Sitter solutions. These will have a cosmological horizon in addition to the ordinary Taub–NUT ones. One can remove all the horizon and Dirac string singularities simultaneously in a limiting case which is CP^2, complex, projective 2-space, with the standard Kaehler metric (Gibbons and Pope, 1978). The action is $-\tfrac{9}{4}\pi\Lambda^{-1}$.

One can also obtain solutions which are the product of two two-dimensional spaces of constant curvature (Gibbons, 1977b). The case of

$S^2 \times S^2$ has already been mentioned, and there is also the trivial flat torus $T^2 \times T^2$. In the other examples the two spaces have genera g_1 and $g_2 > 1$ and the Λ-term has to be negative. The action is $-(2\pi/\Lambda)(g_1 - 1)(g_2 - 1)$.

Finally, to complete this catalogue of known positive-definite solutions on the Einstein equations, one should mention $K3$. This is a compact four-dimensional manifold which can be realized as a quartic surface in CP^3, complex projective 3-space. It can be given a positive-definite metric whose curvature is self-dual and which is therefore a solution of the Einstein equation with $\Lambda = 0$ (Yau, 1977). Moreover $K3$ is, up to identifications, the only compact manifold to admit a self-dual metric. The action is 0.

There are two topological invariants of compact four-dimensional manifolds that can be expressed as integrals of the curvature:

$$\chi = \frac{1}{128\pi^2} \int R_{abcd} R_{efgh} \varepsilon^{abef} \varepsilon^{cdgh} (g)^{1/2} \, \mathrm{d}^4 x, \qquad (15.96)$$

$$\tau = \frac{1}{96\pi^2} \int R_{abcd} R^{ab}{}_{ef} \varepsilon^{cdef} (g)^{1/2} \, \mathrm{d}^4 x. \qquad (15.97)$$

χ is the Euler number of the manifold and is equal to the alternating sum of the Betti numbers:

$$\chi = B_0 - B_1 + B_2 - B_3 + B_4. \qquad (15.98)$$

The pth Betti number, B_p, is the number of independent closed p-surfaces that are not boundaries of some $p + 1$-surface. They are also equal to the number of independent harmonic p-forms. For a closed manifold, $B_p = B_{4-p}$ and $B_1 = B_4 = 1$. If the manifold is simply connected, $B_1 = B_3 = 0$, so $\chi \geq 2$.

The Hirzebruch signature, τ, has the following interpretation. The B_2 harmonic 2-forms can be divided into B_2^+ self-dual and B_2^- anti-self-dual 2-forms. Then $\tau = B_2^+ - B_2^-$. It determines the gravitational contribution to the a :al-current anomaly (Eguchi and Freund, 1976; Hawking, 1977; Hawking and Pope, 1978).

S^4 has $\chi = 2$ and $\tau = 0$; CP^2 has $\chi = 3$, $\tau = 1$; the S^2 bundle over S^2 has $\chi = 4$, $\tau = 0$; $K3$ has $\chi = 24$, $\tau = 16$ and the product of two-dimensional spaces with genera g_1, g_2 has $\chi = 4(g_1 - 1)(g_2 - 1)$, $\tau = 0$.

In the non-compact case there are extra surface terms in the formulae for χ and τ. Euclidean space and the self-dual Taub–NUT solution has $\chi = 1$, $\tau = 0$ and the Schwarzschild solution has $\chi = 2$, $\tau = 0$.

15.8 Gravitational thermodynamics

As explained in section 15.3, the partition function

$$Z = \sum \exp\left(-\beta E_n\right)$$

for a system at temperature $T = \beta^{-1}$, contained in a spherical box of radius r_0, is given by a path integral over all metrics which fit inside the boundary, ∂M, with topology $S^2 \times S^1$, where the S^2 is a sphere of radius r_0 and the S^1 has circumference β. By the stationary-phase approximation described in section 15.5, the dominant contributions will come from metrics near classical solutions g_0 with the given boundary conditions. One such solution is just flat space with the Euclidean time coordinate identified with period β. This has topology $R^3 \times S^1$. The action of the background metric is zero, so it makes no contribution to the logarithm of the partition function. If one neglects small corrections arising from the finite size of the box, the one-loop term also can be evaluated exactly as Z_g

$$\log Z_g = \frac{4\pi^5 r_0^3 T^3}{135}. \tag{15.99}$$

This can be interpreted as the partition function of thermal gravitons on a flat-space background.

The Schwarzschild metric with $M = (8\pi T)^{-1}$ is another solution which fits the boundary conditions. It has topology $R^2 \times S^2$ and action $\hat{I} = \beta^2/16\pi = 4\pi M^2$. The one-loop term has not been computed, but by the scaling arguments given in section 15.6 it must have the form

$$\frac{106}{45} \log\left(\frac{\beta}{\beta_0}\right) + f(r_0 \beta^{-1}) \tag{15.100}$$

where β_0 is related to the normalization constant μ. If $r_0 \beta^{-1}$ is much greater than 1, the box will be much larger than the black hole and one would expect $f(r_0 \beta^{-1})$ to approach the flat-space value (15.99). Thus f should have the form

$$f(r_0 \beta^{-1}) = \frac{4\pi^5 r_0^3}{135\beta^3} + O(r_0^2 \beta^{-2}). \tag{15.101}$$

From the partition function one can calculate the expectation value of the energy

$$\langle E \rangle = \frac{\sum E_n \exp\left(-\beta E_n\right)}{\exp\left(-\beta E_n\right)}$$

$$= -\frac{\partial}{\partial \beta} \log Z. \tag{15.102}$$

Applying this to the contribution $(-\beta^2/16\pi)$ to $\log Z$ from the action of the action of the Schwarzschild solution, one obtains $\langle E \rangle = M$, as one might expect. One can also obtain the entropy, which can be defined to be

$$S = -\sum p_n \log p_n, \tag{15.103}$$

where $p_n = Z^{-1} \exp(-\beta E_n)$ is the probability that the system is in the nth state. Then

$$S = \beta \langle E \rangle + \log Z. \tag{15.104}$$

Applying this to the contribution from the action of the Schwarzschild metric, one obtains

$$S = 4\pi M^2 = \tfrac{1}{4}A, \tag{15.105}$$

where A is the area of the event horizon.

This is a remarkable result because it shows that, in addition to the entropy arising from the one-loop term (which can be regarded as the entropy of thermal gravitons on a Schwarzschild background), black holes have an intrinsic entropy arising from the action of the stationary-phase metric. This intrinsic entropy agrees exactly with that assigned to black holes on the basis of particle-creation calculations on a fixed background and the use of the first law of black hole mechanics (see chapters 6 and 13 by Carter and Gibbons). It shows that the idea that gravity introduces a new level of unpredictability or randomness into physics is supported not only by semi-classical approximation but by a treatment in which the gravitational field is quantized.

One reason why classical solutions in gravity have intrinsic entropy while those in Yang–Mills or $\lambda\phi^4$ do not is that the actions of these theories are scale-invariant, unlike the gravitational action. If g_0 is an asymptotically flat solution with period β and action $\hat{I}[g_0]$, then $k^2 g_0$ is a solution with a period $k\beta$ and action $k^2\hat{I}$. This means that the action, $\hat{I}$, must be of the form $c\beta^2$, where c is a constant which will depend on the topology of the solution. Then $\langle E \rangle = 2c\beta$, $\beta\langle E \rangle = 2c\beta^2$, while $\log Z = -\hat{I} = -c\beta^2$. Thus $S = c\beta^2$. The reason that the action $\hat{I}$ is equal to $\tfrac{1}{2}\beta\langle E \rangle$ and not $\beta\langle E \rangle$, as one would expect for a single state with energy $\langle E \rangle$, is that the topology of the Schwarzschild solution is not the same as that of periodically identified flat space. The fact that the Euler number of the Schwarzschild solution is 2 implies that the time-translation Killing vector, $\partial/\partial\tau$, must be zero on some set (in fact a 2-sphere). Thus the surfaces of a constant τ have two boundaries: one at the spherical box of radius r_0 and the other at the horizon $r = 2M$. Consider now the region of

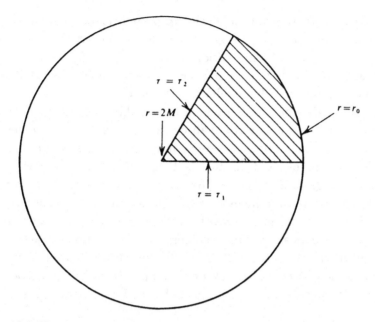

Figure 15.4. The τ–r plane of the Schwarzschild solution. The amplitude $\langle\tau_2|\tau_1\rangle$ to go from the surface τ_1 to the surface $\tau = \tau_2$ is dominated by the action of the shaded portion of the Schwarzschild solution.

the Schwarzschild solution bounded by the surfaces $\tau = \tau_1$, $\tau = \tau_2$ and $r = r_0$ (figure 15.4). The amplitude $\langle\tau_2|\tau_1\rangle$ to go from the surface τ_1 to the surface τ_2 will be given by a path integral over all metrics which fit inside this boundary, with the dominant contribution coming from the stationary-phase metric – which is just the portion of the Schwarzschild solution bounded by these surfaces. The action of this stationary-phase metric will be given by the surface terms because $R = 0$. The surface terms from the surfaces $\tau = \tau_1$ and $\tau = \tau_2$ will cancel out. There will be a contribution of $\frac{1}{2}M(\tau_2 - \tau_1)$ from the surface $r = r_0$. However there will also be a contribution from the 'corner' at $r = 2M$ where the two surfaces $\tau = \tau_1$ and $\tau = \tau_2$ meet, because the second fundamental form, K, of the boundary will have a δ-function behaviour there. By rounding off the corner slightly one can evaluate this contribution, and it turns out to be $\frac{1}{2}M(\tau_2 - \tau_1)$. Thus the total action is $\langle E\rangle (\tau_2 - \tau_1)$ and $\langle\tau_2|\tau_1\rangle = \exp(-\langle E\rangle (\tau_2 - \tau_1))$, as one would expect for a single state with energy $E = \langle E\rangle$. However, if one considers the partition function one simply has the boundary at $r = r_0$ and so the action equals $\frac{1}{2}\beta E$ rather than βE. This difference, which is equal to $\frac{1}{4}A$, gives the entropy of the black hole.

From this one sees that qualitatively new effects arise from the fact that the gravitational field can have different topologies. These effects would not have been found using the canonical approach, because such metrics as the Schwarzschild solution would not have been allowed.

The above derivation of the partition function and entropy of a black hole has been based on the use of the canonical ensemble, in which the system is in equilibrium with an infinite reservoir of energy at temperature T. However the canonical ensemble is unstable when black holes are present because if a hole were to absorb a bit more energy, it would cool down and would continue to absorb more energy than it emitted. This pathology is reflected in the fact that $\langle \Delta E^2 \rangle = \langle E^2 \rangle - \langle E \rangle^2 = (1/Z)(\partial^2 Z/\partial \beta^2) - (\partial \log Z/\partial B)^2 = -1/8\pi$, which is negative. To obtain sensible results with black holes one has to use the micro-canonical ensemble, in which a certain amount of energy E is placed in an insulated box and one considers all possible configurations within that box which have the given energy. Let $N(E)\,dE$ be the number of states of the gravitational field with energies between E and $E + dE$ in a spherical box of radius r_0. The partition function is given by the Laplace transform of $N(E)$,

$$Z(\beta) = \int_0^\infty N(E) \exp(-\beta E)\,dE. \qquad (15.106)$$

Thus, formally, the density of states is given by an inverse Laplace transform,

$$N(E) = \frac{1}{2\pi i} \int_{-i\infty}^{i\infty} Z(\beta) \exp(E\beta)\,d\beta. \qquad (15.107)$$

For large β, the dominant contribution to $Z(\beta)$ comes from the action of the Schwarzschild metric, and is of the form $\exp(-\beta^2/16\pi)$. Thus the right-hand side of (15.107) would diverge if the integral were taken up the imaginary β-axis as it is supposed to be. To obtain a finite value for (15.107) one has to adopt the prescription that the integral be taken along the real β-axis. This is very similar to the procedure used to evaluate the path integral in the stationary-phase approximation, where one rotated the contour of integration for each quadratic term, so that one would obtain a convergent Gaussian integral. With this prescription the factor $1/2\pi i$ in (15.107) would give an imaginary value for the density of states $N(E)$ if the partition function $Z(\beta)$ were real. However, as mentioned in section 5.6, the operator G which governs non-conformal or trace-free perturbations has one negative eigenvalue in the Schwarzschild metric. This contributes a factor i to the one-loop term for Z. Thus the partition function is purely imaginary but the density of space is real. This is what one might expect: the partition function is pathological because the canonical ensemble is not well defined but the density of states is real and positive because the micro-canonical ensemble is well behaved.

It is not appropriate to go beyond the stationary-phase approximation in evaluating the integral in (15.107) because the partition function, Z, has been calculated in this approximation only. If one takes just the contribution $\exp(-\beta^2/16\pi)$ from the action of the background metric, one finds that a black hole of mass M has a density of states $N(M)=2\pi^{-1/2}\exp(4\pi M^2)$. Thus the integral in (15.106) does not converge unless one rotates the contour integration to lie along the imaginary E-axis. If one includes the one-loop term Z_g, the stationary-phase point in the β integration in (15.107) occurs when

$$E = \frac{-\partial \log Z_g}{\partial \beta} \qquad (15.108)$$

for the flat background metric, and

$$E = \frac{\beta}{8\pi} - \frac{\partial \log Z_g}{\partial \beta} \qquad (15.109)$$

for the Schwarzschild background metric. One can interpret these equations as saying that E is equal to the energy of the thermal graviton and the black hole, if present. Using the approximate form of Z_g one finds that if the volume, V, of the box satisfies

$$E^5 < \frac{\pi^2}{15}(8354.5)V, \qquad (15.110)$$

the dominant contribution to N comes from the flat-space background metric. Thus in this case the most probable state of the system is just thermal gravitons and no black hole. If V is less than the inequality (15.110), there are two stationary-phase points for the Schwarzschild background metric. The one with a lower value of β gives a contribution to N which is larger than that of the flat-space background metric. Thus the most probable state of the system is a black hole in equilibrium with thermal gravitons. These results confirm earlier derivations based on the semi-classical approximations (Hawking, 1976; Gibbons and Perry, 1978).

15.9 Beyond one loop

In section 15.5 the action was expanded in a Taylor series around a background field which was a solution of the classical field equations. The path integral over the quadratic terms was evaluated but the higher-order terms were neglected. In renormalizable theories such as quantum electrodynamics, Yang–Mills or $\lambda\phi^4$ one can evaluate these higher or 'interaction' terms with the help of the differential operator A appearing in the quadratic or 'free' part of the action. One can express their effect by Feynman diagrams with two or more closed loops, where the lines in the diagram represent the propagator or Green's function A^{-1} and the vertices correspond to the interaction terms, three lines meeting at a cubic term and so on. In these renormalizable theories the undetermined quantities which arise from regularizing the higher loops turn out to be related to the undetermined normalization quantity, μ, of the single loop. They can thus all be absorbed into a redefinition of the coupling constant and any masses which appear in the theory.

The situation in quantum gravity is very different. The single-loop term about a flat or topologically trivial vacuum metric does not contain the normalization quantity, μ. However, about a topologically non-trivial background one has log Z_g proportional to $(106/45)\chi$ log μ, where Z_g is the one-loop term and χ is the Euler number. One can express this as an addition to the action of an effective topological term $-k(\mu)\chi$, where $k(\mu)$ is a scale-dependent topological coupling constant. One cannot in general provide such a topological interpretation of the μ-dependence of the one-loop term about a background metric which is a solution of the field equations with nonzero matter fields. However one can do it in the special case where the matter fields are related to the gravitational field by local supersymmetry or spinor-dependent gauge transformations. These are the various supergravity and extended supergravity theories (Freedman, Van Nieuwenhuizen and Ferrara, 1976; Deser and Zumino, 1976).

Two loops in supergravity, and maybe also in pure gravity, do not seem to introduce any further undetermined quantities. However it seems likely that, both in supergravity and in pure gravity, further undetermined quantities will arise at three or more loops, though the calculations needed to verify this are so enormous that no-one has attempted them. Even if by some miracle no further undetermined quantities arose from the regularization of the higher loop, one would still not have a good procedure for evaluating the path integral, because the perturbation expansion around a given background field has only a very limited range of validity in gravity, unlike the case in renormalizable theories such as Yang–Mills or $\lambda\phi^4$. In the latter theory the quadratic or 'free' term in the action $\int(\nabla\phi)^2\,\mathrm{d}^4x$ bounds the interaction term $\lambda\int\phi^4\,\mathrm{d}^4x$. This means that one can evaluate the expectation value of the interaction term in the measure $D[\phi] \exp(-\int (\nabla \phi)^2 \, \mathrm{d}^4 x)$ or, in other words, using Feynman diagrams where the lines correspond to the free propagator. Similarly in quantum electrodynamics or Yang–Mills theory, the interaction term is only cubic or quartic and is bounded by the free term. However, in the gravitational case the Taylor expansion about a background metric contains interaction terms of all orders in, and quadratic in derivatives of, the metric perturbations. These interaction terms are not bounded by the free, quadratic term so their expectation values in the measure given by the quadratic term are not defined. In other words, it does not make any sense to represent them by higher-order Feynman diagrams. This should come as no surprise to those who have worked in classical general relativity rather than in quantum field theory. We know that one cannot represent something like a black hole as a perturbation of flat space.

In classical general relativity one can deal with the problem of the limited range of validity of perturbation theory by using matched asymptotic expansions around different background metrics. It would therefore seem natural to try something similar in quantum gravity. In order to ensure gauge-invariance it would seem necessary that these background metrics should be solutions of the classical field equations. As far as we know, in a given topology and with given boundary conditions there is only one solution of the field equations or, at the most, a finite-dimensional family. Thus solutions of a given topology could not be dense in the space of metrics of that topology. However the Einstein action, unlike that of Yang–Mills theory, does not seem to provide any barrier to passing from fields of one topology to another.

One way of seeing this is to use Regge calculus (Regge, 1961). Using this method, one decomposes the spacetime manifold into a simplical complex. Each 4-simplex is taken to be flat and to be determined by its edge (i.e. 1-simplex) lengths. However the angles between the faces (i.e. 2-simplices) are in general such that the 4-simplices could not be joined together in flat four-dimensional space. There is thus a distortion which can be represented as a δ-function in the curvature concentrated on the faces. The total action is $(-1/8\pi) \sum A_i \delta_i$ taken over all 2-simplices, where A_i is the area of the ith 2-simplex and δ_i is the deficit angle at that 2-simplex, i.e. δ_i equals 2π minus the sum of the angles between those 3-simplices which are connected by the given 2-simplex.

A complex in which the action is stationary under small variations of the edge length can be regarded as a discrete approximation to a smooth solution of the Einstein equations. However, one can also regard the

Regge calculus as defining the action of a certain class of metrics without any approximations. This action will remain well defined and finite even if the edge lengths are chosen so that some of the simplices collapse to simplices of lower dimension. For example if a, b, c are the edge lengths of a triangle (a 2-simplex) then they must satisfy the inequalities $a < b + c$ etc. If $a = b + c$, the 2-simplex collapses to a 1-simplex. In general, the simplical complex will not remain a manifold if some of the simplices collapse to lower dimensions. However the action will still be well defined. One can then blow up some of the simplices to obtain a new manifold with a different topology. In this way one can pass continuously from one metric topology to another.

The idea is, therefore, that there can be quantum fluctuations of the metric not only within each topology but from one topology to another. This possibility was first pointed out by Wheeler (1963) who suggested that spacetime might have a 'foam-like' structure on the scale of the Planck length. In the next section I shall attempt to provide a mathematical framework to describe this foam-like structure. The hope is that by considering metrics of all possible topologies one will find that the classical solutions are dense in some sense in the space of all metrics. One could then hope to represent the path integral as a sum of background and one-loop terms from these solutions. One would hope to be able to pick out some finite number of solutions which gave the dominant contributions.

15.10 Spacetime foam

One would like to find which topologies of stationary-phase metrics give the dominant contribution to the path integral. In order to do this it is convenient to consider the path integral over all compact metrics which have a given spacetime volume V. This is not to say that spacetime actually is compact. One is merely using a convenient normalization device, like periodic boundary conditions in ordinary quantum theory: one works with a finite volume in order to have a finite number of states and then considers the values of various quantities per unit volume in the limit that the volume is taken to infinity.

In order to consider path integrals over metrics with a given 4-volume V one introduces into the action a term $\Lambda V/8\pi$, where Λ is to be regarded as a Lagrange multiplier (the factor $1/8\pi$ is chosen for convenience). This term has the same form as a cosmological term in the action but the motivation for it is very different as is its value: observational evidence shows that any cosmological Λ would have to be so small as to be practically negligible whereas the value of the Lagrange multiplier will turn out to be very large, being of the order of one in Planck units.

Let

$$Z[\Lambda] = \int D[g] \exp\left(-\hat{I}[g] - \frac{\Lambda}{8\pi} V[g]\right), \qquad (15.111)$$

where the integral is taken over all metrics on some compact manifold. One can interpret $Z[\Lambda]$ as the 'partition function' for what I shall call the *volume canonical ensemble*, i.e.

$$Z[\Lambda] = \sum_n \left\langle \phi_n \left| \exp - \left(\frac{\Lambda V}{8\pi}\right) \right| \phi_n \right\rangle, \qquad (15.112)$$

where the sum is taken over all states $|\phi_n\rangle$ of the gravitational field. From $Z[\Lambda]$ one can calculate $N(V)\,dV$, the number of the gravitational fields with 4-volumes between V and $V + dV$:

$$N(V) = \frac{1}{16\pi^2 i} \int_{-i\infty}^{i\infty} Z[\Lambda] \exp(\Lambda V)\,d\Lambda \qquad (15.113)$$

In (15.113), the contour of integration should be taken to the right of any singularities in $Z[\Lambda]$ on the imaginary axis.

One wants to compare the contributions to N from different topologies. A convenient measure of the complexity of the topology is the Euler number χ. For simply connected manifolds it seems that χ and the signature τ characterize the manifold up to homotopy and possibly up to homeomorphisms, though this is unproved. In the non-simply connected case there is no possible classification: there is no algorithm for deciding whether two non-simply connected 4-manifolds are homeomorphic or homotopic. This would seem a good reason to restrict attention to simply connected manifolds. Another would be that one could always unwrap a non-simply connected manifold. This might produce a non-compact manifold, but one would expect that one could then close it off at some large volume V with only a small change in the action per unit volume.

By the stationary-phase approximation one would expect the dominant contributions to the path integral Z to come from metrics near solutions of the Einstein equations with a Λ-term. From the scaling behaviour of the action it follows that for such a solution

$$\Lambda = -8\pi c V^{-1/2}, \quad \hat{I} = -\frac{8\pi c^2}{\Lambda} \qquad (15.114)$$

where c is a constant (either positive or negative) which depends on the solution and the topology, and where the action $\hat{I}$ now includes the Λ-term. The constant c has a lower bound of $-(\frac{3}{8})^{1/2}$ which corresponds to its value for S^4. An upper bound can be obtained from (15.96) and (15.97) for χ and τ. For solutions of the Einstein equations with a Λ-term these take the form

$$\chi = \frac{1}{32\pi^2} \int (C_{abcd}C^{abcd} + 2\tfrac{2}{3}\Lambda^2)(g)^{1/2}\, d^4x, \qquad (15.115)$$

$$\tau = \frac{1}{48\pi^2} \int C_{abcd}{}^*C^{abcd}(g)^{1/2}\, d^4x. \qquad (15.116)$$

From (15.115) one sees that there can be a solution only if χ is positive. However this will be the case for simply connected manifolds because then $\chi = 2 + B_2$, where B_2 is the second Betti number. Combining (15.115) and (15.116) one obtains the inequality

$$2\chi - 3|\tau| \geq \frac{32c^2}{3}. \qquad (15.117)$$

From (15.115) one can see that, for large Euler number, at least one of the following must be true:

    (a) c^2 is large

    (b) $\int C_{abcd}C^{abcd}(g)^{1/2}\, d^4x$ is large.

In the former case c must be positive (i.e. Λ must be negative) because there is a lower bound of $-(\frac{3}{8})^{1/2}$ on c. In the latter case the Weyl tensor must be large. As in ordinary general relativity, this will have a converging effect on geodesics similar to that of a positive Ricci tensor. However, between any two points in space there must be a geodesic of minimum length which does not contain conjugate points. Therefore, in order to prevent the Weyl curvature from converging the geodesics too rapidly, one has to put in a negative Ricci tensor or Λ-term of the order of $-C_{abcd}C^{abcd}L^2$, where L is some typical length scale which will be of the order of $V^{1/4}\chi^{-1/4}$, the length per unit of topology. One would then expect the two terms in (15.115) to be of comparable magnitude and c to be of the order of $d\chi^{1/2}$, where $d \leq 3^{1/2}/4$.

This is borne out by a number of examples for which I am grateful to N. Hitchin. For products of two-dimensional manifolds of constant curvature one has $d = \frac{1}{4}$. For algebraic hypersurfaces one has $2^{1/2}/8$.

Hitchin has obtained a whole family of solutions lying between these limits. In addition, if the solution admits a Kähler structure one has the equality

$$3\tau + 2\chi = 32c^2. \tag{15.118}$$

One can interpret these results as saying that one has a collection of the order of χ 'gravitational instantons' each of which has action of the order of L^2, where L is the typical size and is of the order of $V^{1/4}\chi^{-1/4}$. One also has to estimate the dependence of the one-loop curve Z_g on Λ and χ. The dependence on Λ comes from the scaling behaviour and is of the form

$$Z_g \propto \Lambda^{-\gamma},$$

where

$$\gamma = \int \left(\frac{53}{720\pi^2} C_{abcd}C^{abcd} + \frac{763}{540\pi^2} \Lambda^2 \right)(g_0)^{1/2} \, d^4x. \tag{15.119}$$

One can regard γ as the number of extra modes from perturbations about the background metric, over and above those for flat space. From (15.119) one can see that it is of the same order as χ. One can therefore associate a certain number of extra modes with each 'instanton'.

From the above it seems reasonable to make the estimate

$$Z[\Lambda] = \left(\frac{\Lambda}{\Lambda_0} \right)^{-\gamma} \exp{(b\chi\Lambda^{-1})}, \tag{15.120}$$

where $b = 8\pi d^2$ and Λ_0 is related to the normalization constant μ. Using (15.120) in (15.113), one can do the contour integral exactly and obtain

$$N(V) = \Lambda_0^\gamma \left(\frac{8\pi b\chi}{V} \right)^{1-\gamma/2} I_{\gamma-1}\left(\frac{Vb\chi}{2\pi} \right) \tag{15.121}$$

for $V \geq 0$.

However the qualitative dependence on the parameters is seen more clearly by evaluating (15.113) approximately by the stationary-phase method. In fact it is inappropriate to do it more precisely because $Z[\Lambda]$ has been evaluated only in the stationary-phase approximation. The stationary-phase point occurs for

$$\Lambda_s = 4\pi \frac{\gamma \pm (\gamma^2 + Vb\chi/2\pi)^{1/2}}{V}. \tag{15.122}$$

Because the contour should pass to the right of the singularity at $\Lambda = 0$, one should take the positive sign of the square root.

The stationary-phase value of Λ is always positive even though $Z[\Lambda]$ was calculated using background metrics which have negative Λ for large Euler number. This means that one has to analytically continue Z from negative to positive Λ. This analytic continuation is equivalent to multiplying the metric by a purely imaginary conformal factor, which was necessary anyway to make the path integral over conformal factors converge.

From the stationary-phase approximation one has

$$N(V) = Q(\Lambda_s) \equiv \left(\frac{\Lambda_s}{\Lambda_0}\right)^{-\gamma} \exp\left(b\chi\Lambda_s^{-1} + \frac{V\Lambda_s}{8\pi}\right). \qquad (15.123)$$

The dominant contribution to $N(V)$ will come from those topologies for which $dQ/d\chi = 0$. If one assumes $\gamma = a\chi$, where a is constant, one finds that this is satisfied if

$$-a \log\left(\frac{\Lambda_s}{\Lambda_0}\right) + b\Lambda_s^{-1} = 0. \qquad (15.124)$$

If $\Lambda_0 \gtrsim 1$, this will be satisfied by $\Lambda_s \approx \Lambda_0$. If $\Lambda_0 < 1$, $\Lambda_s \approx \Lambda_0^{a/b}$. Equation (15.122) then implies that $\chi = hV$, where the constant of proportionality, h, depends on Λ_0. In other words the dominant contribution to $N(V)$ comes from metrics with one gravitational instanton per volume h^{-1}.

What observable effects this foam-like structure of spacetime would give rise to has yet to be determined, but it might include the gravitational decay of baryons or muons, caused by their falling into gravitational instantons or virtual black holes and coming out again as other species of particles. One would also expect to get non-conservation of the axial-vector current caused by topologies with non-vanishing signature τ.

PHYSICAL REVIEW D VOLUME 28, NUMBER 12 15 DECEMBER 1983

Wave function of the Universe

J. B. Hartle

Enrico Fermi Institute, University of Chicago, Chicago, Illinois 60637
and Institute for Theoretical Physics, University of California, Santa Barbara, California 93106

S. W. Hawking

Department of Applied Mathematics and Theoretical Physics, Silver Street, Cambridge, England
and Institute for Theoretical Physics, University of California, Santa Barbara, California 93106
(Received 29 July 1983)

The quantum state of a spatially closed universe can be described by a wave function which is a functional on the geometries of compact three-manifolds and on the values of the matter fields on these manifolds. The wave function obeys the Wheeler-DeWitt second-order functional differential equation. We put forward a proposal for the wave function of the "ground state" or state of minimum excitation: the ground-state amplitude for a three-geometry is given by a path integral over all compact positive-definite four-geometries which have the three-geometry as a boundary. The requirement that the Hamiltonian be Hermitian then defines the boundary conditions for the Wheeler-DeWitt equation and the spectrum of possible excited states. To illustrate the above, we calculate the ground and excited states in a simple minisuperspace model in which the scale factor is the only gravitational degree of freedom, a conformally invariant scalar field is the only matter degree of freedom and $\Lambda > 0$. The ground state corresponds to de Sitter space in the classical limit. There are excited states which represent universes which expand from zero volume, reach a maximum size, and then recollapse but which have a finite (though very small) probability of tunneling through a potential barrier to a de Sitter-type state of continual expansion. The path-integral approach allows us to handle situations in which the topology of the three-manifold changes. We estimate the probability that the ground state in our minisuperspace model contains more than one connected component of the spacelike surface.

I. INTRODUCTION

In any attempt to apply quantum mechanics to the Universe as a whole the specification of the possible quantum-mechanical states which the Universe can occupy is of central importance. This specification determines the possible dynamical behavior of the Universe. Moreover, if the uniqueness of the present Universe is to find any explanation in quantum gravity it can only come from a restriction on the possible states available.

In quantum mechanics the state of a system is specified by giving its wave function on an appropriate configuration space. The possible wave functions can be constructed from the fundamental quantum-mechanical amplitude for a complete history of the system which may be regarded as the starting point for quantum theory.[1] For example, in the case of a single particle a history is a path $x(t)$ and the amplitude for a particular path is proportional to

$$\exp(iS[x(t)]) , \qquad (1.1)$$

where $S[x(t)]$ is the classical action. From this basic amplitude, the amplitude for more restricted observations can be constructed by superposition. In particular, the amplitude that the particle, having been prepared in a certain way, is located at position x and nowhere else at time t is

$$\psi(x,t)=N \int_C \delta x(t)\exp(iS[x(t)]) . \qquad (1.2)$$

Here, N is a normalizing factor and the sum is over a class

of paths which intersect x at time t and which are weighted in a way that reflects the preparation of the system. $\psi(x,t)$ is the wave function for the state determined by this preparation. As an example, if the particle were previously localized at x' at time t' one would sum over all paths which start at x' at t' and end at x at t thereby obtaining the propagator $\langle x,t \,|\, x',t' \rangle$. The oscillatory integral in Eq. (1.2) is not well defined but can be made so by rotating the time to imaginary values.

An alternative way of calculating quantum dynamics is to use the Schrödinger equation,

$$i\partial\psi/\partial t=H\psi . \qquad (1.3)$$

This follows from Eq. (1.2) by varying the end conditions on the path integral. For a particular state specified by a weighting of paths C, the path integral (1.2) may be looked upon as providing the boundary conditions for the solution of Eq. (1.3).

A state of particular interest in any quantum-mechanical theory is the ground state, or state of minimum excitation. This is naturally defined by the path integral, made definite by a rotation to Euclidean time, over the class of paths which have vanishing action in the far past. Thus, for the ground state at $t=0$ one would write

$$\psi_0(x,0)=N \int \delta x(\tau)\exp(-I[x(\tau)]) , \qquad (1.4)$$

where $I[x(\tau)]$ is the Euclidean action obtained from S by sending $t \rightarrow -i\tau$ and adjusting the sign so that it is positive.

In cases where there is a well-defined time and a corresponding time-independent Hamiltonian, this definition of ground state coincides with the lowest eigenfunction of the Hamiltonian. To see this specialize the path-integral expression for the propagator $\langle x,t | x',t' \rangle$ to $t=0$ and $x'=0$ and insert a complete set of energy eigenstates between the initial and final state. One has

$$\langle x,0 | 0,t' \rangle = \sum_n \psi_n(x)\overline{\psi}_n(0)\exp(iE_n t')$$
$$= \int \delta x(t)\exp(iS[x(t)]) , \qquad (1.5)$$

where $\psi_n(x)$ are the time-independent energy eigenfunctions. Rotate $t' \rightarrow -i\tau'$ in (1.5) and take the limit as $\tau' \rightarrow -\infty$. In the sum only the lowest eigenfunction (normalized to zero energy) survives. The path integral becomes the path integral on the right of (1.4) so that the equality is demonstrated.

The case of quantum fields is a straightforward generalization of quantum particle mechanics. The wave function is a functional of the field configuration on a spacelike surface of constant time, $\Psi = \Psi[\phi(\overline{x}),t]$. The functional Ψ gives the amplitude that a particular field distribution $\phi(\overline{x})$ occurs on this spacelike surface. The rest of the formalism is similarly generalized. For example, for the ground-state wave functional one has

$$\Psi_0[\phi(\overline{x}),0] = N \int \delta\phi(x)\exp(-I[\phi(x)]) , \qquad (1.6)$$

where the integral is over all Euclidean field configurations for $\tau < 0$ which match $\phi(\overline{x})$ on the surface $\tau = 0$ and leave the action finite at Euclidean infinity.

In the case of quantum gravity new features enter. For definiteness and simplicity we shall restrict our attention throughout this paper to spatially closed universes. For these there is no well-defined intrinsic measure of the location of a spacelike surface in the spacetime beyond that contained in the intrinsic or extrinsic geometry of the surface itself. One therefore labels the wave function by the three-metric h_{ij} writing $\Psi = \Psi[h_{ij}]$. Quantum dyanmics is supplied by the functional integral

$$\Psi[h_{ij}] = N \int_C \delta g(x)\exp(iS_E[g]) . \qquad (1.7)$$

S_E is the classical action for gravity including a cosmological constant Λ and the functional integral is over all four-geometries with a spacelike boundary on which the induced metric is h_{ij} and which to the past of that surface satisfy some appropriate condition to define the state. In particular for the amplitude to go from a three-geometry h'_{ij} on an initial spacelike surface to a three-geometry h''_{ij} on a final spacelike surface is

$$\langle h''_{ij} | h'_{ij} \rangle = \int \delta g \exp(iS_E[g]) , \qquad (1.8)$$

where the sum is over all four-geometries which match h'_{ij} on the initial surface and h''_{ij} on the final surface. Here one clearly sees that one cannot specify time in these states. The proper time between the surfaces depends on the four-geometries in the sum.

As in the mechanics of a particle the functional integral (1.7) implies a differential equation on the wave function. This is the Wheeler-DeWitt equation[2] which we shall derive from this point of view in Sec. II. With a simple choice of factor ordering it is

$$\left[-G_{ijkl}\frac{\delta^2}{\delta h_{ij}\delta h_{kl}} - {}^3R(h)h^{1/2} + 2\Lambda h^{1/2} \right]\Psi[h_{ij}] = 0 ,$$
$$(1.9)$$

where G_{ijkl} is the metric on superspace,

$$G_{ijkl} = \tfrac{1}{2}h^{-1/2}(h_{ik}h_{jl} + h_{il}h_{jk} - h_{ij}h_{kl}) \qquad (1.10)$$

and ${}^3R$ is the scalar curvature of the intrinsic geometry of the three-surface. The problem of specifying cosmological states is the same as specifying boundary conditions for the solution of the Wheeler-DeWitt equation. A natural first question to ask is what boundary conditions specify the ground state?

In the quantum mechanics of closed universes we do not expect to find a notion of ground state as a state of lowest energy. There is no natural definition of energy for a closed universe just as there is no independent standard of time. Indeed in a certain sense the total energy for a closed universe is always zero—the gravitational energy canceling the matter energy. It is still reasonable, however, to expect to be able to define a state of minimum excitation corresponding to the classical notion of a geometry of high symmetry. This paper contains a proposal for the definition of such a ground-state wave function for closed universes. The proposal is to extend to gravity the Euclidean-functional-integral construction of nonrelativistic quantum mechanics and field theory [Eqs. (1.4) and (1.6)]. Thus, we write for the ground-state wave function

$$\Psi_0[h_{ij}] = N \int \delta g \exp(-I_E[g]) , \qquad (1.11)$$

where I_E is the Euclidean action for gravity including a cosmological constant Λ. The Euclidean four-geometries summed over must have a boundary on which the induced metric is h_{ij}. The remaining specification of the class of geometries which are summed over determines the ground state. Our proposal is that the sum should be over compact geometries. This means that the Universe does not have any boundaries in space or time (at least in the Euclidean regime) (cf. Ref. 3). There is thus no problem of boundary conditions. One can interpret the functional integral over all compact four-geometries bounded by a given three-geometry as giving the amplitude for that three-geometry to arise from a zero three-geometry, i.e., a single point. In other words, the ground state is the amplitude for the Universe to appear from nothing.[4] In the following we shall elaborate on this construction and show in simple models that it indeed supplies reasonable wave functions for a state of minimum excitation.

The specification of the ground-state wave function is a constraint on the other states allowed in the theory. They must be such, for example, as to make the Wheeler-DeWitt equation Hermitian in an appropriate norm. In analogy with ordinary quantum mechanics one would expect to be able to use these constraints to extrapolate the boundary conditions which determine the excited states of the theory from those fixed for the ground state by Eq. (1.7). Thus, one can in principle determine all the allowed cosmological states.

The wave functions which result from this specification will not vanish on the singular, zero-volume· three-geometries which correspond to the big-bang singularity. This is analogous to the behavior of the wave function of the electron in the hydrogen atom. In a classical treatment, the situation in which the electron is at the proton is singular. However, in a quantum-mechanical treatment the wave function in a state of zero angular momentum is finite and nonzero at the proton. This does not cause any problems in the case of the hydrogen atom. In the case of the Universe we would interpret the fact that the wave function can be finite and nonzero at the zero three-geometry as allowing the possibility of topological fluctuations of the three-geometry. This will be discussed further in Sec. VIII.

After a general discussion of this proposal for the ground-state wave function we shall implement it in a minisuperspace model. The geometrical degrees of freedom in the model are restricted to spatially homogeneous, isotropic, closed universes with S^3 topology, the matter degrees of freedom to a single, homogeneous, conformally invariant scalar field and the cosmological constant is assumed to be positive. A semiclassical evaluation of the functional integral for the ground-state wave function shows that it indeed does possess characteristics appropriate to a "state of minimum excitation."

Extrapolating the boundary conditions which allow the ground state to be extracted from the Wheeler-DeWitt equation, we are able to go further and identify the wave functions in the minisuperspace models corresponding to excited states of the matter field. These wave functions display some interesting features. One has a complete spectrum of excited states which show that a closed universe similar to our own and possessed of a cosmological constant can escape the big crunch and tunnel through to an eternal de Sitter expansion. We are able to calculate the probability for this transition.

In addition to the excited states we make a proposal for the amplitudes that the ground-state three-geometry consists of disconnected three-spheres thus giving a meaning to a gravitational state possessing different topologies.

Our conclusion will be that the Euclidean-functional-integral prescription (1.7) does single out a reasonable candidate for the ground-state wave function for cosmology which when coupled with the Wheeler-DeWitt equation yields a basis for constructing quantum cosmologies.

II. QUANTUM GRAVITY

In this section we shall review the basic principles and machinery of quantum gravity with which we shall explore the wave functions for closed universes. For simplicity we shall represent the matter degrees of freedom by a single scalar field ϕ, more realistic cases being straightforward generalizations. We shall approach this review from the functional-integral point of view although we shall arrive at many canonical results.[5] None of these are new and for different approaches to the same ends the reader is referred to the standard literature.[6]

A. Wave functions

Our starting point is the quantum-mechanical amplitude for the occurrence of a given spacetime and a given field history. This is

$$\exp(iS[g,\phi]) , \qquad (2.1)$$

where $S[g,\phi]$ is the total classical action for gravity coupled to a scalar field. We are envisaging here a fixed manifold although there is no real reason that amplitudes for different manifolds may not be considered provided a rule is given for their relative phases. Just as the interesting observations of a particle are not typically its entire history but rather observations of position at different times, so also the interesting quantum-mechanical questions for gravity correspond to observations of spacetime and field on different spacelike surfaces. Following the general rules of quantum mechanics the amplitudes for these more restricted sets of observations are obtained from (2.1) by summing over the unobserved quantities.

It is easy to understand what is meant by fixing the field on a given spacelike surface. What is meant by fixing the four-geometry is less obvious. Consider all four-geometries in which a given spacelike surface occurs but whose form is free to vary off the surface. By an appropriate choice of gauge near the surface (e.g., Gaussian normal coordinates) all these four-geometries can be expressed so that the only freedom in the four-metric is the specification of the three-metric h_{ij} in the surface. Specifying the three-metric is therefore what we mean by fixing the four-geometry on a spacelike surface. The situation is not unlike gauge theories. There a history is specified by a vector potential $A_\mu(x)$ but by an appropriate gauge transformation $A_0(x)$ can be made to vanish so that the field on a surface can be completely specified by the $A_i(x)$.

As an example of the quantum-mechanical superposition principle the amplitude for the three-geometry and field to be fixed on two spacelike surfaces is

$$\langle h_{ij}'',\phi'' \mid h_{ij}',\phi' \rangle = \int \delta g \, \delta\phi \exp(iS[g,\phi]) , \qquad (2.2)$$

where the integral is over all four-geometries and field configurations which match the given values on the two spacelike surfaces. This is the natural analog of the propagator $\langle x'',t'' \mid x',t' \rangle$ in the quantum mechanics of a single particle. We note again that the proper time between the two surfaces is not specified. Rather it is summed over in the sense that the separation between the surfaces depends on the four-geometry being summed over. It is not that one could not ask for the amplitude to have the three-geometry and field fixed on two surfaces *and* the proper time between them. One could. Such an amplitude, however, would not correspond to fixing observations on just two surfaces but rather would involve a set of intermediate observations to determine the time. It would therefore not be the natural analog of the propagator.

Wave functions Ψ are defined by

$$\Psi[h_{ij},\phi] = \int_C \delta g \, \delta\phi \exp(iS[g,\phi]) . \qquad (2.3)$$

The sum is over a class C of spacetimes with a compact boundary on which the induced metric is h_{ij} and field configurations which match ϕ on the boundary. The remaining specification of the class C is the specification of the state.

If the Universe is in a quantum state specified by a wave function Ψ then that wave function describes the correlations between observables to be expected in that state. For example, in the semiclassical wave function describing a universe like our own, one would expect Ψ to be large when ϕ is big and the spatial volume is small, large when ϕ is small and the spatial volume is big, and small when these quantities are oppositely correlated. This is the only interpretative structure we shall propose or need.

B. Wheeler-DeWitt equation

A differential equation for Ψ can be derived by varying the end conditions on the path integral (2.3) which defines it. To carry out this derivation first recall that the gravitational action appropriate to keeping the three-geometry fixed on a boundary is

$$l^2 S_E = 2 \int_{\partial M} d^3 x \, h^{1/2} K + \int_M d^4 x (-g)^{1/2} (R - 2\Lambda) . \tag{2.4}$$

The second term is integrated over spacetime and the first over its boundary. K is the trace of the extrinsic curvature K_{ij} of the boundary three-surface. If its unit normal is n^i, $K_{ij} = -\nabla_i n_j$ in the usual Lorentzian convention. l is the Planck length $(16\pi G)^{1/2}$ in the units with $\hbar = c = 1$ we use throughout. Introduce coordinates so that the boundary is a constant t surface and write the metric in the standard $3+1$ decomposition:

$$ds^2 = -(N^2 - N_i N^i) dt^2 + 2 N_i dx^i dt + h_{ij} dx^i dx^j . \tag{2.5}$$

The action (2.4) becomes

$$l^2 S_E = \int d^4 x \, h^{1/2} N [K_{ij} K^{ij} - K^2 + {}^3R(h) - 2\Lambda] , \tag{2.6}$$

where explicitly

$$K_{ij} = \frac{1}{N} \left\{ -\frac{1}{2} \frac{\partial h_{ij}}{\partial t} + N_{(i|j)} \right\} \tag{2.7}$$

and a stroke and ${}^3R$ denote the covariant derivative and scalar curvature constructed from the three-metric h_{ij}. The matter action S_M can similarly be expressed as a function of N, N_i, h_{ij}, and the matter field. The functional integral defining the wave function contains an integral over N. By varying N at the surface we push it forward or backward in time. Since the wave function does not depend on time we must have

$$0 = \int \delta g \, \delta \phi \left(\frac{\delta S}{\delta N} \right) \exp(i S[g, \phi]) . \tag{2.8}$$

More precisely, the value of the integral (2.3) should be left unchanged by an infinitesimal translation of the integration variable N. If the measure is invariant under translation this leads to (2.8). If it is not, there will be in addition a divergent contribution to the relation which must be suitably regulated to zero or cancel divergences arising from the calculation of the right-hand side of (2.8).

Classically the field equation $H \equiv \delta S / \delta N = 0$ is the Hamiltonian constraint for general relativity. It is

$$H = h^{1/2} (K^2 - K_{ij} K^{ij} + {}^3R - 2\Lambda - l^2 T_{nn}) = 0 , \tag{2.9}$$

where T_{nn} is the stress-energy tensor of the matter field projected in the direction normal to the surface. Equation (2.8) shows how $H = 0$ is enforced as an operator identity for the wave function. More explicitly one can note that the K_{ij} involve only first-time derivatives of the h_{ij} and therefore may be completely expressed in terms of the momenta π_{ij} conjugate to the h_{ij} which follow from the Lagrangian in (2.6):

$$\pi_{ij} = -h^{1/2} (K_{ij} - h_{ij} K) . \tag{2.10}$$

In a similar manner the energy of the matter field can be expressed in terms of the momentum conjugate to the field π_ϕ and the field itself. Equation (2.8) thus implies the operator identity $H(\pi_{ij}, h_{ij}, \pi_\phi, \phi) \Psi = 0$ with the replacements

$$\pi^{ij} = -i \frac{\delta}{\delta h_{ij}}, \quad \pi_\phi = -i \frac{\delta}{\delta \phi} . \tag{2.11}$$

These replacements may be viewed as arising directly from the functional integral, e.g., from the observation that when the time derivatives in the exponent are written in differenced form

$$-i \frac{\delta}{\delta h_{ij}} \int \delta g \, \delta \phi \, e^{iS} = \int \delta g \, \delta \phi \, \pi^{ij} e^{iS} . \tag{2.12}$$

Alternatively, they are the standard representation of the canonical commutation relations of h_{ij} and π^{ij}.

In translating a classical equation like $\delta S / \delta N = 0$ into an operator identity there is always the question of factor ordering. This will not be important for us so making a convenient choice we obtain

$$\left\{ -G_{ijkl} \frac{\delta^2}{\delta h_{ij} \delta h_{kl}} + h^{1/2} \left[-{}^3R(h) + 2\Lambda + l^2 T_{nn} \left(-i \frac{\delta}{\delta \phi}, \phi \right) \right] \right\}$$

$$\times \Psi[h_{ij}, \phi] = 0 . \tag{2.13}$$

This is the Wheeler-DeWitt equation which wave functions for closed universes must satisfy. There are also the other constraints of the classical theory, but the operator versions of these express the gauge invariance of the wave function rather than any dynamical information.[6]

We should emphasize that the ground-state wave function constructed by a Euclidean functional-integral prescription [(Eq. (1.11)] will satisfy the Wheeler-DeWitt equation in the form (2.13). Indeed, this can be demonstrated explicitly by repeating the steps in the above demonstration starting with the Euclidean functional integral.

C. Boundary conditions

The quantity G_{ijkl} can be viewed as a metric on superspace—the space of all three-geometries (no connection with supersymmetry). It has signature

$(-,+,+,+,+,+)$ and the Wheeler-DeWitt equation is therefore a "hyperbolic" equation on superspace. It would be natural, therefore, to expect to impose boundary conditions on two "spacelike surfaces" in superspace. A convenient choice for the timelike direction is $h^{1/2}$ and we therefore expect to impose boundary conditions at the upper and lower limits of the range of $h^{1/2}$. The upper limit is infinity. The lower limit is zero because if h_{ij} is positive definite or degenerate, $h^{1/2} \geq 0$. Positive-definite metrics are everywhere spacelike surfaces; degenerate metrics may signal topology change. Summarizing the remaining functions of h_{ij} by the conformal metric $\tilde{h}_{ij} = h_{ij}/h^{1/3}$ we may write an important boundary condition on Ψ as

$$\Psi[\tilde{h}_{ij}, h^{1/2}, \phi] = 0, \quad h^{1/2} < 0 . \tag{2.14}$$

Because $h^{1/2}$ has a semidefinite range it is for many purposes convenient to introduce a representation in which $h^{1/2}$ is replaced by its canonically conjugate variable $-\frac{4}{3}Kl^{-2}$ which has an infinite range. The advantages of this representation have been extensively discussed.[7] In the case of pure gravity since $-\frac{4}{3}Kl^{-2}$ and $h^{1/2}$ are conjugate, we can write for the transformation to the representation where $\tilde{h}_{ij}$ and K are definite

$$\Phi[\tilde{h}_{ij}, K] = \int_0^\infty \delta h^{1/2} \exp\left[-i\frac{4}{3}l^{-2} \int d^3x \, h^{1/2}K \right] \Psi[h_{ij}] \tag{2.15}$$

and inversely,

$$\Psi[h_{ij}] = \int_{-\infty}^{+\infty} \delta K \exp\left[+i\frac{4}{3}l^{-2} \int d^3x \, h^{1/2}K \right] \Phi[\tilde{h}_{ij}, K] . \tag{2.16}$$

In each case the functional integrals are over the values of $h^{1/2}$ or K at each point of the spacelike hypersurface and we have indicated limits of integration.

The condition (2.14) implies through (2.15) that $\Phi[\tilde{h}_{ij}, K]$ is analytic in the lower-half K plane. The contour in (2.16) can thus be distorted into the lower-half K plane. Conversely, if we are given $\Phi[\tilde{h}_{ij}, K]$ we can reconstruct the wave function Ψ which satisfies the boundary condition (2.14) by carrying out the integration in (2.16) over a contour which lies below any singularities of $\Phi[\tilde{h}_{ij}, K]$ in K.

In the presence of matter K and $\tilde{h}_{ij}$ remain convenient labels for the wave functional provided the labels for the matter-field amplitudes $\tilde{\phi}$ are chosen so that a multiple of K is canonically conjugate to $h^{1/2}$. In cases where the matter-field action itself involves the scalar curvature this means that the label $\tilde{\phi}$ will be the field amplitude rescaled by some power of $h^{1/2}$. For example, in the case of a conformally invariant scalar field the appropriate label is $\tilde{\phi} = \phi h^{1/6}$. With this understanding we can write for the functionals

$$\Psi = \Psi[h_{ij}, \tilde{\phi}], \quad \Phi = \Phi[\tilde{h}_{ij}, K, \tilde{\phi}] \tag{2.17}$$

and the transformation formulas (2.15) and (2.16) remain unchanged.

D. Hermiticity

The introduction of wave functions as functional integrals [Eq. (2.3)] allows the definition of a scalar product with a simple geometric interpretation in terms of sums over spacetime histories. Consider a wave function Ψ defined by the integral

$$\Psi[h_{ij}, \phi] = N \int_C \delta g \, \delta\phi \exp(iS[g,\phi]) , \tag{2.18}$$

over a class of four-geometries and fields C, and a second wave function Ψ' defined by a similar sum over a class C'. The scalar product

$$(\Psi', \Psi) = \int \delta h \, \delta\phi \, \bar{\Psi}'[h_{ij}, \phi] \Psi[h_{ij}, \phi] \tag{2.19}$$

has the geometric interpretation of a sum over all histories

$$(\Psi', \Psi) = \bar{N}'N \int \delta g \, \delta\phi \exp(iS[g,\phi]) , \tag{2.20}$$

where the sum is over histories which lie in class C to the past of the surface and in the time reversed of class C' to its future.

The scalar product (2.19) is not the product that would be required by canonical theory to define the Hilbert space of physical states. That would presumably involve integration over a hypersurface in the space of all three-geometries rather than over the whole space as in (2.19). Rather, Eq. (2.19) is a mathematical construction made natural by the functional-integral formulation of quantum gravity.

In gravity we expect the field equations to be satisfied as identities. An extension of the argument leading to Eq. (2.8) will give

$$\int \delta g \, \delta\phi \, H(x) \exp(iS[g,\phi]) = 0 \tag{2.21}$$

for any class of geometries summed over and for any intermediate spacelike surface on which $H(x)$ is evaluated. Equation (2.21) can be evaluated for the particular sum which enters Eq. (2.20). $H(x)$ can be interpreted in the scalar product as an operator acting on either Ψ' or Ψ. Thus,

$$(H\Psi', \Psi) = (\Psi', H\Psi) = 0 . \tag{2.22}$$

The Wheeler-DeWitt operator must therefore be Hermitian in the scalar product (2.19).

Since the Wheeler-DeWitt operator is a second-order functional-differential operator, the requirement of Hermiticity will essentially be a requirement that certain surface terms on the boundary of the space of three-metrics vanish and, in particular, at $h^{1/2} = 0$ and $h^{1/2} = \infty$. As in ordinary quantum mechanics these conditions will prove useful in providing boundary conditions for the solution of the equation.

III. GROUND-STATE WAVE FUNCTION

In this section, we shall put forward in detail our proposal for the ground-state wave function for closed cosmologies. The wave function depends on the topology and the three-metric of the spacelike surface and on the values of the matter field on the surface. For simplicity we shall begin by considering only S^3 topology. Other possibilities will be considered in Sec. VIII.

As discussed in the Introduction, the ground-state wave function is to be constructed as a functional integral of the form

$$\Psi_0[h_{ij},\phi]=N\int\delta g\,\delta\phi\exp(-I[g,\phi])\;,\qquad(3.1)$$

where I is the total Euclidean action and the integral is over an appropriate class of Euclidean four-geometries with compact boundary on which the induced metric is h_{ij} and an appropriate class of Euclidean field configurations which match the value given on the boundary. To complete the definition of the ground-state wave function we need to give the class of geometries and fields to be summed over. Our proposal is that the geometries should be compact and that the fields should be regular on these geometries. In the case of a positive cosmological constant Λ any regular Euclidean solution of the field equations is necessarily compact.[8] In particular, the solution of greatest symmetry is the four-sphere of radius $3/\Lambda$, whose metric we write as

$$ds^2=(\sigma/H)^2(d\theta^2+\sin^2\theta\,d\Omega_3^2)\;,\qquad(3.2)$$

where $d\Omega_3^2$ is the metric on the three-sphere. $H^2=\sigma^2\Lambda/3$ and we have introduced the normalization factor $\sigma^2=l^2/24\pi^2$ for later convenience. Thus, it is clear that compact four-geometries are the only reasonable candidates for the class to be summed over when $\Lambda>0$.

If Λ is zero or negative there are noncompact solutions of the field equations. The solutions of greatest symmetry are Euclidean space ($\Lambda=0$) with

$$ds^2=\sigma^2(d\theta^2+\theta^2\,d\Omega_3^2)\qquad(3.3)$$

and Euclidean anti–de Sitter space ($\Lambda<0$) with

$$ds^2=(\sigma/H)^2(d\theta^2+\sinh^2\theta\,d\Omega_3^2)\;.\qquad(3.4)$$

One might therefore feel that the ground state for $\Lambda\leq0$ should be defined by a functional integral over geometries which are asymptotically Euclidean or asymptotically anti–de Sitter. This is indeed appropriate to defining the ground state for scattering problems where one is interested in particles which propagate in from infinity and then out to infinity again.[9] However, in the case of cosmology, one is interested in measurements that are carried out in the interior of the spacetime, whether or not the interior points are connected to some infinite regions does not matter. If one were to use asymptotically Euclidean or anti–de Sitter four-geometries in the functional integral that defines the ground state one could not exclude a contribution from four-geometries that consisted of two disconnected pieces, one of which was compact with the three-geometry as boundary and the other of which was asymptotically Euclidean or anti–de Sitter with no interior boundary. Such disconnected geometries would in fact give the dominant contribution to the ground-state wave function. Thus, one would effectively be back with the prescription given above.

The ground-state wave function obtained by summing over compact four-geometries diverges for large three-geometries in the cases $\Lambda\leq0$ and the wave function cannot be normalized. This is because the Λ in the action damps large four-geometries when $\Lambda>0$, but it enhances

them when $\Lambda<0$. We shall therefore consider only the case $\Lambda>0$ in this paper and shall regard $\Lambda=0$ as a limiting case of $\Lambda>0$.

An equivalent way of describing the ground state is to specify its wave function in the $\tilde{\phi},\tilde{h}_{ij},K$ representation. Here too it can be constructed as a functional integral:

$$\Phi_0[\tilde{h}_{ij},K,\tilde{\phi}]=N\int\delta g\,\delta\phi\exp(-I^K[g,\phi])\;.\qquad(3.5)$$

The sum is over the same class of fields and geometries as before except that now $\tilde{\phi}$, $\tilde{h}_{ij}$, and K are fixed on the boundary rather than ϕ and h_{ij}. The action I^K is therefore the Euclidean action appropriate to holding $\tilde{\phi}$, $\tilde{h}_{ij}$, and K fixed on a boundary. It is a sum of the appropriate pure gravitational action which up to an additive constant is

$$l^2I_E^K[g]=-\tfrac{2}{3}\int_{\partial M}d^3x\,h^{1/2}K-\int_M d^4x\,g^{1/2}(R-2\Lambda)$$

$$(3.6)$$

and a contribution from the matter. The latter is well illustrated by the action of a single conformally invariant scalar field, an example which we shall use exclusively in the rest of this paper. We have

$$I_M^K[g,\phi]=\tfrac{1}{2}\int_M d^4x\,g^{1/2}[(\nabla\phi)^2+\tfrac{1}{6}R\phi^2]\;.\qquad(3.7)$$

These actions differ from the more familiar ones in which ϕ and h_{ij} are fixed only in having different surface terms. Indeed, these surface terms are just those required to ensure the equivalence of (3.1) and (3.5) as a consequence of the transformation formulas (2.15) and (2.16). In the case of the matter action of a conformally invariant scalar field with $\tilde{\phi},h_{ij},K$ fixed the additional surface term conveniently cancels that required in the action when ϕ and h_{ij} are fixed.

It is important to recognize that the functional integral (3.5) does not yield the wave function at the Lorentzian value of K but rather at a Euclidean value of K. For the moment denote the Lorentzian value by K_L. If the hypersurfaces of interest were labeled by a time coordinate t in a coordinate system with zero shift [$N_i=0$ in Eq. (2.5)] then the rotation $t\to i\tau$ and the use of the traditional conventions $K_L=-\nabla\cdot n$ and $K=\nabla\cdot n$ will send $K_L\to-iK$. In terms of the Euclidean K the transformation formulas (2.15) and (2.16) can be rewritten to read

$$\Phi[\tilde{h}_{ij},K,\tilde{\phi}]=\int_0^\infty\delta h^{1/2}\exp\left[-\tfrac{4}{3}l^{-2}\int d^3x\,h^{1/2}K\right]$$

$$\times\Psi[h_{ij},\tilde{\phi}]\;,\qquad(3.8)$$

$$\Psi[h_{ij},\phi]=-\frac{1}{2\pi i}\int_C\delta K\exp\left[\tfrac{4}{3}l^{-2}\int d^3x\,h^{1/2}K\right]$$

$$\times\Phi[\tilde{h}_{ij},K,\tilde{\phi}]\;,\qquad(3.9)$$

where the contour C runs from $-i\infty$ to $+i\infty$. At the risk of some confusion we shall continue to use K in the remainder of this paper to denote the Euclidean K despite having used the same symbol in Secs. I and II for the Lorentzian quantity.

There is one advantage to constructing the ground-state wave function from the functional integral (3.5) rather than (3.1) and it is the following: the integral in Eq. (3.9) will always yield a wave function $\Psi_0[h_{ij},\phi]$ which vanishes for $h^{1/2}<0$ if the contour C is chosen to the right of any singularities of $\Phi_0[\tilde{h}_{ij},K,\tilde{\phi}]$ in K provided Φ does not diverge too strongly in K. The boundary condition (2.14) is thus automatically enforced. This is a considerable advantage when the wave function is only evaluated approximately.

The Euclidean gravitational action [Eq. (3.6)] is not positive definite. The functional integrals in Eqs. (3.1) and (3.5) therefore require careful definition. One way of doing this is to break the integration up into an integral over conformal factors and over geometries in a given conformal equivalence class. By appropriate choice of the contour of integration of the conformal factor the integral can probably be made convergent. If this is the case a properly convergent functional integral can be constructed.

This then is our prescription for the ground state. In the following sections we shall derive some of its properties and demonstrate its reasonableness in a simple minisuperspace model.

IV. SEMICLASSICAL EXPECTATIONS

An important advantage of a functional-integral prescription for the ground-state wave function is that it yields the semiclassical approximation for that wave function directly. In this section, we shall examine the semiclassical approximation to the ground-state wave function defined in Sec. III. For simplicity we shall consider the case of pure gravity. The extension to include matter is straightforward.

The semiclassical approximation is obtained by evaluating the functional integral by the method of steepest descents. If there is only one stationary-phase point the semiclassical approximation is

$$\Psi_0[h_{ij}]=N\Delta^{-1/2}[h_{ij}]\exp(-I_{\text{cl}}[h_{ij}]) \ . \tag{4.1}$$

Here, I_{cl} is the Euclidean gravitational action evaluated at the stationary-phase point, that is, at that solution $g_{\mu\nu}^{\text{cl}}$ of the Euclidean field equations

$$R_{\mu\nu}=\Lambda g_{\mu\nu} \ , \tag{4.2}$$

which induces the metric h_{ij} on the closed three-surface boundary and satisfies the asymptotic conditions discussed in Sec. III. $\Delta^{-1/2}$ is a combination of determinants of the wave operators defining the fluctuations about $g_{\mu\nu}^{\text{cl}}$, including those contributed by the ghosts. We shall focus mainly on the exponent. For further information on Δ in the case without boundary see Ref. 10.

If there is more than one stationary-phase point, it is necessary to consider the contour of integration in the path integral more carefully in order to decide which gives the dominant contribution. In general this will be the stationary-phase point with the lowest value of ReI although it may not be if there are two stationary-phase points which correspond to four-metrics that are conformal to one another. We shall see an example of this in Sec. VI. The ground-state wave function is real. This means that if the stationary-phase points have complex values of the action, there will be equal contributions from

stationary-phase points with complex-conjugate values of the action. If there is no four-geometry which is a stationary-phase point, the wave function will be zero in the semiclassical approximation.

The semiclassical approximation for Ψ_0 can also be obtained by first evaluating the semiclassical approximation to Φ_0 from the functional integral (3.5) and then evaluating the transformation integral (3.9) by steepest descents. This will be more convenient to do when the boundary conditions of fixing $\tilde{h}_{ij}$ and K yield a unique dominant stationary-phase solution to (4.2) but fixing h_{ij} does not.

One can fix the normalization constant N in (4.1) by the requirement

$$\int \delta h \ \bar{\Psi}_0[h_{ij}]\Psi_0[h_{ij}]=1 \ . \tag{4.3}$$

As explained in Sec. II, one can interpret (4.3) geometrically as a path integral over all four-geometries which are compact on both sides of the three-surface with the metric h_{ij}. The semiclassical approximation to this path integral will thus be given by the action of the compact four-geometry without boundary which is the solution of the Einstein field equation. In the case of $\Lambda>0$ the solution with the most negative action is the four-sphere. Thus,

$$N^2=\exp\left[-\frac{2}{3H^2}\right] \ . \tag{4.4}$$

The semiclassical approximation for the wave function gives one considerable insight into the boundary conditions for the Wheeler-DeWitt equation, which are implied by the functional-integral prescription for the wave function. As discussed in Sec. II, these are naturally imposed on three-geometries of very large volumes and vanishing volumes.

Consider the limit of small three-volumes first. If the limiting three-geometry is such that it can be embedded in flat space then the classical solution to (4.2) when $\Lambda>0$ is the four-sphere and remains so as the three-geometry shrinks to zero. The action approaches zero. The value of the wave function is therefore controlled by the behavior of the determinants governing the fluctuations away from the classical solution. These fluctuations are to be computed about a vanishingly small region of a space of constant positive curvature. In this limit one can neglect the curvature and treat the fluctuations as about a region of flat space. The determinant can therefore be evaluated by considering its behavior under a constant conformal rescaling of the four-metric and the boundary three-metric. The change in the determinant under a change of scale is given by the value of the associated ζ function at zero argument.[11]

Regular four-geometries contain many hypersurfaces on which the three-volume vanishes. For example, consider the four-sphere of radius R embedded in a five-dimensional flat space. The three-surfaces which are the intersection of the four-sphere with surfaces of x^5 equals constant have a regular three-metric for $|x^5|<R$. The volume vanishes when $|x^5|=R$ at the north and south poles even though these are perfectly regular points of the four-geometry. One therefore would not expect the wave function to vanish at vanishing three-volume. Indeed, the three-volume will have to vanish somewhere if the topolo-

gy of the four-geometry is not that of a product of a three-surface with the real line or the circle. When the volume does vanish, the topology of the three-geometry will change. One cannot calculate the amplitude for such topology change from the Wheeler-DeWitt equation but one can do so using the Euclidean functional integral. We shall estimate the amplitude in some simple cases in Sec. VIII.

A qualitative discussion of the expected behavior of the wave function at large three-volumes can be given on the basis of the semiclassical approximation when $\Lambda > 0$ as follows. The four-sphere has the largest volume of any real solution to (4.2). As the volume of the three-geometry becomes large one will reach three-geometries which no longer fit anywhere in the four-sphere. We then expect that the stationary-phase geometries become complex. The ground-state wave function will be a real combination of two expressions like (4.1) evaluated at the complex-conjugate stationary-phase four-geometries. We thus expect the wave function to oscillate as the volume of the three-geometry becomes large. If it oscillates without being strongly damped this corresponds to a universe which expands without limit.

The above considerations are only qualitative but do suggest how the behavior of the ground-state wave function determines the boundary conditions for the Wheeler-DeWitt equation. In the following we shall make these considerations concrete in a minisuperspace model.

V. MINISUPERSPACE MODEL

It is particularly straightforward to construct minisuperspace models using the functional-integral approach to quantum gravity. One simply restricts the functional integral to the restricted degrees of freedom to be quantized. In this and the following sections, we shall illustrate the general discussion of those preceding with a particularly simple minisuperspace model. In it we restrict the cosmological constant to be positive and the four-geometries to be spatially homogeneous, isotropic, and closed so that they are characterized by a single scale factor. An explicit metric in a useful coordinate system is

$$ds^2 = \sigma^2[-N^2(t)dt^2 + a^2(t)d\Omega_3{}^2] , \qquad (5.1)$$

where $N(t)$ is the lapse function and $\sigma^2 = l^2/24\pi^2$. For the matter degrees of freedom, we take a single conformally invariant scalar field which, consistent with the geometry, is always spatially homogeneous, $\phi = \phi(t)$. The wave function is then a function of only two variables:

$$\Psi = \Psi(a,\phi), \quad \Phi = \Phi(K,\tilde{\phi}) . \qquad (5.2)$$

Models of this general structure have been considered previously by DeWitt,[12] Isham and Nelson,[13] and Blyth and Isham.[14]

To simplify the subsequent discussion we introduce the following definitions and rescalings of variables:

$$\phi = \frac{\tilde{\phi}}{a} = \frac{\chi}{(2\pi^2\sigma^2)^{1/2}a} , \qquad (5.3)$$

$$\Lambda = 3\lambda/\sigma^2, \quad H^2 = |\lambda| . \qquad (5.4)$$

The Lorentzian action keeping χ and a fixed on the boundaries is

$$S^a = \frac{1}{2} \int dt \left[\frac{N}{a} \right] \left[\left[\frac{a}{N}\frac{da}{dt} \right]^2 + a^2 - \lambda a^4 \right.$$
$$\left. + \left[\frac{a}{N}\frac{d\chi}{dt} \right]^2 - \chi^2 \right] . \qquad (5.5)$$

From this action the momenta π_a and π_χ conjugate to a and χ can be constructed in the usual way. The Hamiltonian constraint then follows by varying the action with respect to the lapse function and expressing the result in terms of a, χ, and their conjugate momenta. One finds

$$\frac{1}{2}(-\pi_a{}^2 - a^2 + \lambda a^4 + \pi_\chi{}^2 + \chi^2) = 0 . \qquad (5.6)$$

The Wheeler-DeWitt equation is the operator expression of this classical constraint. There is the usual operator-ordering problem in passing from classical to quantum relations but its particular resolution will not be central to our subsequent semiclassical considerations. A class wide enough to remind oneself that the issue exists can be encompassed by writing

$$\pi_a{}^2 = -\frac{1}{a^p}\frac{\partial}{\partial a}\left[a^p\frac{\partial}{\partial a} \right] , \qquad (5.7)$$

although this is certainly not the most general form possible. In passing from the classical constraint to its quantum operator form there is also the possibility of a matter-energy renormalization. This will lead to an additive arbitrary constant in the equation. We thus write for the quantum version of Eq. (5.6)

$$\frac{1}{2}\left[\frac{1}{a^p}\frac{\partial}{\partial a}\left[a^p\frac{\partial}{\partial a} \right] - a^2 + \lambda a^4 - \frac{\partial^2}{\partial \chi^2} + \chi^2 - 2\epsilon_0 \right]$$
$$\times \Psi(a,\chi) = 0 . \qquad (5.8)$$

A useful property stemming from the conformal invariance of the scalar field is that this equation separates. If we assume reasonable behavior for the function Ψ in the amplitude of the scalar field we can expand in harmonic-oscillator eigenstates

$$\Psi(a,\chi) = \sum_n c_n(a)u_n(\chi) , \qquad (5.9)$$

where

$$\frac{1}{2}\left[-\frac{d^2}{d\chi^2} + \chi^2 \right]u_n(\chi) = (n + \tfrac{1}{2})u_n(\chi) . \qquad (5.10)$$

The consequent equation for the $c_n(a)$ is

$$\frac{1}{2}\left[-\frac{1}{a^p}\frac{d}{da}\left[a^p\frac{dc_n}{da} \right] + (a^2 - \lambda a^4)c_n \right] = (n + \tfrac{1}{2} - \epsilon_0)c_n . \qquad (5.11)$$

For small a this equation has solutions of the form

$$c_n \approx \text{constant}, \quad c_n \approx a^{1-p} \qquad (5.12)$$

[if p is an integer there may be a $\log(a)$ factor]. For large a the possible behaviors are

$$c_n \sim a^{-(p/2+1)} \exp(\pm \tfrac{1}{3} i H a^3) \ . \tag{5.13}$$

To construct the solution of Eq. (5.11) which corresponds to the ground state of the minisuperspace model we turn to our Euclidean functional-integral prescription. As applied to this minisuperspace model, the prescription of Sec. III for $\Psi_0(a_0, \chi_0)$ would be to sum $\exp(-I[g, \phi])$ over those Euclidean geometries and field configurations which are represented in the minisuperspace and which satisfy the ground-state boundary conditions. The geometrical sum would be over compact geometries of the form

$$ds^2 = \sigma^2[d\tau^2 + a^2(\tau) d\Omega_3{}^2] \tag{5.14}$$

for which $a(\tau)$ matches the prescribed value of a_0 on the hypersurface of interest. The prescription for the matter field would be to sum over homogeneous fields $\chi(\tau)$ which match the prescribed value χ_0 on the surface and which are regular on the compact geometry. Explicitly we could write

$$\Psi_0(a_0, \chi_0) = \int \delta a \, \delta \chi \exp(-I[a, \chi]) \ , \tag{5.15}$$

where, defining $d\eta = d\tau/a$, the action is

$$I = \tfrac{1}{2} \int d\eta \left[-\left(\frac{da}{d\eta} \right)^2 - a^2 + \lambda a^4 + \left(\frac{d\chi}{d\eta} \right)^2 + \chi^2 \right] \ . \tag{5.16}$$

A conformal rotation [in this case of $a(\eta)$] is necessary to make the functional integral in (5.15) converge.[15]

An alternative way of constructing the ground-state wave function for the minisuperspace model is to work in the K representation. Here, introducing

$$k = \sigma K/9 \tag{5.17}$$

as a simplifying measure of K, one would have

$$\Phi_0(k_0, \chi_0) = \int \delta a \, \delta \chi \exp(-I^k[a, \chi]) \ . \tag{5.18}$$

The sum is over the same class of geometries and fields as in (5.15) except they must now assume the given value of k on the bounding three-surface. That is, on the boundary they must satisfy

$$k_0 = \frac{1}{3a} \frac{da}{d\tau} \ . \tag{5.19}$$

The action I^k appropriate for holding k fixed on the boundary is

$$I^k = k_0 a_0{}^3 + I \tag{5.20}$$

[cf. Eq. (3.6)]. Once $\Phi_0(k_0, \chi_0)$ has been computed, the ground-state wave function $\Psi_0(k_0, \chi_0)$ may be recovered by carrying out the contour integral

$$\Psi_0(a_0, \chi_0) = -\frac{1}{2\pi i} \int_C dk \, e^{ka_0{}^3} \Phi_0(k, \chi_0) \ , \tag{5.21}$$

where the contour runs from $-i\infty$ to $+i\infty$ to the right of any singularities of $\Phi_0(k_0, \chi_0)$.

From the general point of view there is no difference between computing $\Psi_0(a_0, \chi_0)$ directly from (5.15) or via the K representation from (5.21). In Sec. VI we shall calculate the semiclassical approximation to $\Psi_0(a_0, \chi_0)$ both

ways with the aim of advancing arguments that the rules of Sec. III define a wave function which may reasonably be considered as the state of minimal excitation and of displaying the boundary conditions under which Eq. (5.11) is to be solved.

VI. GROUND-STATE COSMOLOGICAL WAVE FUNCTION

In this section, we shall evaluate the ground-state wave function for our minisuperspace model and show that it possesses properties appropriate to a state of minimum excitation. We shall first evaluate the wave function in the semiclassical approximation from the steepest-descents approximation to the defining functional integral as described in Sec. IV. We shall then solve the Wheeler-DeWitt equation with the boundary conditions implied by the semiclassical approximation to obtain the precise wave function.

It is the exponent of the semiclassical approximation which will be most important in its interpretation. We shall calculate only this exponent from the extrema of the action and leave the determination of the prefactor [cf. Eq. (4.1)] to the solution of the differential equation. Thus, for example, if there were a single real Euclidean extremum of least action we would write for the semiclassical approximation to the functional integral in Eq. (5.15)

$$\Psi_0(a_0, \chi_0) \approx N e^{-I(a_0, \chi_0)} \ . \tag{6.1}$$

Here, $I(a_0, \chi_0)$ is the action (5.16) evaluated at the extremum configurations $a(\tau)$ and $\chi(\tau)$ which satisfy the ground-state boundary conditions spelled out in Sec. III and which match the arguments of the wave function on a fixed-τ hypersurface.

A. The matter wave function

A considerable simplification in evaluating the ground-state wave function arises from the fact that the energy-momentum tensor of an extremizing conformally invariant field vanishes in the compact geometries summed over as a consequence of the ground-state boundary conditions. One can see this because the compact four-geometries of the class we are considering are conformal to the interior of three-spheres in flat Euclidean space. A constant scalar field is the only solution of the conformally invariant wave equation on flat space which is a constant on the boundary three-sphere. The energy-momentum tensor of this field is zero. This implies that it is zero in any geometry of the class (5.14) because the energy-momentum tensor of a conformally invariant field scales by a power of the conformal factor under a conformal transformation.

More explicitly in the minisuperspace model we can show that the matter and gravitational functional integrals in (5.15) may be evaluated separately. The ground-state boundary conditions imply that geometries in the sum are conformal to half of a Euclidean Einstein-static universe, i.e., that the range of η is $(-\infty, 0)$. The boundary conditions at infinite η are that $\chi(\eta)$ and $a(\eta)$ vanish. The boundary conditions at $\eta = 0$ are that $a(0)$ and $\chi(0)$ match the arguments of the wave function a_0 and χ_0. Thus, not only does the action (5.16) separate into a sum of a gravitational part and a matter part, but the boundary conditions on the $a(\eta)$ and $\chi(\eta)$ summed over do not depend on one another. The matter and gravitational integrals can thus be evaluated separately.

Let us consider the matter integral first. In Eq. (5.16) the matter action is

$$I_M = \tfrac{1}{2} \int d\eta \left[\left| \frac{d\chi}{d\eta} \right|^2 + \chi^2 \right] . \tag{6.2}$$

This is the Euclidean action for the harmonic oscillator. Evaluation of the matter field integral in (5.15) therefore gives

$$\Psi_0(a_0, \chi_0) = e^{-\chi_0^2/2} \psi_0(a_0) . \tag{6.3}$$

Here, $\psi_0(a)$ is the wave function for gravity alone given by

$$\psi_0(a_0) = \int \delta a \exp(-I_E[a]) , \tag{6.4}$$

I_E being the gravitational part of (5.16). Equivalently we can write in the K representation

$$\Phi_0(k_0, \chi_0) = e^{-\chi_0^2/2} \phi_0(k_0) , \tag{6.5}$$

where

$$\phi_0(k_0) = \int \delta a \exp(-I_E^k[a]) . \tag{6.6}$$

$I_E^k[a]$ is related to I_E as in (5.20) and the sum is over $a(\tau)$ which satisfy (5.19) on the boundary. Equation (6.3) shows that as far as the matter field is concerned, $\Psi_0[a_0, \chi_0]$ is reasonably interpreted as the ground-state wave function. The field oscillators are in their state of minimum excitation—the ground state of the harmonic oscillator. We now turn to a semiclassical calculation of the gravitational wave function $\psi_0(a_0)$.

B. The semiclassical ground-state gravitational wave function

The integral in (6.4) is over $a(\tau)$ which represent [through (5.14)] compact geometries with three-sphere boundaries of radius a. The integral in (6.6) is over the same class of geometries except that the three-sphere boundary must possess the given value of k. The compact geometry which extremizes the gravitational action in these cases is a part of the Euclidean four-sphere of radius $1/H$ with an appropriate three-sphere boundary. In the case where the three-sphere radius is fixed on the boundary there are two extremizing geometries. For one the part of the four-sphere bounded by the three-sphere is greater than a hemisphere and for the other it is less. A careful analysis must therefore be made of the functional integral to see which of these extrema contributes to the semiclassical approximation. We shall give such an analysis below but first we show that the correct answer is achieved more directly in the K representation from (6.6) because there is a single extremizing geometry with a prescribed value of k on a three-sphere boundary and thus no ambiguity in constructing the semiclassical approximation to (6.6).

For three-sphere hypersurfaces of the four-sphere with an outward pointing normal, k ranges from approaching $+\infty$ for a surface encompassing a small region about a pole to approaching $-\infty$ for the whole four-sphere (see Fig. 1). More exactly, in the notation of Eq. (3.7)

$$k = \frac{H}{3} \cot\theta . \tag{6.7}$$

The extremum action is constructed through (5.20) with the integral in (5.16) being taken over that part of the four-sphere bounded by the three-sphere of given k. It is

$$I_E^k(k) = -\frac{1}{3H^2} \left[1 - \frac{\kappa}{(\kappa^2+1)^{1/2}} \right] , \tag{6.8}$$

where

$$k = \tfrac{1}{3} \kappa H . \tag{6.9}$$

The semiclassical approximation to (6.6) is now

$$\phi_0(k_0) \approx N \exp[-I_E^k(k_0)] . \tag{6.10}$$

The wave function $\psi_0(a_0)$ in the same approximation can be constructed by carrying out the contour integral

$$\psi_0(a_0) = -\frac{N}{2\pi i} \int_C dk \exp[ka_0^3 - I_E^k(k)] \tag{6.11}$$

by the method of steepest descents. The exponent in the integrand of Eq. (6.11) is minus the Euclidean action for pure gravity with a kept fixed instead of k:

$$I_E(a) = -ka^2 + I_E^k(k) . \tag{6.12}$$

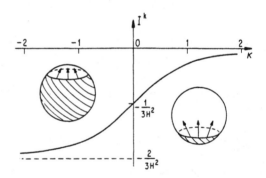

FIG. 1. The action I^k for the Euclidean four-sphere of radius $1/H$. The Euclidean gravitational action for the part of a four-sphere bounded by a three-sphere of definite K is plotted here as a function of κ (a dimensionless measure of K [Eq. (6.9)]). The action is that appropriate for holding K fixed on the boundary. The shaded regions of the inset figures show schematically the part of the four-sphere which fills in the three-sphere of given K used in computing the action. A three-sphere of given K fits in a four-sphere at only one place. Three-spheres with positive K (diverging normals) bound less than a hemisphere of four-sphere while those with negative K (converging normals) bound more than a hemisphere. The action tends to its flat-space value (zero) as K tends to positive infinity. It tends to the Euclidean action for all of de Sitter space as K tends to negative infinity.

To evaluate (6.11) by steepest descents we must find the extrema of Eq. (6.12). There are two cases depending on whether Ha_0 is greater or less than unity.

For $Ha_0 < 1$ the extrema of $I_E(k)$ occur at real values of k which are equal in magnitude and opposite in sign. They are the values of k at which a three-sphere of radius a_0 would fit into the four-sphere of radius $1/H$. That is, they are those values of k for which Eq. (6.7) is satisfied with $a_0{}^2 = (\sin\theta/H)^2$. This is not an accident; it is a consequence of the Hamilton-Jacobi theory. The value of I_E at these extrema is

$$I_\pm = -\frac{1}{3H^2}[1\pm(1-H^2a_0{}^2)^{3/2}] , \qquad (6.13)$$

where the upper sign corresponds to $k<0$ and the lower to $k>0$, i.e., to filling in the three-sphere with greater than a hemisphere of the four-sphere or less than a hemisphere, respectively.

There are complex extrema of I_E but all have actions whose real part is greater than the real extrema described above. The steepest-descents approximation to the integral (6.11) is therefore obtained by distorting the contour into a steepest-descents path (or sequence of them) passing through one or the other of the real extrema. The two real extrema and the corresponding steepest-descents directions are shown in Fig. 2. One can distort the contour

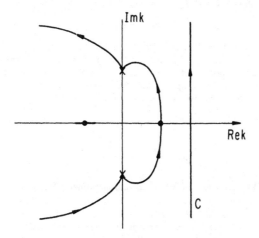

FIG. 2. The integration contour for constructing the semiclassical ground-state wave function of the minisuperspace model in the case $\Lambda > 0$, $Ha_0 < 1$. The figure shows schematically the original integration contour C used in Eq. (6.11) and the steepest-descents contour into which it can be distorted. The branch points of the exponent of Eq. (6.11) at $\kappa = \pm i$ are located by crosses. There are two extrema of the exponent which correspond to filling in the three-sphere of given radius a with greater than a hemisphere of four-sphere or less than a hemisphere. For $Ha_0 < 1$ they lie at the equal and opposite real values of K indicated by dots. The contour C can be distorted into a steepest-descents contour through the extremum with positive K as shown. It cannot be distorted to pass through the extremum with negative K in the steepest-descents direction indicated. The contour integral thus picks out the extremum corresponding to less than a hemisphere of four-sphere (cf. Fig. 1) as the leading term in the semiclassical approximation.

into a steepest-descents path passing through only one of them—the one with positive k as shown. The functional integral thus singles out a unique semiclassical approximation to $\psi_0(a_0)$ which is

$$\psi_0(a_0) \approx N \exp[-I_-(a_0)], \quad Ha_0 < 1 , \qquad (6.14)$$

corresponding to filling in the three-sphere with less than a hemisphere's worth of four-sphere.

From Eq. (4.4) we recover the normalization factor N:

$$N = \exp(-\tfrac{1}{3}H^{-2}) . \qquad (6.15)$$

Thus, for $Ha_0 \ll 1$

$$\psi_0(a_0) = \exp(\tfrac{1}{2}a_0{}^2 - \tfrac{1}{3}H^{-2}) . \qquad (6.16)$$

One might have thought that the extremum I_+, which corresponds to filling in the three-geometry with more than a hemisphere, would provide the dominant contribution to the ground-state wave function as $\exp(-I_+)$ is greater than $\exp(-I_-)$. However, the steepest-descents contour in the integral (6.7) does not pass through the extremum corresponding to I_+. This is related to the fact that the contour of integration of the conformal factor has to be rotated in the complex plane in order to make the path integral converge as we shall show below.

For $Ha_0 > 1$ there are no real extrema because we cannot fit a three-sphere of radius $a_0 > 1/H$ into a four-sphere of radius $1/H$. There are, however, complex extrema of smallest real action located at

$$k = \pm\frac{i}{3}H\left[1 - \frac{1}{H^2a_0{}^2}\right]^{1/2} . \qquad (6.17)$$

It is possible to distort the contour in Eq. (6.11) into a steepest-descents contour passing through both of them as shown in Fig. 3. The resulting wave function has the form

$$\psi_0(a_0) = 2\cos\left[\frac{(H^2a_0{}^2-1)^{3/2}}{3H^2} - \frac{\pi}{4}\right], \quad Ha_0 > 1$$

$$(6.18)$$

or for $Ha_0 \gg 1$

$$\psi_0(a_0) \approx e^{+iHa_0{}^3/3} + e^{-iHa_0{}^3/3} . \qquad (6.19)$$

The semiclassical approximation to the ground-state gravitational wave function $\psi_0(a)$ contained in Eqs. (6.16) and (6.19) may also be obtained directly from the functional integral (6.4) without passing through the k representation. We shall now sketch this derivation. We must consider explicitly the conformal rotation which makes the gravitational part of the action in (5.16) positive definite. The gravitational action is

$$I_E[a] = \tfrac{1}{2}\int d\eta\left[-\left|\frac{da}{d\eta}\right|^2 - a^2 + H^2a^4\right] . \qquad (6.20)$$

If one performed the functional integration

$$\psi_0(a_0) = \int \delta a(\eta)\exp(-I_E[a]) \qquad (6.21)$$

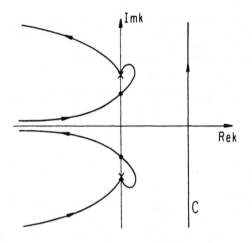

FIG. 3. The integration contour for constructing the semi-classical ground-state wave function of the minisuperspace model in the case $\Lambda > 0$, $Ha_0 > 1$. The figure shows schematically the original contour C used in Eq. (6.11) and the steepest-descents contour into which it can be distorted. The branch points of the exponent of Eq. (6.11) at $\kappa = \pm i$ are located by crosses. There are two complex-conjugate extrema of the exponent as indicated by dots and the contour C can be distorted to pass through both along the steepest-descents directions at 45° to the real axis as shown.

over real values of a, one would obtain a divergent result because the first term in (6.20) is negative definite. One could make the action infinitely negative by choosing a rapidly varying a. The solution to this problem seems to be to integrate the variable a in Eq. (6.21) along a contour that is parallel to the imaginary axis.[15] For each value of η, the contour of integration of a will cross the real axis at some value. Suppose there is some real function $\bar{a}(\eta)$ which maximizes the action. Then if one distorts the contour of integration of a at each value of η so that it crosses the real axis at $\bar{a}(\eta)$, the value of the action at the solution $\bar{a}(\eta)$ will give the saddle-point approximation to the functional integral (6.21), i.e.,

$$\psi_0(a_0) \approx N \exp(-I_E[\bar{a}(\eta)]) . \qquad (6.22)$$

If there were another real function $\hat{a}(\eta)$ which extremized the action but which did not give its maximum value there would be a nearby real function $\hat{a}(\eta) + \delta a(\eta)$ which has a greater action. By choosing the contour of integration in (6.21) to cross the real a axis at $\hat{a}(\eta) + \delta a(\eta)$, one would get a smaller contribution to the ground-state wave function. Thus, the dominant contribution comes from the real function $\bar{a}(\eta)$ with the greatest value of the action.

It may be that there is no real $a(\eta)$ which maximizes the action. In this case the dominant contribution to the ground-state wave function will come from complex functions $a(\eta)$ which extremize the action. These will occur in complex-conjugate pairs because the wave function is real.

In the case of $Ha_0 < 1$, we have already seen that there are two real functions $a(\eta)$ which extremize the action and which correspond to less than or more than a hemisphere of the four-sphere. Their actions are I_- and I_+, respectively, given by (6.13). In fact, I_- is the maximum value of the action for real $a(\eta)$ and therefore gives the dominant contribution to the ground-state wave function. Thus, we again recover Eqs. (6.14) and (6.16). In the case of $Ha_0 > 1$, there is no maximum of the action for real $a(\eta)$. In this case the dominant contribution to the ground-state wave function comes from a pair of complex-conjugate $a(\eta)$ which extremize the action. Thus, we would expect an oscillatory wave function like that given by Eq. (6.19).

C. Ground-state solution of the Wheeler-DeWitt equation

The ground-state wave function must be a solution of the Wheeler-DeWitt equation for the minisuperspace model [Eqs. (5.8) or (5.11)]. The $\exp(-\chi^2/2)$ dependence of the wave function on the matter field deduced in Sec. VI A shows that in fact $\psi_0(a)$ must solve Eq. (5.11) with $n = 0$. There are certainly solutions of this equation which have the large-a combination of exponentials required of the semiclassical approximation by Eq. (6.19) as a glance at Eq. (5.13) shows. In fact the prefactor in these asymptotic behaviors shows that the ground-state wave function will be *normalizable* in the norm

$$(\psi_0', \psi_0) = \int_0^\infty da \, a^p \bar{\psi}_0(a) \psi_0(a) \qquad (6.23)$$

in which the Wheeler-DeWitt operator is Hermitian.

The Wheeler-DeWitt equation enables us to determine the prefactor in the semiclassical approximation from the standard WKB-approximation formulas. With $p = 0$, for example, this would give when $Ha_0 > 1$

$$\psi_0(a_0) = 2(H^2 a_0^4 - a_0^2 + \epsilon_0 + \tfrac{1}{2})^{-1/4}$$
$$\times \cos\left[\frac{(H^2 a_0^2 - 1)^{3/2}}{3H^2} - \frac{\pi}{4} \right] . \qquad (6.24)$$

We could also solve the equation numerically. Figure 4 gives an example when $p = 0$ and $\epsilon_0 = -\tfrac{1}{2}$. There we have assumed that the wave function vanishes at $a = 0$. The dotted lines represent graphs of the prefactor in Eq. (6.24) and show that the semiclassical approximation becomes rapidly more accurate as Ha increases beyond 1. We shall return to an interpretation of these facts below.

D. Correspondence with de Sitter space

Having obtained $\psi_0(a)$, we are now in a position to assess its suitability as the ground-state wave function. Classically the vacuum geometry with the highest symmetry, hence minimum excitation, is de Sitter space—the surface of a Lorentz hyperboloid in a five-dimensional Lorentz-signatured flat spacetime. The properties of the wave function contained in Eqs. (6.16) and (6.19) are those one would expect to be semiclassically associated with this geometry. Sliced into three-spheres de Sitter space contains spheres only with a radius greater than $1/H$. Equation (6.16) shows that the wave function is an exponential-

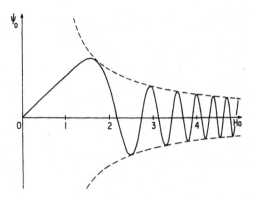

FIG. 4. A numerical solution of the Wheeler-DeWitt equation for the ground-state wave function $\psi_0(a)$. A solution of Eq. (5.11) is shown for $H=1$ in Planck units. We have assumed for definiteness $p=0$, $\epsilon_0=-\frac{1}{2}$, and a vanishing wave function at the origin. The wave function is damped for $Ha<1$ corresponding to the absence of spheres of radii smaller than H^{-1} in Lorentzian de Sitter space. It oscillates for $Ha>1$ decaying only slowly for large a. This reflects the fact that de Sitter space expands without limit. In fact, the envelope represented by the dotted lines is the distribution of three-spheres in Lorentzian de Sitter space: $[Ha(H^2a^2-1)^{1/2}]^{-1}$.

ly decreasing function with decreasing a for radii below that radius. Equation (6.24) shows the spheres of radius larger than $1/H$ are found with an amplitude which varies only slowly with the radius. This is a property expected of de Sitter space which expands both to the past and the future without limit. Indeed, tracing the origin of the two terms in (6.19) back to extrema with different signs of k one sees that one of these terms corresponds to the contracting phase of de Sitter space while the other corresponds to the expanding phase. The slow variation in the amplitude of the ground-state wave function reflects precisely the distribution of three-spheres in classical de Sitter space. Lorentzian de Sitter space is conformal to a finite region of the Einstein static universe

$$ds^2=\sigma^2a^2(\eta)(-d\eta^2+d\Omega_3{}^2) ,\qquad (6.25)$$

where $a(t)=(\cosh Ht)/H$ and $dt=ad\eta$. Three-spheres are evidently distributed uniformly in η in the Einstein static universe. The distribution of spheres in a in Lorentzian de Sitter space is therefore proportional to

$$[a(H^2a^2-1)^{1/2}]^{-1} .\qquad (6.26)$$

This is the envelope of the probability distribution $a^p|\psi(a)|^2$ for spheres of radius a deduced from the semiclassical wave function and shown in Fig. 4. The wave function constructed from the Euclidean prescription of Sec. III appropriately reflects the properties of the classical vacuum solution of highest symmetry and is therefore reasonably called the ground-state wave function.

VII. EXCITED STATES

Our Universe does not correspond to the ground state of the simple minisuperspace model. It might be that the

inclusion of more degrees of freedom in the model would produce a ground state which resembles our Universe more closely or it might be that we do not live in the ground state but in an excited state. Such excited states are not to be calculated by a simple path-integral prescription, but rather by solving the Wheeler-DeWitt equation with the boundary conditions that are required to maintain Hemiticity of the Hamiltonian operator between these states and the ground state. In this section, we shall construct the excited states for the minisuperspace model discussed in Sec. VI.

In the minisuperspace model where the spacelike sections are metric three-spheres all excitations in the gravitational degrees of freedom have been frozen out. We can study, however, excitations in the matter degrees of freedom. These are labeled by the harmonic-oscillator quantum number n as we have already seen [cf. Eq. (5.10)]. The issue then is what solution of Eq. (5.11) for $c_n(a)$ corresponds to this excited state. The equation can be written in the form of a one-dimensional Schrödinger equation

$$-\frac{1}{2a^p}\frac{d}{da}\left[a^p\frac{dc_n}{da}\right]+V(a)c_n=(n+\tfrac{1}{2}+\epsilon_0)c_n , \quad (7.1)$$

where

$$V(a)=\tfrac{1}{2}(a^2-\lambda a^4) .\qquad (7.2)$$

At $a=0$ Eq. (7.1) will in general have two types of solutions one of which is more convergent than the other [cf. Eq. (5.12)]. The behavior for the ground state which corresponds to the functional-integral prescription could be deduced from an evaluation of the determinant in the semiclassical approximation as discussed in Sec. IV. Whatever the result of such an evelation, the solution must be purely of one type or the other in order to ensure the Hermiticity of the Hamiltonian constraint. The same requirement ensures a. similar behavior for the excited-state solutions. In the following by "regular" solutions we shall mean those conforming to the boundary conditions arising from the functional-integral prescription. The exact type will be unimportant to us.

The potential $V(a)$ is a barrier of height $1/(4\lambda)$. At large a, the cosmological-constant part of the potential dominates and one has solutions which are linear combinations of the oscillating functions in (5.13). As we have already seen in the analysis of the ground state, the two possibilities correspond to a de Sitter contraction and a de Sitter expansion. With either of these asymptotic behaviors, a wave packet constructed by superimposing states of different n to produce a wave function with narrow support about some mean value of the scalar field would show this mean value increasing as one moved from large to small a.

Since each of the asymptotic behaviors in (5.13) is physically acceptable there will be solutions of (7.1) for all n. If, however, λ is small and n not too large, there are some values of n which are more important than others. These are the values which make the left-hand side of (7.1) at or close to those values of the energy associated with the metastable states (resonances) of the Schrödinger Hamiltonian on the right-hand side. To make this precise write

$$-\frac{1}{2}\frac{1}{a^p}\frac{d}{da}\left[a^p\frac{dc}{da}\right]+V(a)c=\epsilon c \ . \qquad (7.3)$$

This is the zero angular momentum Schrödinger equation in $d=p+1$ dimensions for single-particle motion in the potential $V(a)$. Classically, for $\epsilon<1/(4\lambda)$ there are two classes of orbits: bound orbits with a maximum value of a and unbound orbits with a minimum value of a. Quantum mechanically there are no bound states. For discrete values of $\epsilon\ll1/(4\lambda)$, however, there are metastable states. They lie near those values of ϵ which would be bound states if $\lambda=0$ and the barrier had infinite height. Since when $\lambda=0$ (7.3) is the zero angular momentum Schrödinger equation for a particle in a "radial" harmonic-oscillator potential in $d=p+1$ dimensions, these values are

$$\epsilon_N=2N+d/2, \ \ N=0,1,2,\ldots \ . \qquad (7.4)$$

For nonzero λ, if the particle has an energy near one of these values and much less than $1/(4\lambda)$ it can execute many oscillations inside the well but eventually it will tunnel out.

For the cosmological problem the classical Hamiltonian corresponding to (7.3) describes the evolution of homogeneous, isotropic, spatially closed cosmologies with radiation and a cosmological constant. The bound orbits correspond to those solutions for which the radiation density is sufficiently high that its attractive effect causes an expanding universe to recollapse before the repulsive effect of the cosmological constant becomes important. By contrast the unbound orbits correspond to de Sitter evolutions in which a collapsing universe never reaches a small enough volume for the increasing density of radiation to reverse the effect of the cosmological constant. There are thus two possible types of classical solutions. Quantum mechanically the Universe can tunnel between the two.

We can calculate the tunneling probability for small λ by using the usual barrier-penetration formulas from ordinary quantum mechanics. Let P be the probability for tunneling from inside the barrier to outside per transversal of the potential inside from minimum to maximum a. Then

$$P\approx e^{-B} \ , \qquad (7.5)$$

where

$$B=2\int_{a_0}^{a_1}da[V(a)-\epsilon]^{1/2} \qquad (7.6)$$

and a_0 and a_1 are the two turning points where $V(a)=\epsilon$. In the limit of $\epsilon\ll1/(4\lambda)$ the barrier-penetration factor becomes

$$B=\frac{2}{3\lambda}=\frac{2}{3H^2} \ . \qquad (7.7)$$

In magnitude this is just the total gravitational action for the Euclidean four-sphere of radius $1/H$ which is the analytic continuation of de Sitter space. This is familiar from general semiclassical results.[16]

Our own Universe corresponds to a highly excited state of the minisuperspace model. We know that the age of the Universe is about 10^{60} Planck times. The maximum

expansion, assuming a radiation dominated model, is therefore at least of order $a_{max}^2\approx10^{120}$. A wave packet describing our Universe would therefore have to be superpositions of states of definite n, with n at least $\approx a_{max}^2\approx10^{120}$. As large as this number is, the dimensionless limit on the inverse cosmological constant is even larger. In order to have such a large radiation dominated Universe λ must be less than 10^{-120}. The probability for our Universe to tunnel quantum mechanically at the moment of its maximum expansion to a de Sitter-type phase rather than recollapse is $P\approx\exp(-10^{120})$. This is a very small number but of interest if only because it is nonzero.

VIII. TOPOLOGY

In the preceding sections we have considered the amplitudes for three-geometries with S^3 topology to occur in the ground state. The functional-integral construction of the ground-state wave function, however, permits a natural extension to calculate the amplitudes for other topologies. We shall illustrate this extension in this section with some simple examples in the semiclassical approximation.

There is no compelling reason for restricting the topologies of the Euclidean four-geometries which enter in the sum defining the ground-state wave function. Whatever one's view on this question, however, there must be a ground-state wave function for every topology of a three-geometry which can be embedded in a four-geometry which enters the sum. In the general case this will mean all possible three-topologies—disconnected as well as connected, multiply connected as well as simply connected. The general ground-state wave function will therefore have N arguments representing the possibility of N compact disconnected three-geometries. The functional-integral prescription for the ground-state wave function in the case of pure gravity would then read

$$\Psi_0[\partial M^{(1)},h_{ij}^{(1)},\ldots,\partial M^{(N)},h_{ij}^{(N)}]=\int\delta g\exp(-I_E[g]) \ ,$$

$$(8.1)$$

where the sum is over all compact Euclidean four-geometries which have N disconnected compact boundaries $\partial M^{(i)}$ on which the induced three-metrics are $h_{ij}^{(i)}$. Since there is nothing in the sum which distinguishes one three-boundary from another the wave function must be symmetric in its arguments.

The wave function defined by (8.1) obeys a type of Wheeler-DeWitt equation in each argument but this is no longer sufficient to determine its form—in particular the correlations between the three-geometries. The functional integral is here the primary computational tool.

It is particularly simple to construct the semiclassical approximations to ground-state wave functions for those three-geometries with topologies which can be embedded in a compact Euclidean solution of the field equations. Consider for example the four-sphere. If the three-geometry has a single connected component and can be embedded in the four-sphere, then the extremal geometry at which the action is evaluated to give the semiclassical approximation is the smaller part of the four-sphere bounded by this three-geometry. The semiclassical ground-state wave function is

$$\Psi_0[h_{ij}] \approx N\Delta^{-1/2}[h_{ij}]$$

$$\times \exp\left[2l^{-2}\left[\int_{\partial M} dx^3 h^{1/2}K + \int_M d^4x\, g^{1/2}\Lambda\right]\right]$$

(8.2)

where M is the smaller part of the four-sphere and K is the trace of the extrinsic curvature of the three-surface computed with outward-pointing normals. Since there is a large variety of topologies of three-surfaces which *can* be embedded in the four-sphere—spheres, toruses, etc.,—we can easily compute their associated wave functions. Of course, these are many interesting three-surfaces which cannot be so embedded and for which the extremal solution defining the semiclassical approximation is not part of the four-sphere. In general one would expect to find wave functions for arbitrary topologies since any three-geometry is cobordant to zero and therefore there is some compact four-manifold which has it as its boundary. The problem of finding solutions of the field equations on these four-manifolds which match the given three-geometry and are compact thus becomes an interesting one.

Similarly, the semiclassical approximation for wave functions representing N disconnected three-geometries are equally easily computed when the geometries can be embedded in the four-sphere. The extremal geometry defining the semiclassical approximation is then simply the four-sphere with the N three-geometries cut out of it. The symmetries of the solution guarantee that as far as the exponent of the semiclassical approximation is concerned, it does not matter where the three-geometries are cut out provided that they do not overlap. To give a specific example, we calculate the amplitude for two disconnected three-spheres of radius $a_{(1)}$ and $a_{(2)}$ assuming $a_{(1)} < a_{(2)} < H^{-1}$. One possible extremal geometry is two disconnected portions of a four-sphere attached to the two three-spheres. This gives a product wave function with no correlation. Another extremal geometry is the smaller half of the four-sphere bounded by the spheres of radius $a_{(2)}$ with the portion interior to a sphere of radius $a_{(1)}$ removed. This gives an additional contribution to the wave function which expresses the correlation between the spheres. The correlated part in the semiclassical approximation is

$$\Psi_0^c(a_{(1)}, a_{(2)}) = N\Delta^{-1/2}(a_{(1)}, a_{(2)})$$

$$\times \exp\left[\frac{1}{3H^2}[-(1-H^2 a_{(2)}^2)^{3/2}\right.$$

$$\left. + (1-H^2 a_{(1)}^2)^{3/2}]\right]. \quad (8.3)$$

While the exponent is simple, the calculation of the determinant is now more complicated—it does not factor.

Equation (8.3) shows that the amplitude to have two correlated three-spheres of radius $a_{(1)} < a_{(2)} < H^{-1}$ is smaller than the amplitude to have a single three-sphere of radius $a_{(2)}$. In this crude sense topological complexity is suppressed. The amplitude for the Universe to bifurcate is of the order $\exp[-1/(3H^2)]$—a very large factor.

IX. CONCLUSIONS

The ground-state wave function for closed universes constructed by the Euclidean functional-integral prescription put forward in this paper can be said to represent a state of minimal excitation for these universes for two reasons. First, it is the natural generalization to gravity of the Euclidean functional integral for the ground-state wave function of flat-spacetime field theories. Second, when the prescription is applied to simple minisuperspace models, it yields a semiclassical wave function which corresponds to the classical solution of Einstein's equations of highest spacetime symmetry and lowest matter excitation.

The advantages of the Euclidean function-integral prescription are many but perhaps three may be singled out. First it is a complete prescription for the wave function. It implies not only the Wheeler-DeWitt equation but also the boundary conditions which determine the ground-state solution. The requirement of Hermiticity of the Wheeler-DeWitt operator extends these boundary conditions to the excited states as well.

A second advantage of this prescription for the ground-state wave function is common to all functional-integral formulations of quantum amplitudes. They permit the direct and explicit calculation of the semiclassical approximation. At the current stage of the development of quantum gravity where qualitative understanding is more important than precise numerical results, this is an important advantage. It is well illustrated by our minisuperspace model in which we were able to calculate semiclassically the probability of tunneling between a universe doomed to end in a big crunch and an eternal de Sitter expansion.

A final advantage of the Euclidean functional-integral prescription for the ground-state wave function is that it naturally generalizes to permit the calculation of amplitudes not usually considered in the canonical theory. In particular, we have been able to provide a functional-integral prescription for amplitudes for the occurrence of three-geometries with multiply connected and disconnected topologies in the ground state. In the semiclassical approximation we have been able to evaluate simple examples of such amplitudes.

The Euclidean functional-integral prescription sheds light on one of the fundamental problems of cosmology: the singularity. In the classical theory the singularity is a place where the field equations, and hence predictability, break down. The situation is improved in the quantum theory. An analogous improvement occurs in the problem of an electron orbiting a proton. In the classical theory there is a singularity and a breakdown of predictability when the electron is at the same position as the proton. However, in the quantum theory there is no singularity or breakdown. In an s-wave state, the amplitude for the electron to coincide with the proton is finite and nonzero, but the electron just carries on to the other side. Similarly, the amplitude for a zero-volume three-sphere in our minisuperspace model is finite and nonzero. One might interpret this as implying that the universe could continue through the singularity to another expansion period, although the classical concept of time would break down so that one could not say that the expansion happened after the contraction.

The ground-state wave function in the simple minisuperspace model that we have considered with a conformally invariant field does not correspond to the quantum state of the Universe that we live in because the matter wave function does not oscillate. However, it seems that this may be a consequence of using only zero rest mass fields and that the ground-state wave function for a universe with a massive scalar field would be much more complicated and might provide a model of quantum state of the observed Universe. If this were the case, one would have solved the problem of the initial boundary conditions of the Universe: the boundary conditions are that it has no boundary.[3]

ACKNOWLEDGMENTS

The authors are grateful for the hospitality of the Institute of Theoretical Physics, Santa Barbara, California, where part of this work was carried out. The research of one of us (J.H.) was supported in part by NSF Grants Nos. PHY81-07384 and PHY80-26043.

[1]See, e.g., R. P. Feynman, Rev. Mod. Phys. 20, 367 (1948); R. P. Feynman and A. R. Hibbs, *Quantum Mechanics and Path Integrals* (McGraw-Hill, New York, 1965) for discussions of quantum mechanics from this point of view.

[2]B. S. DeWitt, Phys. Rev. 160, 1113 (1967); J. A. Wheeler, in *Battelle Rencontres*, edited by C. DeWitt and J. A. Wheeler (Benjamin, New York, 1968).

[3]S. W. Hawking, in *Astrophysical Cosmology*, Pontificia Academiae Scientarium Scripta Varia, 48 (Pontificia Academiae Scientarium, Vatican City, 1982).

[4]For related ideas, see A. Vilenkin, Phys. Lett. 117B, 25 (1982); Phys. Rev. D 27, 2848 (1983).

[5]The connection between the canonical and functional-integral approaches to quantum gravity has been extensively discussed. See, in particular, H. Leutwyler, Phys. Rev. 134, B1155 (1964); L. Faddeev and V. Popov, Usp. Fiz. Nauk. 111, 427 (1973) [Sov. Phys. Usp. 16, 777 (1974)]; E. S. Fradkin and G. A. Vilkovisky, CERN Report No. TH-2332, 1977 (unpublished).

[6]For reviews of the canonical theory, see K. Kuchar, in *Quantum Gravity 2*, edited by C. Isham, R. Penrose, and D. W. Sciama (Clarendon, Oxford, 1981); A Hanson, T. Regge, and C. Teitelboim, *Constrained Hamiltonian Systems* (Academia Nazionale dei Lincei, Rome, 1976); K. Kuchar, in *Relativity, Astrophysics and Cosmology*, edited by W. Israel (Reidel, Dordrecht, 1973).

[7]J. M. York, Phys. Rev. Lett. 28, 1082 (1972).

[8]J. Milnor, *Morse Theory* (Princeton University Press, Princeton, New Jersey, 1962).

[9]For example, J. B. Hartle (unpublished).

[10]See, e.g., G. W. Gibbons and M. J. Perry, Nucl. Phys. B146, 90 (1978).

[11]S. W. Hawking, Commun. Math. Phys. 55, 133 (1977).

[12]B. DeWitt, Phys. Rev. 160, 1113 (1967).

[13]C. Isham and J. E. Nelson, Phys. Rev. D 10, 3226 (1974).

[14]W. E. Blyth and C. Isham, Phys. Rev. D 11, 768 (1975).

[15]G. W. Gibbons, S. W. Hawking, and M. J. Perry, Nucl. Phys. B138, 141 (1978).

[16]See, e.g., S. Coleman, Phys. Rev. D 15, 2929 (1977).

14

Quantum cosmology

S. W. HAWKING

14.1 Introduction

A few years ago I received a reprint request from an Institute of Quantum Oceanography somewhere in the Soviet Far East. I thought: What could be more ridiculous? Oceanography is a subject that is pre-eminently classical because it describes the behaviour of very large systems. Moreover, oceanography is based on the Navier–Stokes equation, which is a classical effective theory describing how large numbers of particles interact according to a more basic theory, quantum electrodynamics. Presumably, any quantum effects would have to be calculated in the underlying theory.

Why is quantum cosmology any less ridiculous than quantum oceanography? After all, the universe is an even bigger and more classical system than the oceans. Further, general relativity, which we use to describe the universe, may be only a low energy effective theory which approximates some more basic theory, such as string theory.

The answer to the first objection is that the spacetime structure of the universe is certainly classical today, to a very good approximation. However, there are problems with a large or infinite universe, as Newton realised. One would expect the gravitational attraction between all the different bodies in the universe to cause them to accelerate towards each other. Newton argued that this would indeed happen in a large but finite universe. However, he claimed that in an infinite universe the bodies would not all come together because there would not be a central point for them to fall to. This is a fallacious argument because in an infinite universe any point can be regarded as the centre. A correct treatment shows that an infinite universe can not remain in a stationary state if gravity is attractive. Yet so firmly held was the belief in an unchanging universe that when Einstein first proposed general relativity he added a cosmological constant in order to

223

obtain a static solution for the universe, thus missing a golden opportunity to predict that the universe should be expanding or contracting. I shall discuss later why it should be that we observe it to be expanding and not contracting.

If one traces the expansion back in time, one finds that all the galaxies would have been on top of each other about 15 thousand million years ago. At first it was thought that there was an earlier contracting phase and that the particles in the universe would come very close to each other but would miss each other. The universe would reach a high but finite density and would then re-expand (Lifshitz and Khalatnikov, 1963). However, a series of theorems (Hawking and Penrose, 1970; Hawking and Ellis, 1973) showed that if classical general relativity were correct, there would inevitably be a singularity at which all physical laws would break down. Thus classical cosmology predicts its own downfall. In order to determine how the classical evolution of the universe began one has to appeal to quantum cosmology and study the early quantum era.

But what about the second objection? Is general relativity the fundamental underlying theory of gravity or is it just a low energy approximation to some more basic theory? The fact that pure general relativity is not finite at two loops (Goroff and Sagnotti, 1985) suggests it is not the ultimate theory. It is an open question whether supergravity, the supersymmetric extention of general relativity, is finite at three loops and beyond but no-one is prepared to do the calculation. Recently, however, people have begun to consider seriously the possibility that general relativity may be just a low energy approximation to some theory such as superstrings, although the evidence that superstrings are finite is not, at the moment, any better than that for supergravity.

Even if general relativity is only a low energy effective theory it may yet be sufficient to answer the key question in cosmology: Why did the classical evolution phase of the universe start off the way it did? An indication that this is indeed the case is provided by the fact that many of the features of the universe that we observe can be explained by supposing that there was a phase of exponential 'inflationary' expansion in the early universe. This is described in more detail in the articles by Linde, and Blau and Guth (Chapters 13, 12, this volume). In order not to generate fluctuations in the microwave background bigger than the observational upper limit of 10^{-4}, the energy density in the inflationary era cannot have been greater than about $10^{-10} m_p^4$ (Rubakov *et al.*, 1982; Hawking, 1984a). This would put the inflationary era well inside the regime in which general relativity should be a good approximation. It would also be well inside the region in which any possible extra dimensions were compactified. Thus it might be reasonable to hope that the saddle point or semi-classical approximation to the quantum mechanical path integral for general relativity in four dimensions would give a reasonable indication of how the universe began. In what follows I shall assume that the lowest-order term in the action for a spacetime metric is the Einstein one, as it must be for agreement with ordinary, low energy, observations. However, I shall bear in mind the possibilities of higher-order terms and extra dimensions.

14.2 The quantum state of the universe

I shall use the Euclidean path integral approach. The basic assumption of this is that the 'probability' in some sense of a positive definite spacetime metric $g_{\mu\nu}$ and matter fields ϕ on a manifold M is proportional to $\exp(-\tilde{I})$ where $\tilde{I}$ is the Euclidean action. In general relativity

$$\tilde{I} = -\frac{1}{16\pi} \int_M R(g)^{1/2} \, \mathrm{d}^d x + \frac{1}{8\pi} \int_{\partial M} K(h)^{1/2} \, \mathrm{d}^{d-1} x - \int L_{\mathrm{m}}[\phi](g)^{1/2} \, \mathrm{d}^d x,$$

where h and K are respectively the determinant of the first fundamental form and the trace of the second fundamental form of the boundary ∂M of M. In string theory the action $\tilde{I}$ of a metric $g_{\mu\nu}$, antisymmetric tensor field $B_{\mu\nu}$ and dilaton field ϕ is given by the log of the path integral of the string action over all maps of string world sheets into the given space. For most fields the path integral will not be conformally invariant. This will mean that the path integral diverges and $\tilde{I}$ will be infinite. Such fields will be suppressed by an infinite factor. However, the path integral over maps into certain background fields will be conformally invariant. The action for these fields will be that of general relativity plus higher-order terms.

The probability of an observable O having the value A can be found by summing the projection operator Π_A over the basic probability over all Euclidean metrics and fields belonging to some class C.

$$P_O(A) = \int_C \mathrm{d}[g_{\mu\nu}] \, \mathrm{d}[\phi] \Pi_A \exp(-\tilde{I}),$$

where $\Pi_A = 1$ if the value of O is A and zero otherwise. From such probabilities and the conditional probability, the probability of A given B,

$$P(A|B) = \frac{P(A, B)}{P(B)},$$

where $P(A, B)$ is the joint probability of A and B, one can calculate the outcome of all allowable measurements.

The choice of the class C of metrics and fields on which one considers the probability measure $\exp(-\tilde{I})$ determines the quantum state of the universe. C is usually specified by the asymptotic behaviour of the metric and matter fields, just as the state of the universe in classical general relativity can be specified by the asymptotic behaviour of these fields. For instance, one could demand that C consist of all metrics that approach the metric of Euclidean flat space outside some compact region and all matter fields that go to zero at infinity. The quantum state so defined is the vacuum state used in S matrix calculations. In these one considers incoming and outgoing states that differ from Euclidean flat space and zero matter fields at infinity in certain ways. The path integral over all such fields gives the amplitude to go from the initial to the final state.

In these S matrix calculations one considers only measurements at infinity and does not ask questions about what happens in the middle of the spacetime. However, this is not much help for cosmology: it is unlikely that the universe is asymptotically flat, and, even if it were, we are not really interested in what happens at infinity but in events in some finite region surrounding us. Suppose we took the class C of metrics and matter fields that defines the quantum state of the universe to be the class described above of asymptotically Euclidean metrics and fields. Then the path integral to calculate the probability of a value of an observable O would receive contributions from two kinds of metrics. There would be connected asymptotically Euclidean metrics and there would be a disconnected metric which consisted of a compact component that contained the observable O and a separate asymptotically Euclidean component. One can not exclude disconnected metrics from the class C because any disconnected metric can be approximated arbitrarily closely by a connected metric in which the different components are joined by thin tubes with negligible action. It turns out that for observables that depend only on a compact region the dominant contribution to the path integral comes from the compact regions of disconnected metrics. Thus, as far as cosmology is concerned, the probabilities of observables would be almost the same if one took the class C to consist of compact metrics and matter fields that are regular on them.

In fact, this seems a much more natural choice for the class C that defines the quantum state of the universe. It does not refer to any unobserved asymptotic region and it does not involve any boundary or edge to spacetime at infinity or a singularity where one would have to appeal to some outside agency to set the boundary conditions. It would mean that spacetime would be completely self contained and would be determined completely by the laws of physics: there would not be any points where the laws broke down and there would not be any edge of spacetime at which unpredictable influences could enter the universe. This choice of boundary conditions for the class C can be paraphrased as: 'The boundary condition of the universe is that it has no boundary' (Hawking, 1982; Hartle and Hawking, 1983; Hawking, 1984b).

This choice of the quantum state of the universe is very analogous to the vacuum state in string theory which is defined by all maps of closed string world sheets without boundary into Euclidean flat space. More generally, one can define a 'ground' state of no string excitations about any set of background fields that satisfy certain conditions by all maps of closed string world sheets into the background. Thus one can regard the 'no boundary' quantum state for the universe as a 'ground' state (Hartle and Hawking, 1983). It is, however, different from other ground states. In other quantum theories non-trivial field configurations have positive energy. They therefore cannot appear in the zero energy ground state except as quantum fluctuations. In the case of gravity it is also true that any asymptotically flat metric has positive energy, except flat space, which has zero energy. However, in a closed, non-asymptotically flat universe there is no infinity at which to define the energy of the field configuration. In a sense the total energy of a closed universe is zero: the positive energy of the matter fields and gravitational waves is exactly balanced by the negative potential energy which arises because gravity is attractive. It is this negative potential energy that allows non-trivial gravitational fields to appear in the 'ground' state of the universe.

Unfortunately, this negative energy also causes the Euclidean action $\tilde{I}$ for general relativity to be unbounded below (Gibbons et al., 1978), thus causing $\exp(-\tilde{I})$ not to be a good probability measure on the space C of field configurations. In certain cases it may be possible to deal with this difficulty by rotating the contour of integration of the conformal factor in the path integral from real values to be parallel to the imaginary axis. However, there does not seem to be a general prescription that will guarantee that the path integral converges. This difficulty might be overcome in string theory where the string action is positive in Euclidean backgrounds. It may be, however, that the difficulty in making the path integral converge is fundamental to the fact that the 'ground' state of the universe seems to be highly non-trivial. In any event it would seem reasonable to expect that the main contribution to the path integral would come from fields that are near stationary points of the action $\tilde{I}$, that is, near solutions of the field equations.

It should be emphasised that the 'no boundary' condition on the metrics in the class C that defines the quantum state of the universe is just a proposal: it cannot be proved from something else. It is quite possible that the universe is in some different quantum state though it would be difficult to think of one that was defined in a natural manner. The 'no boundary' proposal does have the great advantage, however, that it provides a definite basis on which to calculate the probabilities of observable quantities and compare them with what we see. This basis seems to be lacking in many other approaches to quantum cosmology in which the assumptions on the quantum state of the universe are not clearly stated. For instance, Vilenkin (1986) defines the quantum state in a toy minisuperspace model by requiring that a certain current on minisuperspace be ingoing at one point of the boundary of minisuperspace (corresponding to 'creation from nothing') and outgoing elsewhere on the boundary (annihilation into nothing?). However, he does not seem to have a general prescription that would define the quantum state except in simple minisuperspace cases. Moreover, his state is not CPT invariant, which is a property that one might think the quantum state of the universe should have. Similarly, Linde (1985; Chapter 13, this volume) does not give a definition of the quantum state of the universe. He also suggests that the Wick rotation for the Euclidean action of the gravitational field should be in the opposite direction to that for other fields. This would be equivalent to changing the sign of the gravitational constant and making gravity repulsive instead of attractive.

14.3 The density matrix

One thinks of a quantum system as being described by its state at one time. In the case of cosmology, 'at one time' can be interpreted as on a spacelike surface S. One can therefore ask for the probability that the metric and matter fields have given values on a $d-1$ surface S. In fact, it is meaningful to ask questions only about the $d-1$ metric h_{ij} induced on S by the d metric $g_{\mu\nu}$ on M because the components $n^\mu g_{\mu\nu}$ of $g_{\mu\nu}$ that lie out of S can be given any values by a diffeomorphism of M that leaves S fixed. Thus the probability that the surface S has the induced metric h_{ij} and matter fields ϕ_0 is

$$P(h_{ij}, \phi_0) = \int_C d[g_{\mu\nu}]\, d[\phi_0] \exp(-\tilde{I})\Pi_{(h_{ij}, \phi_0)},$$

where $\Pi_{(h_{ij}, \phi_0)}$ is the projection operator which has value 1 if the induced metric and matter fields on S have the given values and is zero otherwise.

One can cut the manifold M at the surface S to obtain a new manifold $\tilde{M}$

228

bounded by two copies $\tilde{S}$ and $\tilde{S}'$ of S. One can then define $\rho(h_{ij}, \phi_0; h'_{ij}, \phi'_0)$ to be the path integral over all metrics and matter fields on $\tilde{M}$ which agree with the given values h_{ij}, ϕ_0 on $\tilde{S}$ and h'_{ij}, ϕ'_0 on $\tilde{S}$. The quantity ρ can be regarded as a density matrix describing the quantum state of the universe as seen from a single spacelike surface for the following reasons:

(i) The diagonal elements of ρ, that is, when $h_{ij} = h'_{ij}$ and $\phi_0 = \phi'_0$, give the probability of finding a surface S with the metric h_{ij} and matter fields ϕ_0.

(ii) If S divides M into two parts, the manifold $\tilde{M}$ will consist of two disconnected parts, $\tilde{M}_+$ and $\tilde{M}_-$. The path integral for ρ will factorise:

$$\rho(h_{ij}, \phi_0; h'_{ij}, \phi'_0) = \Psi_+(h_{ij}, \phi_0)\Psi_-(h'_{ij}, \phi'_0),$$

where the wave functions Ψ_+ and Ψ_- are given by the path integral over all metrics and matter fields on $\tilde{M}_+$ and $\tilde{M}_-$ respectively which have the given values on S and S'. If the matter fields ϕ are CP invariant, $\Psi_+ = \Psi_-$ and both are real (Hawking, 1985). Ψ is known as 'The Wave Function Of The Universe'. A density matrix which factorises can be interpreted as corresponding to a pure quantum state.

(iii) If the surface S does not divide M into two parts, the manifold $\tilde{M}$ will be connected. In this case the path integral for ρ will not factorise into the product of two wave functions. This means that ρ will correspond to the density matrix of a mixed quantum state, rather than a pure state for which the density matrix would factorise (Page, 1986; Hawking, 1987).

One can think of the density matrices which do not factorise in the following way: Imagine a set of surfaces T_i which, together with S, divide the spacetime manifold M into two parts. One can take the disjoint union of the T_i and S as the surface which is used to define ρ (there is no reason why this surface has to be connected). In this case the manifold $\tilde{M}$ will be disconnected and the path integral for ρ will factorise into the product of two wave functions which will depend on the metrics and matter fields on two sets of surfaces, S, T_i and S', T'_i. The quantity ρ will therefore be the density matrix for a pure quantum state. However, an observer will be able to measure the metric and matter fields only on one connected component of the surface (say, S) and will not know anything about their values on the other components, T_i, or even if any other components are required to divide the spacetime manifold into two parts. The observer will therefore have to sum over all possible metrics and matter fields on the surfaces $\tilde{T}_i$. This summation or trace over the fields on the $\tilde{T}_i$ will reduce ρ to a density matrix corresponding to a mixed state in the fields on the remaining surfaces $\tilde{S}$ and

$\tilde{S}'$. It is like when you have a system consisting of two parts A and B. Suppose the system is in a pure quantum state but that you can observe only part A. Then, as you have no knowledge about B, you have to sum over all possibilities for B, with equal weight. This reduces the density matrix for the system from a pure state to a mixed state.

The summation over all fields on the surfaces $\tilde{T}_i$ is equivalent to joining the surfaces $\tilde{T}_i$ to $\tilde{T}'_i$ and doing the path integral over all metrics and matter fields on a manifold $\tilde{M}$ whose only boundaries are the surfaces $\tilde{S}$ and $\tilde{S}'$. There is an overcounting because, as well as summing over all metrics and matter fields, one is summing over all positions of the surfaces T_i in $\tilde{M}$. However, the path integral over these extra degrees of freedom can be factored out by introducing ghosts. The reduced path integral is then the same as that for the density matrix ρ for a single pair of surfaces $\tilde{S}$ and $\tilde{S}'$. Thus one can see that the reason that the density matrix for S corresponds to a mixed state is that one is observing the state of the universe on a single spacelike surface and ignoring the possibility that spacetime may be not simply connected and so require other surfaces T_i as well as S to divide M into two parts.

14.4 The Wheeler–DeWitt equation

In a neighbourhood of the boundary surface $\tilde{S}$ of the manifold $\tilde{M}$, one can write the metric $g_{\mu\nu}$ in the $(d-1)+1$ form:

$$ds^2 = (N^2 + N^i N_i)\,dt^2 + 2N_i\,dx^i\,dt + h_{ij}\,dx^i\,dx^j,$$

where $\tilde{S}$ is the surface $t=0$. The Euclidean action can then be written in the Hamiltonian form:

$$\tilde{I} = -\int dt\,d^{d-1}x(\pi^{ij}\dot{h}_{ij} + \pi_\phi\dot{\phi} - NH^0 - N_iH^i),$$

where $\pi^{ij} = -(h^{1/2}/16\pi)(K^{ij} - h^{ij}K)$ is the Euclidean momentum conjugate to h_{ij}, K_{ij} is the second fundamental form of $\tilde{S}$,

$$H_0 = 16\pi G_{ijkl}\pi^{ij}\pi^{kl} - \frac{1}{16\pi}h^{1/2\ d-1}R + T^{00}$$

$$H^i = -2\pi^{ij}_{\ \ ;j} + T^{0i}$$

$$G_{ijkl} = \tfrac{1}{2}h^{-1/2}(h_{ik}h_{jl} + h_{il}h_{jk} - h_{ij}h_{kl}).$$

As was stated above, the components of $g_{\mu\nu}$ that lie out of the surface $\tilde{S}$ can be given any values by a diffeomorphism of $\tilde{M}$ that leaves $\tilde{S}$ fixed. This means that the variational derivative of the path integral for ρ with respect to N and N_i on $\tilde{S}$ must be zero:

$$\frac{\delta\rho}{\delta N_i} = -\int d[g_{\mu\nu}]\, d[\phi_0]\, \frac{\delta \tilde{I}}{\delta N_i}\, \exp(-\tilde{I}) = \tilde{H}_i\rho = 0$$

$$\frac{\delta\rho}{\delta N} = -\int d[g_{\mu\nu}]\, d[\phi_0]\, \frac{\delta \tilde{I}}{\delta N}\, \exp(-\tilde{I}) = \tilde{H}\rho = 0,$$

where the operators $\tilde{H}$ and $\tilde{H}_i$ are obtained from the corresponding classical expressions by replacing the Euclidean momentum π^{ij} by $-\delta/\delta h_{ij}$ and π_ϕ by $-\delta/\delta\phi$.

The first equation is called the momentum constraint. It is a first-order equation for ρ on superspace, the space W of all metrics h_{ij} and matter fields ϕ on a surface S. It implies that ρ is the same for metrics and matter fields which can be obtained from each other by coordinate transformations in S. The second equation is called the Wheeler–DeWitt equation. It holds at each point of superspace, except where $h_{ij} = h'_{ij}$ and $\phi_0 = \phi'_0$. When this is true, the separation between $\tilde{S}$ and $\tilde{S}'$ in the metric $g_{\mu\nu}$ on the manifold $\tilde{M}$ may be zero. In this case, it is no longer true that the variation of ρ with respect to N is zero. There is an infinite dimensional delta function on the right-hand side of the Wheeler–DeWitt equation. Thus, the Wheeler–DeWitt equation is like the equation for the propagator, $G(x, x') = \langle\phi(x)\phi(x')\rangle$:

$$(-\Box + m^2)G(x, x') = \delta(x, x').$$

As the point x tends towards x', the propagator diverges like r^{2-d}, where r is the distance between x and x'. Thus $G(x, x')$ will be infinite. Similarly, $\rho(h_{ij}, \phi_0; h_{ij}, \phi_0)$, the diagonal elements of the density matrix, will be infinite. This infinity arises from Euclidean geometries of the form $S \times S^1$, where the S^1 is of very short radius. However, we are interested really only in the probabilities for Lorentzian geometries, because we live in a Lorentzian universe, not a Euclidean one. One can recognise the part of the density matrix ρ that corresponds to Lorentzian geometries by the fact that it will oscillate rapidly as a function of the scale factor of the metrics h_{ij} and h'_{ij} (Hawking, 1984b). One therefore wants to subtract out the infinite, Euclidean, component and leave a finite, Lorentzian, component. One way of doing this is to consider only spacetime manifolds M which the surface S divides into two parts. The density matrix from such geometries will be of the factorised form:

$$\rho(h_{ij}, \phi_0; h'_{ij}, \phi'_0) = \Psi(h_{ij}, \phi_0)\Psi(h'_{ij}, \phi'_0),$$

where the wave function Ψ obeys the Wheeler–DeWitt equation with no delta function on the right-hand side. This part of the density matrix will therefore remain finite when $h_{ij} = h'_{ij}$ and $\phi_0 = \phi'_0$. In a supersymmetric theory, such as supergravity or superstrings, the infinity at the diagonal in the density matrix would probably be cancelled by the fermions.

14.5 Minisuperspace

The Wheeler–DeWitt equation can be regarded as a second-order differential equation for ρ or Ψ on superspace, the infinite-dimensional space of all metrics and matter fields on S. It is hard to solve such an equation. Instead, progress has been made by using finite dimensional approximations to superspace, called minisuperspaces, first introduced by Misner (1970). In other words, one reduces the infinite number of degrees of freedom of the gravitational and matter fields and of the gauge to a finite number and solves the Wheeler–DeWitt equation on a finite-dimensional space.

14.5.1 de Sitter model

The simplest example is a homogeneous isotropic four-dimensional universe with a cosmological constant and metric

$$ds^2 = \sigma^2 [N^2 \, dt^2 + a^2 \, d\Omega_3^2].$$

The action is

$$\tilde{I} = -\frac{1}{2} \int dt N a \left[\frac{1}{N^2} \left[\frac{da}{dt} \right]^2 + 1 - \lambda a^2 \right],$$

where $\sigma^2 = \frac{2}{3}\pi m_p^2$, a is the radius of the 3-sphere space-like surfaces and $\lambda = \frac{1}{3}\sigma^2 \Lambda$. One can choose $N = a$. The first two terms in the Euclidean action are negative definite. This means that the path integral over a does not converge. However, one can make the path integral converge by taking a to be imaginary. This corresponds to integrating the conformal factor over a contour parallel to the imaginary axis (Gibbons *et al.*, 1978).

With a imaginary, the action is the same as that of the anharmonic oscillator. The density matrix $\rho(a, a')$ is given by a path integral over all values of a on a manifold $\tilde{M}$ bounded by surfaces with radii a and a'. There are two kinds of such manifold: ones that have two disconnected components, which correspond to spacetimes that are divided in two by S, and connected ones, which correspond to non-simply connected spacetimes that S does not divide.

Consider first the case in which S divides M in two. The density matrix from these geometries that S divides into two is the product of wave functions:

$$\rho(a, a') = \Psi(a)\Psi(a'),$$

where the wave function Ψ is given by a path integral over compact 4-geometries bounded by a 3-sphere of radius a or a'. One would expect this path integral to be approximately $A \exp(-B)$, where B is the action of a solution of the classical Euclidean field equations with the given boundary conditions and the prefactor A is given by a path integral over small fluctuations about the solution of the classical field equations. The compact homogeneous isotropic solution of the Euclidean field equations is a 4-sphere of radius $\lambda^{-1/2}$. A 3-sphere of radius $a < \lambda^{-1/2}$ can fit into such a 4-sphere in two positions: it can bound more or less than half the 4-sphere. The action B of both these solutions of the classical equations is negative, with the action of more than half the 4-sphere being the more negative. One might therefore expect that this solution would provide the dominant contribution to the path integral. However, if one takes the scale factor a to be imaginary, in order to make the path integral converge, and then analytically continues back to real a, one finds that the dominant contribution comes from the solution that corresponds to less than half the 4-sphere, rather than the other solution which corresponds to more than half the 4-sphere, as one might have expected. This conclusion also follows from an analysis of the path integral in the K representation (Hartle and Hawking, 1983).

In terms of the gauge choice $N = a$, used above, the path integral is over a with a the given value at $t = 0$ and $a = 0$ at $t = \pm\infty$. This path integral is the same as that for the propagator for the anharmonic oscillator from ia at $t = 0$ to 0 at $t = \infty$. But this gives the ground state wave function. Thus

$$\Psi(a) \propto \mathrm{Re}(A_0(i\,a)),$$

where $A_0(x)$ is the ground state wave function of the anharmonic oscillator.

For small x, $A_0(x)$ behaves like $\exp(-\frac{1}{2}x^2)$. Thus $\Psi(a)$ behaves exponentially like $\exp(\frac{1}{2}x^2)$. This agrees with the estimates from the action of less than half the 4-sphere, as above. However, for $a > \lambda^{-1/2}$, there is no Euclidean solution of the classical field equations for a compact homogeneous isotropic 4-space bounded by a 3-sphere of radius a. Instead there are complex metrics which are solutions of the field equations with the required properties. Near the 3-sphere of radius a, one can take a section through the complexified spacetime manifold on which the metric is real and Lorentzian. This is reflected in the fact that $A_0(i\,a)$ will oscillate for $a > \lambda^{-1/2}$: exponential wave functions correspond to Euclidean 4-geometries and oscillating wave functions correspond to Lorentzian 4-geometries (Hawking, 1984*b*).

For large a, $\Psi(a)$ behaves like $a^{-1}\cos(\lambda^{1/2}a^3)$. One can interpret this by writing the wave function in the WKB form: $C(\exp(\mathrm{i}\,S)+\exp(-\mathrm{i}\,S))$, where S is a rapidly varying phase factor and C is a slowly varying amplitude. The wave function will satisfy the Wheeler–DeWitt equation to leading order if the phase factor S obeys the classical Hamilton–Jacobi equation. Thus, an oscillating wave function will correspond in the classical limit to an $(n-1)$-dimensional family of solutions of the classical Lorentzian field equations, where n is the dimension of the minisuperspace.

In the example above, $n=1$. The oscillating part of the wave function corresponds to the classical de Sitter solution which collapses from infinite radius to a minimum radius $a=\lambda^{-1/2}$ and then expands again exponentially to infinite radius. The classical Lorentzian solution does not go below a radius of $\lambda^{-1/2}$, so one can interpret the exponentially damped wave function below that radius as corresponding to a Euclidean geometry in the classically forbidden region. Note that, for this explanation to make sense, the wave function has to decrease with decreasing a, and not increase as authors such as Linde and Vilenkin have argued on the analogy of tunnelling 'from nothing'. Anyway, if one believes that the quantum state of the universe is determined by a path integral over compact geometries, one has no freedom of choice of the solution of the Wheeler–DeWitt equation: it has to be the one that increases exponentially with increasing a.

Another feature of the wave function that is worth remarking on is that it is real. This means that, in the oscillating region, the WKB ansatz is $C(\exp(\mathrm{i}\,S)+\exp(-\mathrm{i}\,S))$. One can regard the first term as representing an expanding universe and the second a contracting universe. More generally, if the wave function represents some history of the universe, it also represents the *CPT* image of that history (Hawking, 1985). This should be contrasted with the approach of Vilenkin and others, who try to choose a solution of the Wheeler–DeWitt equation which corresponds only to expanding universes. The fallacy of this attempt is that the direction of the time coordinate has no intrinsic meaning: it can be changed by a coordinate transformation. The physically meaningful question is: how does the entropy or degree of disorder behave during the histories of the universe that are described by the wave function? The minisuperspace models considered here are too simple to answer this but it will be discussed for models with the full number of degrees of freedom in Section 14.7.

The contribution to the density matrix from geometries that S does not divide into two parts is given by a path integral with a fixed at the given values at $t=0$ and $t=t_1$ for some Euclidean time interval t_1. But this is equal to the real part of the propagator $K(\mathrm{i}\,a,0;\,\mathrm{i}\,a',t_1)$ for the anharmonic oscillator from $\mathrm{i}\,a$ at $t=0$ to $\mathrm{i}\,a'$ at $t=t_1$.

$$K(\mathrm{i}\,a,0;\,\mathrm{i}\,a',t_1)=\sum_n A_n(\mathrm{i}\,a)A_n(\mathrm{i}\,a')\exp(-E_n t_1),$$

where $A_n(x)$ are the wave functions of the excited states of the anharmonic oscillator and E_n are the energy levels. To obtain the density matrix one has to integrate over all values of t_1 because the two surfaces can have any time separation:

$$\rho(a,a')=\mathrm{Re}\int_0^\infty K(\mathrm{i}\,a,0;\,\mathrm{i}\,a',t_1)\,\mathrm{d}t_1=\mathrm{Re}\sum_n\frac{A_n(\mathrm{i}\,a)A_n(\mathrm{i}\,a')}{E_n}.$$

One can interpret this as saying that the universe is in the state specified by the wave function $\mathrm{Re}(A_n(\mathrm{i}\,a))$ with the relative probability $(E_n)^{-1}$. Note that the universe need not be 'on shell' in the sense that the Wheeler–DeWitt operator acting on A_n is not 0, but E_n. This term in the Wheeler–DeWitt equation acts as if the universe contained a certain amount of negative energy radiation. It will cause the classical solution corresponding to A_n by the WKB approximation to bounce at a larger radius than $\lambda^{-1/2}$. Thus, the effect of the universe being in a mixed quantum state might be observable. However, at large values of a, the effect of the negative energy radiation would be very small and the universe would expand exponentially, like the de Sitter solution.

14.5.2 The massive scalar field model

The de Sitter model was interesting because it showed that the 'no boundary' proposal for the quantum state of the universe leads to inflation if there is some process which gives rise to an effective cosmological constant in the early universe. However, the universe is not expanding exponentially at the present time, so there has to be some way in which the cosmological 'constant' can reduce to zero at late times. One mechanism, and possibly the only one, for generating such a decaying effective cosmological constant is a scalar field with a potential which has a minimum at zero and which is exponentially bounded. I shall consider the simplest example, a massive scalar field.

The action of a homogeneous isotropic universe of radius a with a massive scalar field ϕ that is constant on the surfaces of homogeneity is

$$\tilde{I} = -\frac{1}{2}\int dt\, Na\left[\frac{1}{N^2}\left[\left[\frac{da}{dt}\right]^2 - a^2\left[\frac{d\phi}{dt}\right]^2\right] + 1 - a^2 m^2 \phi^2\right].$$

Unfortunately, in this case, there does not seem to be any simple prescription for making the Euclidean action positive definite. Taking a imaginary leaves the kinetic term for ϕ negative, while taking ϕ imaginary would cure this problem but would make the mass term negative. One could, however, make the action positive in this manner if the potential was pure ϕ^4. On physical grounds, one would not expect that there would be a qualitative difference between the behaviour of a universe in which the scalar potential was ϕ^2 and one in which it was ϕ^4.

In the case that the surface S divides the spacetime into two parts, the wave function will obey the Wheeler–DeWitt equation

$$\frac{1}{2}\left[\frac{1}{a^p}\frac{\partial}{\partial a}a^p\frac{\partial}{\partial a} - \frac{1}{a^2}\frac{\partial^2}{\partial\phi^2} - a^2 + a^4 m^2 \phi^2\right]\Psi[a,\phi] = 0,$$

where p reflects some of the uncertainty in the factor ordering of the operators in the Wheeler–DeWitt equation. It is thought that the value of p does not have much effect, so it is usual to take $p = 1$, because this simplifies the equation. One can introduce new coordinates:

$$x = a\sinh\phi, \quad y = a\cosh\phi.$$

In these coordinates, the Wheeler–DeWitt equation becomes

$$\left[\frac{\partial^2}{\partial y^2} - \frac{\partial^2}{\partial x^2} + V\right]\Psi(x, y) = 0,$$

where $V = (y^2 - x^2)[-1 + (y^2 - x^2)m^2(\operatorname{arctanh} x/y)^2]$.

For small values of a, one can expect that Ψ is approximately $A\exp(-B)$, where B is the action of a solution of the Euclidean field equations. If $\phi \gg 1$ and $a < 1/m\phi$, the value of ϕ will not vary much over the solution and the $m^2\phi^2$ term in the action will act as an effective cosmological constant. One would therefore expect B to be the action of the smaller part of a 4-sphere of radius $1/m\phi$, bounded by a 3-sphere of radius a. From the de Sitter model, one would expect the wave function to oscillate for $a > 1/m\phi$ and the phase factor S to be $\frac{1}{3}m\phi a^3$, the analytic continuation of B. Such a wave function is a solution to the Wheeler–DeWitt equation to leading order.

One can interpret the oscillating part of the wave function as corresponding to a complex compact metric which is a solution of the field equations and which is bounded by the surface S. In a neighbourhood of S one can take a section through the complexified spacetime manifold on which the metric is nearly real and Lorentzian. This solution will have a minimum radius of order $1/m\phi$ and will expand exponentially with ϕ slowly decreasing. It will be a quantum realisation of the 'chaotic inflation' model proposed by Linde (1983).

After an exponential expansion of the universe by a factor of order $\exp(\frac{1}{2}\phi^2)$, the scalar field will start to oscillate with frequency m. The energy momentum tensor of the scalar field will change from that of an effective cosmological constant to that of pressure-free matter. The universe will change from an exponential expansion to a matter-dominated one. In a model with other matter fields, one would expect the energy in the massive scalar field oscillations to be converted into zero rest mass particles. The universe would then expand as a radiation-dominated model.

The universe would expand to a maximum radius and then recollapse. One would expect that if such complex, almost Lorentzian, geometries contributed to the wave function in their expanding phase, they would also contribute in their contracting phase. However, although a few solutions will bounce at small radius and expand again (Hawking, 1984b; Page, 1985a,b), most solutions will collapse to a singularity. They will give an oscillating contribution to the wave function, even in the region $a < 1/m\phi$ of superspace where the dominant contribution is exponential. It will also mean that the boundary condition for the Wheeler–DeWitt equation on the light cone $x = \pm y$ is not exactly $\Psi = 1$, as was assumed in some earlier papers (Hawking and Wu, 1985; Moss and Wright, 1983).

The density matrix from geometries that S does not divide into two parts has not been calculated yet. By analogy with the de Sitter model, one might expect that the part which corresponds to Lorentzian geometries would behave like solutions with a massive scalar field and negative energy radiation. One would not expect the negative energy to prevent collapse to a singularity.

To summarise, in this model, the universe begins its expansion from a non-singular state. It expands in an inflationary manner, goes over to a matter or radiation-dominated expansion, reaches a maximum radius and recollapses to a singularity. This will be discussed further in Sections 14.7 and 14.8.

14.6 Beyond minisuperspace

The minisuperspace models were useful because they showed that the 'no boundary' proposal for the quantum state of the universe can lead to a universe like the one that we observe, at least in its large scale features.

However, ultimately one would like to know the density matrix or wave function on the whole of superspace, not just a finite-dimensional subspace. This is a bit of a tall order but one can use a 'midisuperspace' approximation in which one takes the action to all orders in a finite number of degrees of freedom and to second order in the remaining degrees of freedom.

A treatment of the massive scalar field model on these lines has been given by Halliwell and Hawking (1985). The two degrees of freedom of the model described above are treated exactly, and the rest as perturbations on the background determined by the two-dimensional minisuperspace model. As in the model above, the oscillating part of the background wave function corresponds by the WKB approximation to a universe which starts at a minimum radius, expands in an inflationary and then a matter-dominated manner, reaches a maximum radius and recollapses to a singularity.

From the 'no boundary' condition the behaviour of the perturbations is determined by a path integral of the perturbation modes over the compact geometries represented by the background wave function. In the case of Euclidean geometries that are part of a 4-sphere or of complex geometries that are near such a Euclidean geometry, one can use an adiabatic approximation to show that the perturbation modes are in their ground state, with the minimum excitation compatible ·with the uncertainty principle. This means that the Lorentzian geometries that correspond to the oscillating part of the wave function start off at the minimum radius with all the perturbation modes in the ground state. As the universe inflates, the adiabatic approximation remains good and the perturbation modes remain in their ground states until their wavelength becomes longer than the horizon size or, in other words, their frequency is red shifted to less than the expansion time scale. After this, the wave functions of the perturbation modes freeze and do not relax adiabatically to remain in the ground state as the frequency of the modes changes.

The perturbation modes remain frozen until the wavelength of the modes becomes less than the horizon size again during the matter- or radiation-dominated expansion. Because they have not been able to relax adiabatically, they will then be in a highly excited state. After this, they will evolve like classical perturbations of a Friedmann universe. They will have a 'scale free' spectrum, that is, their rms amplitude at the time the wavelength equals the horizon size will be independent of the wavelength. The amplitude will be roughly $10(m/m_p)$, where m is the mass of the scalar field. Thus they would have the right amplitude of about 10^{-4} to account for galaxy formation if m is about 10^{14} GeV.

In order to generate sufficient inflation, the initial value of the scalar field ϕ has to be greater than about 8. However, with $m = 10^{-5}m_{\rm P}$, the energy density of the scalar field will still be a lot less than the Planck density. Thus it may be reasonable in quantum cosmology to ignore higher-order terms and extra dimensions.

In the recollapse phase the perturbations will continue to grow classically. They will not return to their ground state when the universe becomes small again, as I suggested (Hawking, 1985). The reason is that when they start expanding, the background compact geometry bounded by the surface S is near to the Euclidean geometry of half a 4-sphere. On such a background the adiabatic approximation will hold for the perturbation modes, so they will be in their ground state. However, when the universe recollapses, the background geometry will be near a Lorentzian solution which expands and recontracts. The adiabatic approximation will not hold on such a background. Thus the perturbation modes will not be in their ground state when the universe recollapses, but will be highly excited.

14.7 The direction of time

The quantum state defined by the 'no boundary' proposal is CPT invariant (Hawking, 1985), though this is not true of other quantum states, such as that proposed by Vilenkin (1986). Yet the observed universe shows a pronounced asymmetry between the future and the past. We remember events in the past but we have to predict events in the future. Imagine a tall building which is destroyed by an explosion and collapses to a pile of rubble and dust. If one took a film of this and ran it backwards, one would see the rubble and dust gather themselves together and jump back into their places in the building. One would easily recognise that the film was being shown backwards because this kind of behaviour is never observed: we do not see tower blocks jumping up. Yet it is not forbidden by the laws of physics. These are CPT invariant. In fact, the laws that are important for the structure of buildings are invariant under C and P separately. Thus, they must be invariant under T alone. In other words, if a building can collapse, it can also resurrect itself.

The explanation that is usually given as to why we do not see buildings jumping up is that the second law of thermodynamics says that entropy or disorder must always increase with time, and that an erect building is in a much more ordered state than a pile of rubble and dust. However, this law has a rather different status from other laws, such as Newton's law of gravity. First, it is not an absolute law that is always obeyed: rather it is a statistical law that says what will probably happen. Second, it is not a local law like other laws of physics: it is a statement about boundary conditions. It says that if a system starts off in a state of high order, it is likely to be found in a disordered state at a later time, simply because there are many more disordered states than ordered ones.

The reason that entropy and disorder increase with time and buildings fall down rather than jump up is that the universe seems to have started out in a state of high order in the past. On the other hand, if, for some reason, the universe obeyed the boundary condition that it was in a state of high order at late times, then at earlier times it would be likely to be in a disordered state and disorder would decrease with time. However, human beings are governed by the second law and the boundary conditions, just like everything else in the universe. Our subjective sense of the direction of time is determined by the direction in which disorder increases because to record information in our memories requires the expenditure of free energy and increases the entropy and disorder of the universe. Thus, if disorder decreased with time, our subjective sense of time would also be reversed and we would still say that entropy and disorder increased with time. The second law is almost a tautology: entropy and disorder increase with time because we measure time in the direction in which disorder increases.

However, there remains the question of why should the universe have been in a state of high order at one end of time? Why was it not in a state of complete disorder or thermal equilibrium at all times? After all, that might seem more probable as there are many more disorder states than order ones. And why does the direction of time in which disorder increases coincide with that in which the universe expands? Put it another way: why do we say that the universe is expanding, and not contracting?

These questions can be answered only by some assumption on the boundary conditions of the universe or, equivalently, on the class of spacetime geometries in the path integral. As we have seen, the 'no boundary' condition implies that the universe would have started off in a smooth and ordered state with all the inhomogeneous perturbations in their ground state of minimum excitation. As the universe expanded, the perturbations would have grown and the universe would have become more inhomogeneous and disordered. This would answer the questions above.

But what would happen if the universe, or some region of it, stopped expanding and began to collapse? At first I thought (Hawking, 1985) that entropy and disorder would have to decrease in the contracting phase so that the universe would get back to a smooth state when it was small again. This was because I thought that at small values of the radius a, the wave function would be given just by a path integral over small Euclidean geometries. This would imply $\Psi = 1$ on the light cone $x = \pm y$ in the model described above and that the adiabatic approximation would hold for the perturbation modes, which would therefore be in their ground state. However, Page (1985b) pointed out that there would also be a contribution to the wave function from compact, complex, almost Lorentzian geometries that represented universes that started at a minimum radius, expanded to a maximum and recollapsed, as described above. This was supported by work by Laflamme (1987), who investigated a minisuperspace model in which the surfaces S had topology $S^1 \times S^2$. He also found almost Lorentzian solutions which started in a non-singular manner but recollapsed to a singularity. The adiabatic approximation for the perturbation modes would not hold in the recollapse. Thus they would not return to their ground states, but would get even more excited as the collapse continued. The universe would get more and more inhomogeneous and disorder would continue to increase with time.

There remains the question of why we observe that the direction of time in which disorder increases is also the direction in which the universe is expanding. Because the 'no boundary' quantum state is CPT invariant, there will also be histories of the universe that are the CPT reverses of that described above. However, intelligent beings in these histories would have the opposite subjective sense of time. They would therefore describe the universe in the same way as above: it would start in a smooth state, expand and collapse to a very inhomogeneous state. The question therefore becomes: why do we live in the expanding phase? If we lived in the contracting phase, we would observe entropy to increase in the opposite direction of time to that in which the universe was expanding. To answer this, I think one has to appeal to the weak anthropic principle. The probability is that the universe will not recollapse for a very long time (Hawking and Page, 1986). By that time, the stars would all have burnt out and the baryons would have decayed. The conditions would therefore not be suitable for the existence of beings like us. It is only in the expanding phase that intelligent beings can exist to ask the question: why is entropy increasing in the same direction of time as that in which the universe is expanding?

14.8 The origin and fate of the universe

Does the universe have a beginning and/or end?

If the 'no boundary' proposal for the quantum state is correct, spacetime is compact. On a compact space, any time coordinate will have a minimum and a maximum. Thus, in this sense, the universe will have a beginning and an end.

Will the beginning and end be singularities?

Here one must distinguish between two different questions: whether there are singularities in the geometries over which the path integral is taken, and whether there are singularities in the Lorentzian geometries that correspond to the density matrix by the WKB approximation. A singularity cannot really be regarded as belonging to spacetime because the laws of physics would not hold there. Thus, the requirement of the 'no boundary' proposal that the path integral is over compact geometries only rules out the existence of any singularities in this sense. Of course, one will have to allow compact metrics that are not smooth in the path integral, just as in the integral over particle histories one has to allow particle paths that are not smooth but satisfy a Hölder continuity condition. However, one can approximate such paths by smooth paths. Similarly, in the path integral for the universe, it must be possible to approximate the non-smooth metrics in a suitable topology by sequences of smooth metrics because otherwise one could not define the action of such metrics. Thus, in this sense, the geometries in the path integral are non-singular.

On the other hand, the Lorentzian geometries that correspond to the density matrix by the WKB approximation can and do have singularities. In the minisuperspace model described above, the Lorentzian geometries began at a non-singular minimum radius or 'bounce' and evolve to a singularity in general, in the direction of time defined by entropy increase. I would conjecture that this is a general feature: oscillating wave functions and Lorentzian geometries arise only when one has a massive scalar field which gives rise to an effective cosmological constant and Euclidean solutions which are like the 4-sphere. The Lorentzian solutions will be the analytic continuation of the Euclidean solutions. They will start in a smooth non-singular state at a minimum radius equal to the radius of the 4-sphere and will expand and become more irregular. When and if they collapse, it will be to a singularity.

One could say that the universe was 'created from nothing' at the minimum radius (Vilenkin, 1982). However, the use of the word 'create' would seem to imply that there was some concept of time in which the universe did not exist before a certain instant and then came into being. But time is defined only within the universe, and does not exist outside it, as was pointed out by Saint Augustine (400): 'What did God do before He made Heaven and Earth? I do not answer as one did merrily: He was preparing Hell for those that ask such questions. For at no time had God not made anything because time itself was made by God.'

The modern view is very similar. In general relativity, time is just a coordinate that labels events in the universe. It does not have any meaning outside the spacetime manifold. To ask what happened before the universe began is like asking for a point on the Earth at 91° north latitude; it just is not defined. Instead of talking about the universe being created, and maybe coming to an end, one should just say: The universe is.

References

Gibbons, G. W., Hawking, S. W. and Perry, M. J. (1978). *Nucl. Phys.*, **B138**, 141.

Goroff, M. H. and Sagnotti, A. (1985). *Phys. Lett.*, **160B**, 81.

Halliwell, J. J. and Hawking, S. W. (1985). *Phys. Rev.*, **D31**, 1777.

Hartle, J. B. and Hawking, S. W. (1983). *Phys. Rev.*, **D28**, 2960.

Hawking, S. W. (1982). In *Astrophysical Cosmology*. Proceedings of the Study Week on Cosmology and Fundamental Physics, ed. H. A. Bruck, G. V. Coyne and M. S. Longair. Pontificia Academiae Scientarium: Vatican City.

Hawking, S. W. and Penrose, R. (1970). *Proc. Roy. Soc. Lon.*, **A314**, 529.

Hawking, S. W. and Ellis, G. F. R. (1973). *The Large Scale Structure of Space-Time*. Cambridge University Press: Cambridge.

Hawking, S. W. (1984a). *Phys. Lett.*, **150B**, 339.

Hawking, S. W. (1984b). *Nucl. Phys.*, **B239**, 257.

Hawking, S. W. and Wu, Z. C. (1985). *Phys. Lett.*, **151B**, 15.

Hawking, S. W. (1985). *Phys. Rev.*, **D32**, 2489.

Hawking, S. W. and Page, D. N. (1986). *Nucl. Phys.*, **B264**, 185.

Hawking, S. W. (1987). *Physica Scripta* (in press).

Laflamme, R. (1987). The wave function of a $S^1 \times S^2$ universe. Preprint, to be published.

Lifshitz, E. M. and Khalatnikov, I. M. (1963). *Adv. Phys.*, **12**, 185.

Linde, A. D. (1983). *Phys. Lett.*, **129B**, 177.

Linde, A. D. (1985). *Phys. Lett.*, **162B**, 281.

Misner, C. W. (1970). In *Magic without Magic*, ed. J. R. Klauder. Freeman: San Francisco.

Moss, I. and Wright, W. (1983). *Phys. Rev.*, **D29**, 1067.

Page, D. N. (1985a). *Class. & Q.G.*, **1**, 417.

Page, D. N. (1985b). *Phys. Rev.*, **D32**, 2496.

Page, D. N. (1986). *Phys. Rev.*, **D34**, 2267.

Rubakov, V. A., Sazhin, M. V. and Veryaskin, A. V. (1982). *Phys. Lett.*, **115B**, 189.

Saint Augustine (400). *Confessions*. Re-edited in *Encyclopedia Britannica* (1952).

Vilenkin, A. (1982). *Phys. Lett.*, **117B**, 25.

Vilenkin, A. (1986). *Phys. Rev.*, **D33**, 3560.

PHYSICAL REVIEW D

PARTICLES AND FIELDS

THIRD SERIES, VOLUME 31, NUMBER 8

15 APRIL 1985

Origin of structure in the Universe

J. J. Halliwell and S. W. Hawking

Department of Applied Mathematics and Theoretical Physics, Silver Street, Cambridge CB3 9EW, United Kingdom
and Max Planck Institut for Physics and Astrophysics, Foehringer Ring 6, Munich, Federal Republic of Germany
(Received 17 December 1984)

It is assumed that the Universe is in the quantum state defined by a path integral over compact four-metrics. This can be regarded as a boundary condition for the wave function of the Universe on superspace, the space of all three-metrics and matter field configurations on a three-surface. We extend previous work on finite-dimensional approximations to superspace to the full infinite-dimensional space. We treat the two homogeneous and isotropic degrees of freedom exactly and the others to second order. We justify this approximation by showing that the inhomogeneous or anisotropic modes start off in their ground state. We derive time-dependent Schrödinger equations for each mode. The modes remain in their ground state until their wavelength exceeds the horizon size in the period of exponential expansion. The ground-state fluctuations are then amplified by the subsequent expansion and the modes reenter the horizon in the matter- or radiation-dominated era in a highly excited state. We obtain a scale-free spectrum of density perturbations which could account for the origin of galaxies and all other structure in the Universe. The fluctuations would be compatible with observations of the microwave background if the mass of the scalar field that drives the inflation is 10^{14} GeV or less.

I. INTRODUCTION

Observations of the microwave background indicate that the Universe is very close to homogeneity and isotropy on a large scale. Yet we know that the early Universe cannot have been completely homogeneous and isotropic because in that case galaxies and stars would not have formed. In the standard hot big-bang model the density perturbations required to produce these structures have to be assumed as initial conditions. However, in the inflationary model of the Universe[1−4] it was possible to show that the ground-state fluctuations of the scalar field that causes the exponential expansion would lead to a spectrum of density perturbations that was almost scale free.[5−7] In the simplest grand-unified-theory (GUT) inflationary model the amplitude of the density perturbations was too large but an amplitude that was consistent with observation could be obtained in other models with a different potential for the scalar field.[8] Similarly, ground-state fluctuations of the gravitational-wave modes would lead to a spectrum of long-wavelength gravitational waves that would be consistent with observation provided that the Hubble constant H in the inflationary period was not more than about 10^{-4} of the Planck mass.[9]

One cannot regard these results as a completely satisfactory explanation of the origin of structure in the Universe because the inflationary model does not make any assumption about the initial or boundary conditions of the Universe. In particular, it does not guarantee that there should be a period of exponential expansion in which the scalar field and the gravitational-wave modes would be in the ground state. In the absence of some assumption about the boundary conditions of the Universe, any present state would be possible: one could pick an arbitrary state for the Universe at the present time and evolve it backward in time to see what initial conditions it arose from. It has recently been proposed[10−13] that the boundary conditions of the Universe are that it has no boundary. In other words, the quantum state of the Universe is defined by a path integral over compact four-metrics without boundary. The quantum state can be described by a wave function Ψ which is a function on the infinite-dimensional space W called superspace which consists of all three-metrics h_{ij} and matter field configurations Φ_0 on a three-surface S. Because the wave function does not depend on time explicitly, it obeys a system of zero-energy Schrödinger equations, one for each choice of the shift N_i and the lapse N on S. The Schrödinger equations can be decomposed into the momentum constraints, which imply that the wave function is the same at all points of W that are related by coordinate transformations, and the Wheeler-DeWitt equations, which can be

31   1777

regarded as a system of second-order differential equations for Ψ on W. The requirement that the wave function be given by a path integral over compact four-metrics then becomes a set of boundary conditions for the Wheeler-DeWitt equations which determines a unique solution for Ψ.

It is difficult to solve differential equations on an infinite-dimensional manifold. Attention has therefore been concentrated on finite-dimensional approximations to W, called "minisuperspaces." In other words, one restricts the number of gravitational and matter degrees of freedom to a finite number and then solves the Wheeler-DeWitt equations on a finite-dimensional manifold with boundary conditions that reflect the fact that the wave function is given by a path integral over compact four-metrics. In particular,[12-15] it has been shown that in the case of a homogeneous isotropic closed universe of radius a with a massive scalar field ϕ the wave function corresponds in the classical limit to a family of classical solutions which have a long period of exponential or "inflationary" expansion and then go over to a matter-dominated expansion, reach a maximum radius, and then collapse in a time-symmetric manner. This model would be in agreement with observation but, because it is so restricted, the only prediction it can make is that the observed value of the density parameter Ω should be exactly one.[15] The aim of this paper is to extend this minisuperspace model to the full number of degrees of freedom of the gravitational and scalar fields. We treat the 2 degrees of freedom of the minisuperspace model exactly and we expand the other inhomogeneous and anisotropic degrees of freedom to second order in the Hamiltonian. In the region of W in which Ψ oscillates rapidly, one can use the WKB approximation to relate the wave function to a family of classical solutions and so introduce a concept of time. As in the minisuperspace case, the family includes solutions with a long period of exponential expansion. We show that the gravitational-wave and density-perturbation modes obey decoupled time-dependent Schrödinger equations with respect to the time parameter of the classical solution. The boundary conditions imply that these modes start off in the ground state. While they remain within the horizon of the exponentially expanding phase, they can relax adiabatically and so they remain in the ground state. However, when they expand outside the horizon of the inflationary period, they become "frozen" until they reenter the horizon in the matter-dominated era. They then give rise to gravitational waves and a scale-free spectrum of density perturbations. These would be consistent with the observations of the microwave background and could be large enough to explain the origins of galaxies if the mass of the scalar field were about 10^{-5} of the Planck mass. Thus the proposal that the quantum state of the Universe is defined by a path integral over compact four-metrics seems to be able to account for the origin of structure in the Universe: it arises, not from arbitrary initial conditions, but from the ground-state fluctuations that have to be present by the Heisenberg uncertainty principle.

In Sec. II we review the Hamiltonian formalism of classical general relativity, and in Sec. III we show how this leads to the canonical treatment of the quantum theory. In Sec. IV we summarize earlier work[13] on a homogeneous isotropic minisuperspace model with a massive scalar field. We extend this to all the matter and gravitational degrees of freedom in Sec. V, treating the inhomogeneous modes to second order in the Hamiltonian. In Sec. VI we decompose the wave function into a background term which obeys an equation similar to that of the unperturbed minisuperspace model, and perturbation terms which obey time-dependent Schrödinger equations. We use the path-integral expression for the wave function in Sec. VII to show that the perturbation wave functions start out in their ground states. Their subsequent evolution is described in Sec. VIII. In Sec. IX we calculate the anisotropy that these perturbations would produce in the microwave background and compare with observation. In Sec. X we summarize the paper and conclude that the proposed quantum state could account not only for the large-scale homogeneity and isotropy but also for the structure on smaller scales.

II. CANONICAL FORMULATION OF GENERAL RELATIVITY

We consider a compact three-surface S which divides the four-manifold M into two parts. In a neighborhood of S one can introduce a coordinate t such that S is the surface $t=0$ and coordinates x^i ($i=1,2,3$). The metric takes the form

$$ds^2 = -(N^2 - N_i N^i)dt^2 + 2N_i dx^i dt + h_{ij} dx^i dx^j . \tag{2.1}$$

N is called the lapse function. It measure the proper-time separation of surfaces of constant t. N_i is called the shift vector. It measures the deviation of the lines of constant x^i from the normal to the surface S. The action is

$$I = \int (L_g + L_m) d^3x \, dt , \tag{2.2}$$

where

$$L_g = \frac{m_P^2}{16\pi} N (G^{ijkl} K_{ij} K_{kl} + h^{1/2\,3}R) , \tag{2.3}$$

$$K_{ij} = \frac{1}{2N} \left[-\frac{\partial h_{ij}}{\partial t} + 2N_{(i\,|j)} \right] , \tag{2.4}$$

is the second fundamental form of S, and

$$G^{ijkl} = \frac{1}{2} h^{1/2} (h^{ik} h^{jl} + h^{il} h^{jk} - 2h^{ij} h^{kl}) . \tag{2.5}$$

In the case of a massive scalar field Φ

$$L_m = \frac{1}{2} N h^{1/2} \left[N^{-2} \left(\frac{\partial \Phi}{\partial t} \right)^2 - 2\frac{N^i}{N^2} \frac{\partial \Phi}{\partial t} \frac{\partial \Phi}{\partial x^i} \right.$$
$$\left. - \left[h^{ij} - \frac{N^i N^j}{N^2} \right] \frac{\partial \Phi \partial \Phi}{\partial x^i \partial x^j} - m^2 \Phi^2 \right] . \tag{2.6}$$

In the Hamiltonian treatment of general relativity one regards the components h_{ij} of the three-metric and the field Φ as the canonical coordinates. The canonically conjugate momenta are

$$\pi^{ij}=\frac{\partial L_g}{\partial \dot{h}_{ij}}=-\frac{h^{1/2}m_P^2}{16\pi}(K^{ij}-h^{ij}K) , \qquad (2.7)$$

$$\pi_\Phi=\frac{\partial L_m}{\partial \dot{\Phi}}=N^{-1}h^{1/2}\left[\dot{\Phi}-N^i\frac{\partial \Phi}{\partial x^i}\right] . \qquad (2.8)$$

The Hamiltonian is

$$H=\int (\pi^{ij}\dot{h}_{ij}+\pi_\Phi\dot{\Phi}-L_g-L_m)d^3x$$

$$=\int (NH_0+N_iH^i)d^3x , \qquad (2.9)$$

where

$$H_0=16\pi m_P^{-2}G_{ijkl}\pi^{ij}\pi^{kl}-\frac{m_P^2}{16\pi}h^{1/2}\,^3R$$

$$+\tfrac{1}{2}h^{1/2}\left[\frac{\pi_\Phi^2}{h}+h^{ij}\frac{\partial \Phi\partial \Phi}{\partial x^i\partial x^j}+m^2\Phi^2\right] , \qquad (2.10)$$

$$H^i=-2\pi^{ij}_{\ |j}+h^{ij}\frac{\partial \Phi}{\partial x^j}\pi_\Phi , \qquad (2.11)$$

and

$$G_{ijkl}=\tfrac{1}{2}h^{-1/2}(h_{ik}h_{jl}+h_{il}h_{jk}-h_{ij}h_{kl}) . \qquad (2.12)$$

The quantities N and N_i are regarded as Lagrange multipliers. Thus the solution obeys the momentum constraint

$$H^i=0 \qquad (2.13)$$

and the Hamiltonian constraint

$$H_0=0 . \qquad (2.14)$$

For given fields N and N^i on S the equations of motion are

$$\dot{h}_{ij}=\frac{\partial H}{\partial \pi^{ij}}, \quad \dot{\pi}^{ij}=-\frac{\partial H}{\partial h_{ij}} ,$$

$$\dot{\Phi}=\frac{\partial H}{\partial \pi_\Phi}, \quad \dot{\pi}_\Phi=-\frac{\partial H}{\partial \Phi} . \qquad (2.15)$$

III. QUANTIZATION

The quantum state of the Universe can be described by a wave function Ψ which is a function on the infinite-dimensional manifold W of all three-metrics h_{ij} and matter fields Φ on S. A tangent vector to W is a pair of fields (γ_{ij},μ) on S where γ_{ij} can be regarded as a small change of the metric h_{ij} and μ can be regarded as a small change of Φ. For each choice of $N>0$ on S there is a natural metric $\Gamma(N)$ on W:[15]

$$ds^2=\int N^{-1}\left[\frac{m_P^2}{32\pi}G^{ijkl}\gamma_{ij}\gamma_{kl}+\tfrac{1}{2}h^{1/2}\mu^2\right]d^3x . \qquad (3.1)$$

The wave function Ψ does not depend explicitly on the time t because t is just a coordinate which can be given arbitrary values by different choices of the undetermined multipliers N and N_i. This means that Ψ obeys the zero-energy Schrödinger equation:

$$H\Psi=0 . \qquad (3.2)$$

The Hamiltonian operator H is the classical Hamiltonian with the usual substitutions:

$$\pi^{ij}(x)\rightarrow -i\frac{\delta}{\delta h_{ij}(x)}, \quad \pi_\phi(x)\rightarrow -i\frac{\delta}{\delta \phi(x)} . \qquad (3.3)$$

Because N and N_i are regarded as independent Lagrange multipliers, the Schrödinger equation can be decomposed into two parts. There is the momentum constraint

$$H_-\Psi\equiv \int N_iH^id^3x\ \Psi$$

$$=\int h^{1/2}N_i\left[2\left[\frac{\delta}{\delta h_{ij}(x)}\right]_{|j}-h^{ij}\frac{\partial \Phi}{\partial x^j}\frac{\delta}{\delta \Phi(x)}\right]d^3x\ \Psi$$

$$=0 . \qquad (3.4)$$

This implies that Ψ is the same on three-metrics and matter field configurations that are related by coordinate transformations in S. The other part of the Schrödinger equation, corresponding to $H_|\Psi=0$, where $H_|=\int NH_0d^3x$ is called the Wheeler-DeWitt equation. There is one Wheeler-DeWitt equation for each choice of N on S. One can regard them as a system of second-order partial differential equations for Ψ on W. There is some ambiguity in the choice of operator ordering in these equations but this will not affect the results of this paper. We shall assume that $H_|$ has the form[15]

$$(-\tfrac{1}{2}\nabla^2+\xi R+V)\Psi=0 , \qquad (3.5)$$

where ∇^2 is the Laplacian in the metric $\Gamma(N)$. R is the curvature scalar of this metric and the potential V is

$$V=\int h^{1/2}N\left[-\frac{m_P^2}{16\pi}{}^3R+\epsilon+U\right]d^3x , \qquad (3.6)$$

where $U=T^{00}-\tfrac{1}{2}\pi_\Phi^2$. The constant ϵ can be regarded as a renormalization of the cosmological constant Λ. We shall assume that the renormalized Λ is zero. We shall also assume that the coefficient ξ of the scalar curvature R of W is zero.

Any wave function Ψ which satisfies the momentum constraint and the Wheeler-DeWitt equation for each choice of N and N_i on S describes a possible quantum state of the Universe. We shall be concerned with the particular solution which represents the quantum state defined by a path integral over compact four-metrics without boundary. In this case[11-13]

$$\Psi=\int d[g_{\mu\nu}]d[\Phi]\exp[-\hat{I}(g_{\mu\nu},\Phi)] , \qquad (3.7)$$

where $\hat{I}$ is the Euclidean action obtained by setting N negative imaginary and the path integral is taken over all compact four-metrics $g_{\mu\nu}$ and matter fields Φ which are bounded by S on which the three-metric is h_{ij} and the matter field is Φ. One can regard (3.7) as a boundary condition on the Wheeler-DeWitt equations. It implies that Ψ tends to a constant, which can be normalized to one, as h_{ij} goes to zero.

IV. UNPERTURBED FRIEDMANN MODEL

References 12–14 considered the minisuperspace model which consisted of a Friedmann model with metric

$$ds^2 = \sigma^2(-N^2 dt^2 + a^2 d\Omega_3^2) , \qquad (4.1)$$

where $d\Omega_3^2$ is the metric of the unit three-sphere. The normalization factor $\sigma^2 = 2/3\pi m_P^2$ has been included for convenience. The model contains a scalar field $(2^{1/2}\pi\sigma)^{-1}\phi$ with mass $\sigma^{-1}m$ which is constant on surfaces of constant t. One can easily generalize this to the case of a scalar field with a potential $V(\phi)$. Such generalizations include models with higher-derivative quantum corrections.[16] The action is

$$I = -\tfrac{1}{2} \int dt \, N a^3 \left[\frac{1}{N^2 a^2} \left[\frac{da}{dt} \right]^2 - \frac{1}{a^2} \right. $$
$$\left. - \frac{1}{N^2} \left[\frac{d\phi}{dt} \right]^2 + m^2\phi^2 \right] . \qquad (4.2)$$

The classical Hamiltonian is

$$H = \tfrac{1}{2} N (-a^{-1}\pi_a^2 + a^{-3}\pi_\phi^2 - a + a^3 m^2 \phi^2) , \qquad (4.3)$$

where

$$\pi_a = -\frac{a \, da}{N \, dt} , \quad \pi_\phi = \frac{a^3 d\phi}{N \, dt} . \qquad (4.4)$$

The classical Hamiltonian constraint is $H = 0$. The classical field equations are

$$N \frac{d}{dt} \left[\frac{1}{N} \frac{d\phi}{dt} \right] + \frac{3}{a} \frac{da}{dt} \frac{d\phi}{dt} + N^2 m^2 \phi = 0 , \qquad (4.5)$$

$$N \frac{d}{dt} \left[\frac{1}{N} \frac{da}{dt} \right] = N^2 a m^2 \phi^2 - 2a \left[\frac{d\phi}{dt} \right]^2 . \qquad (4.6)$$

The Wheeler-DeWitt equation is

$$\tfrac{1}{2} N e^{-3\alpha} \left[\frac{\partial^2}{\partial\alpha^2} - \frac{\partial^2}{\partial\phi^2} + 2V \right] \Psi(\alpha,\phi) = 0 , \qquad (4.7)$$

where

$$V = \tfrac{1}{2} (e^{6\alpha} m^2 \phi^2 - e^{4\alpha}) \qquad (4.8)$$

and $\alpha = \ln a$. One can regard Eq. (4.7) as a hyperbolic equation for Ψ in the flat space with coordinates (α,ϕ) with α as the time coordinate. The boundary condition that gives the quantum state defined by a path integral over compact four-metrics is $\Psi \to 1$ as $\alpha \to -\infty$. If one integrates Eq. (4.7) with this boundary condition, one finds that the wave function starts oscillating in the region $V > 0$, $|\phi| > 1$ (this has been confirmed numerically[14]). One can interpret the oscillatory component of the wave function by the WKB approximation:

$$\Psi = \mathrm{Re}(C \, e^{iS}) , \qquad (4.9)$$

where C is a slowly varying amplitude and S is a rapidly varying phase. One chooses S to satisfy the classical Hamilton-Jacobi equation:

$$H(\pi_\alpha, \pi_\phi, \alpha, \phi) = 0 , \qquad (4.10)$$

where

$$\pi_\alpha = \frac{\partial S}{\partial\alpha} , \quad \pi_\phi = \frac{\partial S}{\partial\phi} . \qquad (4.11)$$

One can write (4.10) in the form

$$\tfrac{1}{2} f^{ab} \frac{\partial S \partial S}{\partial q^a \partial q^b} + e^{-3\alpha} V = 0 , \qquad (4.12)$$

where f^{ab} is the inverse to the metric $\Gamma(1)$:

$$f^{ab} = e^{-3\alpha} \mathrm{diag}(-1,1) . \qquad (4.13)$$

The wave function (4.9) will then satisfy the Wheeler-DeWitt equation if

$$\nabla^2 C + 2i f^{ab} \frac{\partial C \partial S}{\partial q^a \partial q^b} + iC \nabla^2 S = 0 , \qquad (4.14)$$

where ∇^2 is the Laplacian in the metric f_{ab}. One can ignore the first term in Eq. (4.14) and can integrate the equation along the trajectories of the vector field $X^a = dq^a/dt = f^{ab}\partial S/\partial q^b$ and so determine the amplitude C. These trajectories correspond to classical solutions of the field equations. They are parametrized by the coordinate time t of the classical solutions.

The solutions that correspond to the oscillating part of the wave function of the minisuperspace model start out at $V = 0$, $|\phi| > 1$ with $d\alpha/dt = d\phi/dt = 0$. They expand exponentially with

$$S = -\tfrac{1}{3} e^{3\alpha} m |\phi| (1 - m^{-2} e^{-2\alpha} \phi^{-2})$$

$$\approx -\tfrac{1}{3} e^{3\alpha} m |\phi| , \qquad (4.15)$$

$$\frac{d\alpha}{dt} = m |\phi| , \quad \frac{d|\phi|}{dt} = -\tfrac{1}{3} m . \qquad (4.16)$$

After a time of order $3m^{-1}(|\phi_1| - 1)$, where ϕ_1 is the initial value of ϕ, the field ϕ starts to oscillate with frequency m. The solution then becomes matter dominated and expands with e^α proportional to $t^{2/3}$. If there were other fields present, the massive scalar particles would decay into light particles and then the solution would expand with e^α proportional to $t^{1/2}$. Eventually the solution would reach a maximum radius of order $\exp(9\phi_1^2/2)$ or $\exp(9\phi_1^2)$ depending on whether it is radiation or matter dominated for most of the expansion. The solution would then recollapse in a similar manner.

V. THE PERTURBED FRIEDMANN MODEL

We assume that the metric is of the form (2.1) except the right hand side has been multiplied by a normalization factor σ^2. The three-metric h_{ij} has the form

$$h_{ij} = a^2(\Omega_{ij} + \epsilon_{ij}) , \qquad (5.1)$$

where Ω_{ij} is the metric on the unit three-sphere and ϵ_{ij} is a perturbation on this metric and may be expanded in harmonics:

$$\epsilon_{ij} = \sum_{n,l,m} [6^{1/2}a_{nlm}\tfrac{1}{3}\Omega_{ij}Q^n_{lm} + 6^{1/2}b_{nlm}(P_{ij})^n_{lm} + 2^{1/2}c^0_{nlm}(S^0_{ij})^n_{lm} + 2^{1/2}c^e_{nlm}(S^e_{ij})^n_{lm} + 2d^0_{nlm}(G^0_{ij})^n_{lm} + 2d^e_{nlm}(G^e_{ij})^n_{lm}] \ . \tag{5.2}$$

The coefficients $a_{nlm},b_{nlm},c^0_{nlm},c^e_{nlm},d^0_{nlm},d^e_{nlm}$ are functions of the time coordinate t but not the three spatial coordinates x^i.

The $Q(x^i)$ are the standard scalar harmonics on the three-sphere. The $P_{ij}(x^i)$ are given by (suppressing all but the i,j indices)

$$P_{ij} = \frac{1}{(n^2-1)}Q_{|ij} + \tfrac{1}{3}\Omega_{ij}Q \ . \tag{5.3}$$

They are traceless, $P_i{}^i = 0$. The S_{ij} are defined by

$$S_{ij} = S_{i|j} + S_{j|i} \ , \tag{5.4}$$

where S_i are the transverse vector harmonics, $S_i{}^{|i}=0$. The G_{ij} are the transverse traceless tensor harmonics $G_i{}^i = G_{ij}{}^{|j} = 0$. Further details about the harmonics and their normalization can be found in Appendix A.

The lapse, shift, and the scalar field $\Phi(x^i,t)$ can be expanded in terms of harmonics:

$$N = N_0\left[1 + 6^{-1/2}\sum_{n,l,m}g_{nlm}Q^n_{lm}\right] , \tag{5.5}$$

$$N_i = e^\alpha \sum_{n,l,m}[6^{-1/2}k_{nlm}(P_i)^n_{lm} + 2^{1/2}j_{nlm}(S_i)^n_{lm}] , \tag{5.6}$$

$$\Phi = \sigma^{-1}\left[\frac{1}{2^{1/2}\pi}\phi(t) + \sum_{n,l,m}f_{nlm}Q^n_{lm}\right] , \tag{5.7}$$

where $P_i = [1/(n^2-1)]Q_{|i}$. Hereafter, the labels n, l, m, o, and e will be denoted simply by n. One can then expand the action to all orders in terms of the "background" quantities a,ϕ,N_0 but only to second order in the "perturbations" $a_n,b_n,c_n,d_n,f_n,g_n,k_n,j_n$:

$$I = I_0(a,\phi,N_0) + \sum_n I_n \ , \tag{5.8}$$

where I_0 is the action of the unperturbed model (4.2) and I_n is quadratic in the perturbations and is given in Appendix B.

One can define conjugate momenta in the usual manner. They are

$$\pi_\alpha = -N_0^{-1}e^{3\alpha}\dot{\alpha} + \text{quadratic terms} , \tag{5.9}$$

$$\pi_\phi = N_0^{-1}e^{3\alpha}\dot{\phi} + \text{quadratic terms} , \tag{5.10}$$

$$\pi_{a_n} = -N_0^{-1}e^{3\alpha}[\dot{a}_n + \dot{\alpha}(a_n - g_n) + \tfrac{1}{3}e^{-\alpha}k_n] , \tag{5.11}$$

$$\pi_{b_n} = N_0^{-1}e^{3\alpha}\frac{(n^2-4)}{(n^2-1)}(\dot{b}_n + 4\dot{\alpha}b_n - \tfrac{1}{3}e^{-\alpha}k_n) , \tag{5.12}$$

$$\pi_{c_n} = N_0^{-1}e^{3\alpha}(n^2-4)(\dot{c}_n + 4\dot{\alpha}c_n - e^{-\alpha}j_n) , \tag{5.13}$$

$$\pi_{d_n} = N_0^{-1}e^{3\alpha}(\dot{d}_n + 4\dot{\alpha}d_n) , \tag{5.14}$$

$$\pi_{f_n} = N_0^{-1}e^{3\alpha}[\dot{f}_n + \dot{\phi}(3a_n - g_n)] \ . \tag{5.15}$$

The quadratic terms in Eqs. (5.9) and (5.10) are given in Appendix B. The Hamiltonian can then be expressed in terms of these momenta and the other quantities:

$$H = N_0\left[H_{|0} + \sum_n H^n_{|2} + \sum_n g_n H^n_{|1}\right]$$
$$+ \sum_n (k_n{}^S H^n_{-1} + j_n{}^V H^n_{-1}) \ . \tag{5.16}$$

The subscripts 0,1,2 on the $H_|$ and H_- denote the orders of the quantities in the perturbations and S and V denote the scalar and vector parts of the shift part of the Hamiltonian. $H_{|0}$ is the Hamiltonian of the unperturbed model with $N=1$:

$$H_{|0} = \tfrac{1}{2}e^{-3\alpha}(-\pi_\alpha^2 + \pi_\phi^2 + e^{6\alpha}m^2\phi^2 - e^{4\alpha}) \ . \tag{5.17}$$

The second-order Hamiltonian is given by

$$H_{|2} = \sum_n H^n_{|2} = \sum_n ({}^S H^n_{|2} + {}^V H^n_{|2} + {}^T H^n_{|2}) \ ,$$

where

$$\begin{aligned}
{}^S H^n_{|2} = \tfrac{1}{2}e^{-3\alpha}&\left[\left[\tfrac{1}{2}a_n^2 + \frac{10(n^2-4)}{(n^2-1)}b_n^2\right]\pi_\alpha^2 + \left[\tfrac{15}{2}a_n^2 + \frac{6(n^2-4)}{(n^2-1)}b_n^2\right]\pi_\phi^2 \right. \\
&-\pi_{a_n}^2 + \frac{(n^2-1)}{(n^2-4)}\pi_{b_n}^2 + \pi_{f_n}^2 + 2a_n\pi_{a_n}\pi_\alpha + 8b_n\pi_{b_n}\pi_\alpha - 6a_n\pi_{f_n}\pi_\phi \\
&-e^{4\alpha}\left[\tfrac{1}{3}(n^2-\tfrac{3}{2})a_n^2 + \frac{(n^2-7)}{3}\frac{(n^2-4)}{(n^2-1)}b_n^2 + \tfrac{2}{3}(n^2-4)a_nb_n - (n^2-1)f_n^2\right] \\
&\left.+ e^{6\alpha}m^2(f_n^2 + 6a_nf_n\phi) + e^{6\alpha}m^2\phi^2\left[\tfrac{3}{2}a_n^2 - \frac{6(n^2-4)}{(n^2-1)}b_n^2\right]\right] , \tag{5.18}
\end{aligned}$$

$$\begin{aligned}
{}^V H^n_{|2} = \tfrac{1}{2}e^{-3\alpha}&\left[(n^2-4)c_n^2(10\pi_\alpha^2 + 6\pi_\phi^2) + \frac{1}{(n^2-4)}\pi_{c_n}^2 + 8c_n\pi_{c_n}\pi_\alpha + (n^2-4)c_n^2(2e^{4\alpha} - 6e^{6\alpha}m^2\phi^2)\right] , \tag{5.19}
\end{aligned}$$

$${}^T H^n_{|2} = \tfrac{1}{2}e^{-3\alpha}\{d_n^2(10\pi_\alpha^2 + 6\pi_\phi^2) + \pi_{d_n}^2 + 8d_n\pi_{d_n}\pi_\alpha + d_n^2[(n^2+1)e^{4\alpha} - 6e^{6\alpha}m^2\phi^2]\} \ . \tag{5.20}$$

The first-order Hamiltonians are

$$H^n_{|1} = \tfrac{1}{2} e^{-3\alpha} \{ -a_n(\pi_\alpha{}^2 + 3\pi_\phi{}^2) + 2(\pi_\phi \pi_{f_n} - \pi_\alpha \pi_{a_n}) + m^2 e^{6\alpha}(2f_n\phi + 3a_n\phi^2) - \tfrac{2}{3} e^{4\alpha}[(n^2-4)b_n + (n^2 + \tfrac{1}{2})a_n]\} \; . \tag{5.21}$$

The shift parts of the Hamiltonian are

$$^S H^n_{-1} = \tfrac{1}{3} e^{-3\alpha} \left[-\pi_{a_n} + \pi_{b_n} + \left[a_n + \frac{4(n^2-4)}{(n^2-1)} b_n \right] \pi_\alpha + 3f_n \pi_\phi \right] \; , \tag{5.22}$$

$$^V H^n_{-1} = e^{-\alpha}[\pi_{c_n} + 4(n^2-4)c_n \pi_\alpha] \; . \tag{5.23}$$

The classical field equations are given in Appendix B.

Because the Lagrange multipliers N_0, g_n, k_n, j_n are independent, the zero energy Schrödinger equation

$$H\Psi = 0 \tag{5.24}$$

can be decomposed as before into momentum constraints and Wheeler-DeWitt equations. As the momentum constraints are linear in the momenta, there is no ambiguity in the operator ordering. One therefore has

$$^S H^n_{-1} \Psi = -\tfrac{1}{3} e^{-3\alpha} \left[\frac{\partial}{\partial a_n} - \left[a_n + \frac{4(n^2-4)}{(n^2-1)} b_n \right] \frac{\partial}{\partial \alpha} - \frac{\partial}{\partial b_n} - 3f_n \frac{\partial}{\partial \phi} \right] \Psi = 0 \; , \tag{5.25}$$

$$^V H^n_{-1} \Psi = e^{-\alpha} \left[\frac{\partial}{\partial c_n} + 4(n^2-4)c_n \frac{\partial}{\partial \alpha} \right] \Psi = 0 \; . \tag{5.26}$$

The first-order Hamiltonians $H^n_{|1}$ give a series of finite dimensional second-order differential equations, one for each n. In the order of approximation that we are using, the ambiguity in the operator ordering will consist of the possible addition of terms linear in $\partial/\partial\alpha$. The effect of such terms can be compensated for by multiplying the wave function by powers of e^α. This will not affect the relative probabilities of different observations at a given value of α. We shall therefore ignore such ambiguities and terms:

$$\tfrac{1}{2} e^{-3\alpha} \left[a_n \left[\frac{\partial^2}{\partial\alpha^2} + 3\frac{\partial^2}{\partial\phi^2} \right] - 2 \left[\frac{\partial^2}{\partial f_n \partial\phi} - \frac{\partial^2}{\partial a_n \partial\alpha} \right] + m^2 e^{6\alpha}[2\phi f_n + 3a_n\phi^2] - \tfrac{2}{3} e^{4\alpha}[(n^2-4)b_n + (n^2+\tfrac{1}{2})a_n] \right] \Psi = 0 \; . \tag{5.27}$$

Finally, one has an infinite-dimensional second-order differential equation

$$\left[H_{|0} + \sum_n (^S H^n_{|2} + {}^V H^n_{|2} + {}^T H^n_{|2}) \right] \Psi = 0 \; , \tag{5.28}$$

where $H_{|0}$ is the operator in the Wheeler-DeWitt equation of the unperturbed Friedmann minisuperspace model:

$$H_{|0} = \tfrac{1}{2} e^{-3\alpha} \left[\frac{\partial^2}{\partial\alpha^2} - \frac{\partial^2}{\partial\phi^2} + e^{6\alpha} m^2 \phi^2 - e^{4\alpha} \right] \tag{5.29}$$

and

$$\begin{aligned}
^S H^n_{|2} = \tfrac{1}{2} e^{-3\alpha} \Bigg[& -\left[\tfrac{1}{2} a_n{}^2 + \frac{10(n^2-4)}{(n^2-1)} b_n{}^2 \right] \frac{\partial^2}{\partial\alpha^2} - \left[\tfrac{15}{2} a_n{}^2 + \frac{6(n^2-4)}{(n^2-1)} b_n{}^2 \right] \frac{\partial^2}{\partial\phi^2} \\
& + \frac{\partial^2}{\partial a_n{}^2} - \frac{(n^2-1)}{(n^2-4)} \frac{\partial^2}{\partial b_n{}^2} - \frac{\partial^2}{\partial f_n{}^2} - 2a_n \frac{\partial^2}{\partial a_n \partial\alpha} - 8b_n \frac{\partial^2}{\partial b_n \partial\alpha} + 6a_n \frac{\partial^2}{\partial f_n \partial\phi} \\
& - e^{4\alpha} \left[\tfrac{1}{3}(n^2-\tfrac{5}{2})a_n{}^2 + \frac{(n^2-7)}{3} \frac{(n^2-4)}{(n^2-1)} b_n{}^2 + \tfrac{2}{3}(n^2-4)a_n b_n - (n^2-1)f_n{}^2 \right] \\
& + e^{6\alpha} m^2(f_n{}^2 + 6a_n f_n \phi) + e^{6\alpha} m^2 \phi^2 \left[\tfrac{3}{2} a_n{}^2 - \frac{6(n^2-4)}{(n^2-1)} b_n{}^2 \right] \Bigg] \; ,
\end{aligned} \tag{5.30}$$

$$^V H^n_{|2} = \tfrac{1}{2} e^{-3\alpha} \left[-(n^2-4)c_n{}^2 \left[10\frac{\partial^2}{\partial\alpha^2} + 6\frac{\partial^2}{\partial\phi^2} \right] - \frac{1}{(n^2-4)} \frac{\partial^2}{\partial c_n{}^2} - 8c_n \frac{\partial^2}{\partial c_n \partial\alpha} + (n^2-4)c_n{}^2(2e^{4\alpha} - 6e^{6\alpha} m^2 \phi^2) \right] \; , \tag{5.31}$$

$$^T H^n_{|2} = \tfrac{1}{2} e^{-3\alpha} \left[-d_n{}^2 \left[10\frac{\partial^2}{\partial\alpha^2} + 6\frac{\partial^2}{\partial\phi^2} \right] - \frac{\partial^2}{\partial d_n{}^2} - 8d_n \frac{\partial^2}{\partial d_n \partial\alpha} + d_n{}^2[(n^2+1)e^{4\alpha} - 6e^{6\alpha} m^2 \phi^2] \right] \; . \tag{5.32}$$

We shall call Eq. (5.28) the master equation. It is not hyperbolic because, as well as the positive second derivatives $\partial^2/\partial\alpha^2$ in $H_{|0}$, there are the positive second derivatives $\partial^2/\partial a_n{}^2$ in each ${}^S H^n{}_{|2}$. However, one can use the momentum constraint (5.25) to substitute for the partial derivatives with respect to a_n and then solve the resultant differential equation on $a_n = 0$. Similarly, one can use the momentum constraint (5.26) to substitute for the partial derivatives with respect to c_n and then solve on $c_n = 0$. One thus obtains a modified equation which is hyperbolic for small f_n. If one knows the wave function on $a_n = 0 = c_n$, one can use the momentum constraints to calculate the wave function at other values of a_n and c_n.

VI. THE WAVE FUNCTION

Because the perturbation modes are not coupled to each other, the wave function can be expressed as a sum of terms of the form

$$\Psi = \mathrm{Re}\left[\Psi_0(\alpha,\phi)\prod_n \Psi^{(n)}(\alpha,\phi,a_n,b_n,c_n,d_n,f_n)\right]$$

$$= \mathrm{Re}(C\,e^{iS})\,, \qquad (6.1)$$

where S is a rapidly varying function of α and ϕ and C is a slowly varying function of all the variables. If one substitutes (6.1) into the master equation and divides by Ψ, one obtains

$$-\frac{\nabla_2^2 \Psi_0}{2\Psi_0} - \sum_n \frac{\nabla_2^2 \Psi^{(n)}}{2\Psi^{(n)}} - \sum_{n\lesssim m}\frac{(\nabla_2\Psi^{(n)})\cdot(\nabla_2\Psi^{(m)})}{2\Psi^{(n)}\Psi^{(m)}}$$

$$-\frac{(\nabla_2\Psi_0)}{\Psi_0}\cdot\left[\sum_n \frac{\nabla_2\Psi^{(n)}}{\Psi^{(n)}}\right]$$

$$+\sum_n \frac{H^n{}_{|2}\Psi}{\Psi} + e^{-3\alpha}V(\alpha,\phi) = 0\,, \quad (6.2)$$

where ∇_2^2 is the Laplacian in the minisuperspace metric $f_{ab} = e^{3\alpha}\mathrm{diag}(-1,1)$ and the dot product is with respect to this metric.

An individual perturbation mode does not contribute a significant fraction of the sums in the third and fourth terms in Eq. (6.2). Thus these terms can be replaced by

$$-\frac{(\nabla_2\Psi)}{\Psi}\cdot\sum_n \frac{(\nabla_2\Psi^{(n)})}{\Psi^{(n)}} + \frac{1}{2}\left[\sum_n \frac{\nabla_2\Psi^{(n)}}{\Psi^{(n)}}\right]^2$$

$$\approx -i(\nabla_2 S)\cdot\sum_n \frac{(\nabla_2\Psi^{(n)})}{\Psi^{(n)}} + \frac{1}{2}\left[\sum_n \frac{\nabla_2\Psi^{(n)}}{\Psi^{(n)}}\right]^2\,. \quad (6.3)$$

In order that the ansatz (6.1) be valid, the terms in (6.2) that depend on a_n, b_n, c_n, d_n, f_n have to cancel out. This implies

$$\frac{(\nabla_2\Psi)}{\Psi}\cdot(\nabla_2\Psi^{(n)}) + \frac{1}{2}\nabla_2^2\Psi^{(n)} = \frac{H^n{}_{|2}\Psi}{\Psi}\Psi^{(n)}\,, \quad (6.4)$$

$$(-\tfrac{1}{2}\nabla_2^2 + e^{-3\alpha}V + \tfrac{1}{2}J\cdot J)\Psi_0 = 0\,, \quad (6.5)$$

where

$$J = \sum_n \frac{\nabla_2\Psi^{(n)}}{\Psi^{(n)}}\,.$$

In regions in which the phase S is a rapidly varying function of α and ϕ, one can neglect the second term in (6.4) in comparison with the first term. One can also replace the π_α and π_ϕ which appear in $H^n{}_{|2}$ by $\partial S/\partial\alpha$ and $\partial S/\partial\phi$, respectively. The vector $X^a = f^{ab}\partial S/\partial q^b$ obtained by raising the covector $\nabla_2 S$ by the inverse minisuperspace metric f^{ab} can be regarded as $\partial/\partial t$ where t is the time parameter of the classical Friedmann metric that corresponds to Ψ by the WKB approximation. One then obtains a time dependent Schrödinger equation for each mode along a trajectory of the vector field X^a:

$$i\frac{\partial \Psi^{(n)}}{\partial t} = H^n{}_{|2}\Psi^{(n)}\,. \quad (6.6)$$

Equation (6.5) can be interpreted as the Wheeler-DeWitt equation for a two-dimensional minisuperspace model with an extra term $\frac{1}{2}J\cdot J$ arising from the perturbations. In order to make J finite, one will have to make subtractions. Subtracting out the ground-state energies of the $H^n{}_{|2}$ corresponds to a renormalization of the cosmological constant Λ. There is a second subtraction which corresponds to a renormalization of the Planck mass m_P and a third one which corresponds to a curvature-squared counterterm. The effect of such higher-derivative terms in the action has been considered elsewhere.[16]

One can write $\Psi^{(n)}$ as

$$\Psi^{(n)} = {}^S\Psi^{(n)}(\alpha,\phi,a_n,b_n,f_n)\,{}^V\Psi^{(n)}(\alpha,\phi,c_n)\,{}^T\Psi^{(n)}(\alpha,\phi,d_n)\,, \qquad (6.7)$$

where ${}^S\Psi^{(n)}$, ${}^V\Psi^{(n)}$, and ${}^T\Psi^{(n)}$ obey independent Schrödinger equations with ${}^S H^n{}_{|2}$, ${}^V H^n{}_{|2}$, and ${}^T H^n{}_{|2}$, respectively.

VII. THE BOUNDARY CONDITIONS

We want to find the solution of the master equation that corresponds to

$$\Psi[h_{ij},\Phi] = \int d[g_{\mu\nu}]d[\Phi]\exp(-\hat{I})\,, \quad (7.1)$$

where the integral is taken over all compact four-metrics and matter fields which are bounded by the three-surface S. If one takes the scale parameter α to be very negative but keeps the other parameters fixed, the Euclidean action $\hat{I}$ tends to zero like $e^{2\alpha}$. Thus one would expect Ψ to tend to one as α tends to minus infinity.

One can estimate the form of the scalar, vector, and tensor parts ${}^S\Psi^{(n)}$, ${}^V\Psi^{(n)}$, ${}^T\Psi^{(n)}$ of the perturbation $\Psi^{(n)}$ from the path integral (7.1) One takes the four-metric $g_{\mu\nu}$ and the scalar field Φ to be of the background form

$$ds^2 = \sigma^2(-N^2 dt^2 + e^{2\alpha(t)}d\Omega_3{}^2)\,, \quad (7.2)$$

and $\phi(t)$, respectively, plus a small perturbation described by the variables (a_n,b_n,f_n), c_n, and d_n as functions of t. In order for the background four-metric to be compact, it has to be Euclidean when $\alpha = -\infty$, i.e., N has to be purely negative imaginary at $\alpha = -\infty$, which we shall take to be $t = 0$. In regions in which the metric is Lorentzian, N

will be real and positive. In order to allow a smooth transition from Euclidean to Lorentzian, we shall take N to be of the form $-i e^{i\mu}$ where $\mu = 0$ at $t = 0$. In order that the four-metric and the scalar field be regular at $t = 0, a_n, b_n, c_n, d_n, f_n$ have to vanish there.

The tensor perturbations d_n have the Euclidean action

$$^T\hat{I}_n = \tfrac{1}{2} \int dt\, d_n{}^T D d_n + \text{boundary term} , \qquad (7.3)$$

where

$$^T D = \left[-\frac{d}{dt} \left[\frac{e^{3\alpha} d}{iN_0 dt} \right] + iN_0 e^{\alpha}(n^2 - 1) \right] + 4iN_0 e^{3\alpha} \left[+\tfrac{1}{2} e^{-2\alpha} - \tfrac{3}{2} m^2 \phi^2 - \frac{3\dot{\phi}^2}{2(iN_0)^2} - \frac{3\dot{\alpha}^2}{2(iN_0)^2} - \frac{1}{iN_0} \frac{d}{dt} \left[\frac{\dot{\alpha}}{iN_0} \right] \right] . \qquad (7.4)$$

The last term in (7.4) vanishes if the background metric satisfies the background field equations. The action is extremized when d_n satisfies the equation

$$^T D d_n = 0 . \qquad (7.5)$$

For a d_n that satisfies (7.5), the action is just the boundary term

$$^T\hat{I}{}^{\text{cl}}_n = \frac{1}{2iN_0} e^{3\alpha}(d_n \dot{d}_n + 4\dot{\alpha} d_n{}^2) . \qquad (7.6)$$

The path integral over d_n will be

$$\int d[d_n] \exp(-{}^T\hat{I}_n) = (\det{}^T D)^{-1/2} \exp(-{}^T\hat{I}{}^{\text{cl}}_n) . \qquad (7.7)$$

One now has to integrate (7.7) over different background metrics to obtain the wave function $^T\Psi^{(n)}$. One expects the dominant contribution to come from background metrics that are near a solution of the classical background field equations. For such metrics one can employ the adiabatic approximation in which one regards α to be a slowly varying function of t. Then the solution of (7.5) which obeys the boundary condition $d_n = 0$ at $t = 0$ is

$$d_n = A(e^{\nu\tau} - e^{-\nu\tau}) , \qquad (7.8)$$

where $\nu = e^{-\alpha}(n^2 - 1)^{1/2}$ and $\tau = \int iN_0 dt$. This approximation will be valid for background fields which are near a solution of the background field equations and for which

$$\left| \frac{\dot{\alpha}}{N_0} \right| \ll n e^{-\alpha} . \qquad (7.9)$$

For a regular Euclidean metric, $|\dot{\alpha}/N_0| = e^{-\alpha}$ near $t = 0$. If the metric is a Euclidean solution of the background field equations, then $|\dot{\alpha}/N_0| < e^{-\alpha}$. Thus the adiabatic approximation should hold for large values of n into the region in which the solution of the background field equations becomes Lorentzian and the WKB approximation can be used. The wave function $^T\Psi^{(n)}$ will then be

$$^T\Psi^{(n)} = B \exp \left[- \left[\tfrac{1}{2} n\, e^{2\alpha} \coth(\nu\tau) + \frac{2}{iN_0} \dot{\alpha} e^{3\alpha} \right] d_n{}^2 \right] . \qquad (7.10)$$

In the Euclidean region, τ will be real and positive. For large values of n, $\coth(\nu\tau) \approx 1$. In the Lorentzian region where the WKB approximation applies, τ will be complex but it will still have a positive real part and $\coth(\nu\tau)$ will still be approximately 1 for large n. Thus

$$^T\Psi^{(n)} = B \exp \left[-2i \frac{\partial S}{\partial \alpha} d_n{}^2 - \tfrac{1}{2} n\, e^{2\alpha} d_n{}^2 \right] . \qquad (7.11)$$

The normalization constant B can be chosen to be 1. Thus, apart from a phase factor, the gravitational-wave modes enter the WKB region in their ground state.

We now consider the vector part $^V\Psi^{(n)}$ of the wave function. This is pure gauge as the quantities c_n can be given any value by gauge transformations parametrized by the j_n. The freedom to make gauge transformations is reflected quantum mechanically in the constraint

$$e^{-\alpha} \left[\frac{\partial}{\partial c_n} + 4(n^2 - 4)c_n \frac{\partial}{\partial \alpha} \right] \Psi = 0 . \qquad (7.12)$$

One can integrate (7.12) to give

$$\Psi(\alpha, \{c_n\}) = \Psi \left[\alpha - 2 \sum_n (n^2 - 4)c_n{}^2, 0 \right] , \qquad (7.13)$$

where the dependence on the other variables has been suppressed. One can also replace $\partial\Psi/\partial\alpha$ by $i(\partial S/\partial\alpha)\Psi$. One can then solve for $^V\Psi^{(n)}$:

$$^V\Psi^{(n)} = \exp \left[2i(n^2 - 4)c_n{}^2 \frac{\partial S}{\partial \alpha} \right] . \qquad (7.14)$$

The scalar perturbation modes a_n, b_n, and f_n involve a combination of the behavior of the tensor and vector perturbations. The scalar part of the action is given in Appendix B. The action is extremized by solutions of the classical equations

$$N_0 \frac{d}{dt} \left[e^{3\alpha} \frac{\dot{a}_n}{N_0} \right] + \tfrac{1}{3}(n^2 - 4)N_0{}^2 e^{\alpha}(a_n + b_n) + 3 e^{3\alpha}(\dot{\phi} \dot{f}_n - N_0{}^2 m^2 \phi f_n)$$

$$= N_0{}^2 [3 e^{3\alpha} m^2 \phi^2 - \tfrac{1}{3}(n^2 + 2)e^{\alpha}]g_n + e^{3\alpha} \dot{\alpha} \dot{g}_n - \tfrac{1}{3} N_0 \frac{d}{dt} \left[e^{2\alpha} \frac{k_n}{N_0} \right] \qquad (7.15)$$

$$N_0 \frac{d}{dt} \left[e^{3\alpha} \frac{\dot{b}_n}{N_0} \right] - \frac{1}{3}(n^2-1)N_0^2 e^\alpha(a_n+b_n) = \frac{1}{3}(n^2-1)N_0^2 e^\alpha g_n + \frac{1}{3}N_0 \frac{d}{dt} \left[e^{2\alpha} \frac{k_n}{N_0} \right] , \tag{7.16}$$

$$N_0 \frac{d}{dt} \left[e^{3\alpha} \frac{\dot{f}_n}{N_0} \right] + 3e^{3\alpha}\dot{\phi}\dot{a}_n + N_0^2[m^2 e^{3\alpha}+(n^2-1)e^\alpha]f_n = e^{3\alpha}(-2N_0^2 m^2\phi g_n + \dot{\phi}\dot{g}_n - e^{-\alpha}\dot{\phi}k_n) . \tag{7.17}$$

There is a three-parameter family of solutions to (7.15)–(7.17) which obey the boundary condition $a_n=b_n=f_n=0$ at $t=0$. There are however, two constraint equations:

$$\dot{a}_n + \frac{(n^2-4)}{(n^2-1)}\dot{b}_n + 3f_n\dot{\phi} = \dot{\alpha}g_n - \frac{e^{-\alpha}}{(n^2-1)}k_n , \tag{7.18}$$

$$3a_n(-\dot{\alpha}^2+\dot{\phi}^2) + 2(\dot{\phi}\dot{f}_n - \dot{\alpha}\dot{a}_n) + N_0^2 m^2(2f_n\phi+3a_n\phi^2) - \frac{2}{3}N_0^2 e^{-2\alpha}[(n^2-4)b_n+(n^2+\frac{1}{2})a_n]$$
$$= \frac{2}{3}\dot{\alpha}e^{-\alpha}k_n + 2g_n(-\dot{\alpha}^2+\dot{\phi}^2) . \tag{7.19}$$

These correspond to the two gauge degrees of freedom parametrized by k_n and g_n, respectively. The Euclidean action for a solution to Eqs. (7.15)–(7.19) is

$$S\hat{T}_n^{cl} = \frac{1}{2iN_0} e^{3\alpha} \left[-a_n\dot{a}_n + \frac{(n^2-4)}{(n^2-1)}b_n\dot{b}_n + f_n\dot{f}_n + \dot{\alpha}\left(-a_n^2 + \frac{4(n^2-4)}{(n^2-1)}b_n^2\right) + 3\dot{\phi}a_n f_n + g_n(\dot{\alpha}a_n \bar{-}\dot{\phi}f_n) \right.$$
$$\left. - \frac{1}{3}e^{-\alpha}k_n\left[a_n + \frac{(n^2-4)}{(n^2-1)}b_n\right]\right] , \tag{7.20}$$

where the background field equations have been used.

In many ways the simplest gauge to work in is that with $g_n=k_n=0$. However, this gauge does not allow one to find a compact four-metric which is bounded by a three-surface with arbitrary values of a_n, b_n, and f_n and which is a solution of the Eqs. (7.15)–(7.17) and the constraint equations. Instead, we shall use the gauge $a_n=b_n=0$ and shall solve the constraint Eqs. (7.18) and (7.19) to find g_n and k_n:

$$g_n = 3\frac{(n^2-1)\dot{\alpha}\dot{\phi}f_n + \dot{\phi}\dot{f}_n + N_0^2 m^2\phi f_n}{(n^2-4)\dot{\alpha}^2 + 3\dot{\phi}^2} , \tag{7.21}$$

$$k_n = 3(n^2-1)e^\alpha\frac{\dot{\alpha}\dot{\phi}\dot{f}_n + N_0^2 m^2\phi f_n\dot{\alpha} - 3f_n\dot{\phi}(-\dot{\alpha}^2+\dot{\phi}^2)}{(n^2-4)\dot{\alpha}^2+3\dot{\phi}^2} . \tag{7.22}$$

With these substituted, (7.17) becomes a second-order equation for f_n,

$$N_0 \frac{d}{dt} \left[e^{3\alpha} \frac{\dot{f}_n}{N_0} \right] + N_0^2[m^2 e^{3\alpha}+(n^2-1)e^\alpha]f_n = e^{3\alpha}(-2N_0^2 m^2\phi g_n + \dot{\phi}\dot{g}_n - e^{-\alpha}\dot{\phi}k_n) . \tag{7.23}$$

For large n we can again use the adiabatic approximation to estimate the solution of (7.23) when $|\phi| > 1$:

$$f_n = A\sinh(\nu\tau) , \tag{7.24}$$

where $\nu^2 = e^{-2\alpha}(n^2-1)$. Thus for these modes

$$S_{\Psi^{(n)}}(\alpha,\phi,0,0,f_n) \approx \exp\left[-\frac{1}{2}n e^{2\alpha}f_n^2 - \frac{1}{2}i\frac{\partial S}{\partial\phi}g_n f_n\right] . \tag{7.25}$$

This is of the ground-state form apart from a small phase factor. The value of $S_{\Psi^{(n)}}$ at nonzero values of a_n and b_n can be found by integrating the constraint equations (5.25) and (5.27).

The tensor and scalar modes start off in their ground states, apart possibly from the modes at low n. The vector modes are pure gauge and can be neglected. Thus the total energy

$$E = \sum_n \frac{H_{|2}^{(n)}\Psi^{(n)}}{\Psi^{(n)}}$$

of the perturbations will be small when the ground-state energies are subtracted. But $E = i(\nabla_2 S)\cdot J$ where $J = \sum_n \nabla_2\Psi^{(n)}/\Psi^{(n)}$. Thus J is small. This means that the wave function Ψ_0 will obey the Wheeler-DeWitt equation of the unperturbed minisuperspace model and the phase factor S will be approximately $-i\ln\Psi_0$. However the homogeneous scalar field mode ϕ will not start out in its ground state. There are two reasons for this: first, regularity at $t=0$ requires $a_n=b_n=c_n=d_n=f_n=0$, but does not require $\phi = 0$. Second, the classical field equation for ϕ is of the form of a damped harmonic oscillator with a constant frequency m rather than a decreasing frequency $e^{-\alpha}n$. This means that the adiabatic approximation is not valid at small t and that the solution of the classical field equation is ϕ approximately constant. The action of such solutions is small, so large values of $|\phi|$ are not damped as they are for the other variables. Thus the WKB trajectories which start out from large values of $|\phi|$ have high probability. They will correspond to classical solutions which have a long inflationary period and then go over to a matter-dominated expansion. In a realistic model which included other fields of low rest mass, the matter energy in the oscillations of the massive scalar field would decay into light particles with a thermal spectrum. The model would then expand as a radiation-dominated universe.

VIII. GROWTH OF PERTURBATIONS

The tensor modes will obey the Schrödinger equation

$$i\frac{\partial^T\Psi^{(n)}}{\partial t} = {}^TH^n_{\,|2}\,{}^T\Psi^{(n)} \tag{8.1}$$

$$= \tfrac{1}{2}e^{-3\alpha}\left\{ + d_n^{\,2}\left[10\left[\frac{\partial S}{\partial \alpha}\right]^2 + 6\left[\frac{\partial S}{\partial \phi}\right]^2\right]\right.$$

$$-\frac{\partial^2}{\partial d_n^{\,2}} - 8d_n i\frac{\partial S}{\partial \alpha}\frac{\partial}{\partial d_n}$$

$$\left. + d_n^{\,2}[(n^2+1)e^{4\alpha} - 6e^{6\alpha}m^2\phi^2]\right\}. \tag{8.2}$$

One can write

$$^T\Psi^{(n)} = \exp(-2\alpha)\exp\left[-2i\frac{\partial S}{\partial \alpha}d_n^{\,2}\right]{}^T\Psi_0^{(n)}, \tag{8.3}$$

then

$$i\frac{\partial^T\Psi_0^{(n)}}{\partial t} = \tfrac{1}{2}e^{-3\alpha}\left[-\frac{\partial^2}{\partial d_n^{\,2}} + d_n^{\,2}(n^2-1)e^{4\alpha}\right]{}^T\Psi_0^{(n)}. \tag{8.4}$$

The WKB approximation to the background Wheeler-DeWitt equation has been used in deriving (8.4). Then (8.4) has the form of the Schrödinger equation for an oscillator with a time-dependent frequency $\nu = (n^2-1)^{1/2}e^{-\alpha}$. Initially the wave function $^T\Psi_0^{(n)}$ will be in the ground state (apart from a normalization factor) and the frequency ν will be large compared to $\dot\alpha$. In this case one can use the adiabatic approximation to show that $^T\Psi_0^{(n)}$ remains in the ground state

$$^T\Psi_0^{(n)} \approx \exp(-\tfrac{1}{2}n\,e^{2\alpha}d_n^{\,2}). \tag{8.5}$$

The adiabatic approximation will break down when $\nu \approx \dot\alpha$, i.e., the wave length of the gravitational mode becomes equal to the horizon scale in the inflationary period. The wave function $^T\Psi_0^{(n)}$ will then freeze

$$^T\Psi_0^{(n)} \approx \exp(-\tfrac{1}{2}n\,e^{2\alpha_*}d_n^{\,2}), \tag{8.6}$$

where α_* is the value of α at which the mode goes outside the horizon. The wave function $^T\Psi_0^{(n)}$ will remain of the form (8.6) until the mode reenters the horizon in the matter- or radiation-dominated era at the much greater value α_e of α. One can then apply the adiabatic approximation again to (8.4) but $^T\Psi_0^{(n)}$ will no longer be in the ground state; it will be a superposition of a number of highly excited states. This is the phenomenon of the amplification of the ground-state fluctuations in the gravitational-wave modes that was discussed in Refs. 9, 17, and 18.

The behavior of the scalar modes is rather similar but their description is more complicated because of the gauge degrees of freedom. In the previous section we evaluated the wave function $^S\Psi^{(n)}$ on $a_n = b_n = 0$ by the path-integral prescription. The ground-state form (in f_n) that we found will be valid until the adiabatic approximation breaks down, i.e., until the wavelength of the mode exceeds the horizon distance during the inflationary period. In order to discuss the subsequent behavior of the wave function. It is convenient to use the first-order Hamiltonian constraint (5.27) to evaluate $^S\Psi^{(n)}$ on $a_n \neq 0, b_n = f_n = 0$. One finds that

$$^S\Psi^{(n)}(\alpha,\phi,a_n,0,0) = B\exp[iCa_n^{\,2}]\,{}^S\Psi_0^{(n)}(\alpha,\phi,a_n). \tag{8.7}$$

The normalization and phase factors B and C depend on α and ϕ but not a_n:

$$C = \frac{1}{2}\left[\frac{\partial S}{\partial \alpha}\right]^{-1}\left[\left[\frac{\partial S}{\partial \alpha}\right]^2 - \tfrac{1}{3}(n^2-4)e^{4\alpha}\right]. \tag{8.8}$$

At the time the wavelength of the mode equals the horizon distance during the inflationary period, the wave function $^S\Psi_0^{(n)}$ has the form

$$^S\Psi_0^{(n)} = \exp(-\tfrac{1}{2}ny_*^{\,-2}e^{2\alpha_*}a_n^{\,2}), \tag{8.9}$$

where y_* is the value of $y = (\partial S/\partial\alpha)[\partial S/\partial\phi]^{-1}$ when the mode leaves the horizon, $y_* = 3\phi_*$. More generally, in the case of a scalar field with a potential $V(\phi)$, $y = 6V(\partial V/\partial\phi)^{-1}$.

One can obtain a Schrödinger equation for $^S\Psi_0^{(n)}$ by putting $b_n = f_n = 0$ in the scalar Hamiltonian $^SH^n_{\,|2}$ and substituting for $\partial/\partial b_n$ and $\partial/\partial f_n$ from the momentum constraint (5.25) and the first-order Hamiltonian constraint (5.27), respectively. This gives

$$i\frac{\partial^S\Psi_0^{(n)}}{\partial t} = \tfrac{1}{2}e^{-3\alpha}\left\{-y^2\frac{\partial^2}{\partial a_n^{\,2}} + e^{4\alpha}(n^2-4)\right.$$

$$\left. \times\left[\frac{1}{y^2} - \tfrac{1}{3}e^{4\alpha}\left[\frac{\partial S}{\partial \alpha}\right]^{-2}\right]a_n^{\,2}\right\}{}^S\Psi_0^{(n)}, \tag{8.10}$$

where terms of order $1/n^2$ have been neglected. The term $e^{4\alpha}[\partial S/\partial\alpha]^{-2}$ will be small compared to $1/y^2$ except near the time of maximum radius of the background solution. The Schrödinger equation for $^S\Psi_0^{(n)}(a_n)$ is very similar to the equation for $^T\Psi_0^{(n)}(d_n)$, (8.4), except that the kinetic term is multiplied by a factor y^2 and the potential term is divided by a factor y^2. One would therefore expect that for wavelengths within the horizon. ${}^{S}\Psi_0^{(n)}$ would have the ground-state form $\exp(-\frac{1}{2}ny^{-2}e^{2\alpha}a_n{}^2)$ and this is borne out by (8.9). On the other hand, when the wavelength becomes larger than the horizon, the Schrödinger equation (8.10) indicates that ${}^{T}\Psi_0^{(n)}$ will freeze in the form (8.9) until the mode reenters the horizon in the matter-dominated era. Even if the equation of state of the Universe changes to radiation dominated during the period that the wavelength of the mode is greater than the horizon size, it will still be true that ${}^{S}\Psi_0^{(n)}$ is frozen in the form (8.9). The ground-state fluctuations in the scalar modes will therefore be amplified in a similar manner to the tensor modes. At the time of reentry of the horizon the rms fluctuation in the scalar modes, in the gauge in which $b_n = f_n = 0$, will be greater by the factor y_* than the rms fluctuation in the tensor modes of the same wavelength.

IX. COMPARISON WITH OBSERVATION

From a knowledge of ${}^{T}\Psi_0^{(n)}$ and ${}^{S}\Psi_0^{(n)}$ one can calculate the relative probabilities of observing different values of d_n and a_n at a given point on a trajectory of the vector field X^i, i.e., at a given value of α and ϕ in a background metric which is a solution of the classical field equations. In fact, the dependence on ϕ will be unimportant and we shall neglect it. One can then calculate the probabilities of observing different amounts of anisotropy in the microwave background and can compare these predictions with the upper limits set by observation.

The tensor and scalar perturbation modes will be in highly excited states at large values of α. This means that we can treat their development as an ensemble evolving according to the classical equations of motion with initial distributions in d_n and a_n proportional to $|{}^{T}\Psi_0^{(n)}|^2$ and $|{}^{S}\Psi_0^{(n)}|^2$, respectively. The initial distributions in $\dot{d}_n$ and $\dot{a}_n$ will be proportional to $|{}^{T}\Psi_0^{(n)}\pi_{d_n}{}^{T}\Psi_0^{(n)}|$ and $|{}^{S}\Psi_0^{(n)}\pi_{a_n}{}^{S}\Psi_0^{(n)}|$, respectively. In fact, at the time that the modes reenter the horizon, the distributions will be concentrated at $\dot{d}_n = \dot{a}_n = 0$.

The surfaces with $b_n = f_n = 0$ will be surfaces of constant energy density in the classical solution during the inflationary period. By local conservation of energy, they will remain surfaces of constant energy density in the era after the inflationary period when the energy is dominated by the coherent oscillations of the homogeneous background scalar field ϕ. If the scalar particles decay into light particles and heat up the Universe, the surfaces with $b_n = f_n = 0$ will be surfaces of constant temperature. The surface of last scattering of the microwave background will be such a surface with temperature T_s. The microwave radiation can be considered to have propagated freely to us from this surface. Thus the observed temperature will be

$$T_0 = \frac{T_s}{1+z} , \qquad (9.1)$$

where z is the red-shift of the surface of last scattering. Variations in the observed temperature will arise from variations in z in different directions of observation.

These are given by

$$1 + z = l^{\mu}n_{\mu} \qquad (9.2)$$

evaluated at the surface of last scattering where n_{μ} is the unit normal to the surfaces of constant t in the gauge $g_n = k_n = j_n = 0$ and $b_n = f_n = 0$ on the surface of last scattering and l^{μ} is the parallel propagated tangent vector to the null geodesic from the observer normalized by $l^{\mu}n_{\mu} = 1$ at the present time. One can calculate the evolution of $l^{\mu}n_{\mu}$ down the past light cone of the observer:

$$\frac{d}{d\lambda}[l^{\mu}n_{\mu}] = n_{\mu;\nu}l^{\mu}l^{\nu} , \qquad (9.3)$$

where λ is the affine parameter on the null geodesic. The only nonzero components of $n_{\mu;\nu}$ are

$$n_{i;j} = e^{2\alpha}\left[\dot{\alpha}\Omega_{ij} + \sum_n (\dot{a}_n + \dot{\alpha}a_n)\frac{1}{3}\Omega_{ij}Q \right.$$
$$\left. + \sum_n (\dot{b}_n + \dot{\alpha}b_n)P_{ij} + \sum_n (\dot{d}_n + \dot{\alpha}d_n)G_{ij} \right] .$$
$$(9.4)$$

In the gauge that we are using, the dominant anisotropic terms in (9.4) on the scale of the horizon, will be those involving $\dot{\alpha}a_n$ and $\dot{\alpha}d_n$. These will give temperature anisotropies of the form

$$\langle(\Delta T/T)^2\rangle \approx \langle a_n{}^2\rangle \text{ or } \approx \langle d_n{}^2\rangle . \qquad (9.5)$$

The number of modes that contribute to anisotropies on the scale of the horizon is of the order of n^3. From the results of the last section

$$\langle a_n{}^2\rangle = y_*{}^2 n^{-1}e^{-2\alpha_*} , \qquad (9.6)$$

$$\langle d_n{}^2\rangle = n^{-1}e^{-2\alpha_*} . \qquad (9.7)$$

The dominant contribution comes from the scalar modes which give

$$\langle(\Delta T/T)^2\rangle \approx y_*{}^2 n^2 e^{-2\alpha_*} . \qquad (9.8)$$

But $n e^{-\alpha_*} \approx \dot{\alpha}_*$, the value of the Hubble constant at the time that the present horizon size left the horizon during the inflationary period. The observational upper limit of about 10^{-8} on $\langle(\Delta T/T)^2\rangle$ restricts this Hubble constant to be less than about $5\times10^{-5}m_P$ (Ref. 8) which in turn restricts the mass of the scalar field to be less than 10^{14} GeV.

X. CONCLUSION AND SUMMARY

We started from the proposal that the quantum state of the Universe is defined by a path integral over compact four-metrics. This can be regarded as a boundary condition for the Wheeler-DeWitt equation for the wave function of the Universe on the infinite-dimensional manifold, superspace, the space of all three-metrics and matter field configurations on a three-surface S. Previous papers had considered finite-dimensional approximations to superspace and had shown that the boundary condition led to a wave function which could be interpreted as corresponding to a family of classical solutions which were homogeneous and isotropic and which had a period of exponential or inflationary expansion. In the present paper we extended this work to the full superspace without restrictions. We treated the two basic homogeneous and isotropic degrees of freedom exactly and the other degrees of freedom to second order. We justified this approximation by showing that the inhomogeneous or anisotropic modes started out in their ground states.

We derived time-dependent Schrödinger equations for each mode. We showed that they remained in the ground state until their wavelength exceeded the horizon size during the inflationary period. In the subsequent expansion the ground-state fluctuations got frozen until the wavelength reentered the horizon during the radiation- or matter-dominated era. This part of the calculation is similar to earlier work on the development of gravitational waves[9] and density perturbations[5,6] in the inflationary Universe but it has the advantage that the assumptions of a period of exponential expansion and of an initial ground state for the perturbations are justified. The perturbations would be compatible with the upper limits set by observations of the microwave background if the scalar field that drives the inflation has a mass of 10^{14} GeV or less.

In Sec. VIII we calculated the scalar perturbations in a gauge in which the surfaces of constant time are surfaces of constant density. There are thus no density fluctuations in this gauge. However, one can make a transformation to a gauge in which $a_n = b_n = 0$. In this gauge the density fluctuation at the time that the wavelength comes within the horizon is

$$\langle (\Delta \rho / \rho)^2 \rangle \approx y^2 \frac{\dot{\rho}_e^{\,2}}{\dot{\alpha}_e^{\,2} \rho_e^{\,2}} \dot{\alpha}_*^{\,2} \; . \tag{10.1}$$

Because y and $\dot{\alpha}_*$ depend only logarithmically on the wavelength of the perturbations, this gives an almost scale-free spectrum of density fluctuations. These fluctuations can evolve according to the classical field equations to give rise to the formation of galaxies and all the other structure that we observe in the Universe. Thus all the complexities of the present state of the Universe have their origin in the ground-state fluctuations in the inhomogeneous modes and so arise from the Heisenberg uncertainty principle.

APPENDIX A: HARMONICS ON THE THREE-SPHERE

In this appendix we describe the properties of the scalar, vector, and tensor harmonics on the three-sphere S^3. The metric on S^3 is Ω_{ij} and so the line element is

$$dl^2 = \Omega_{ij} dx^i dx^j$$
$$= d\chi^2 + \sin^2\chi (d\theta^2 + \sin^2\theta \, d\phi^2) \; . \tag{A1}$$

A vertical bar will denote covariant differentiation with respect to the metric Ω_{ij}. Indices i,j,k are raised and lowered using Ω_{ij}.

Scalar harmonics

The scalar spherical harmonics $Q_{lm}^n(\chi,\theta,\phi)$ are scalar eigenfunctions of the Laplacian operator on S^3. Thus,

they satisfy the eigenvalue equation

$$Q^{(n)}{}_{|k}{}^{|k} = -(n^2-1)Q^{(n)}, \quad n=1,2,3,\dots . \tag{A2}$$

The most general solution to (A2), for given n, is a sum of solutions

$$Q^{(n)}(\chi,\theta,\phi) = \sum_{l=0}^{n-1} \sum_{m=-l}^{l} A_{lm}^n Q_{lm}^n(\chi,\theta,\phi) \; , \tag{A3}$$

where A_{lm}^n are a set of arbitrary constants. The Q_{lm}^n are given explicitly by

$$Q_{lm}^n(\chi,\theta,\phi) = \Pi_l^n(\chi) Y_{lm}(\theta,\phi) \; , \tag{A4}$$

where $Y_{lm}(\theta,\phi)$ are the usual harmonics on the two-sphere, S^2, and $\Pi_l^n(\chi)$ are the Fock harmonics.[19,20] The spherical harmonics Q_{lm}^n constitute a complete orthogonal set for the expansion of any scalar field on S^3.

Vector harmonics

The transverse vector harmonics $(S_i)_{lm}^n(\chi,\theta,\phi)$ are vector eigenfunctions of the Laplacian operator on S^3 which are transverse. That is, they satisfy the eigenvalue equation

$$S_i^{(n)}{}_{|k}{}^{|k} = -(n^2-2)S_i^{(n)}, \quad n=2,3,4,\dots \tag{A5}$$

and the transverse condition

$$S_i^{(n)|i} = 0 \; . \tag{A6}$$

The most general solution to (A5) and (A6) is a sum of solutions

$$S_i^{(n)}(\chi,\theta,\phi) = \sum_{l=1}^{n-1} \sum_{m=-l}^{l} B_{lm}^n (S_i)_{lm}^n(\chi,\theta,\phi) \; , \tag{A7}$$

where B_{lm}^n are a set of arbitrary constants. Explicit expressions for the $(S_i)_{lm}^n$ are given in Ref. 20 where it is also explained how they are classified as odd (o) or even (e) using a parity transformation. We thus have two linearly independent transverse vector harmonics S_i^o and S_i^e (n,l,m suppressed).

Using the scalar harmonics Q_{lm}^n we may construct a third vector harmonics $(P_i)_{lm}^n$ defined by (n,l,m suppressed)

$$P_i = \frac{1}{(n^2-1)} Q_{|i}, \quad n=2,3,4,\dots . \tag{A8}$$

It may be shown to satisfy

$$P_{i|k}{}^{|k} = -(n^2-3)P_i \quad \text{and} \quad P_i{}^{|i} = -Q \; . \tag{A9}$$

The three vector harmonics S_i^o, S_i^e, and P_i constitute a complete orthogonal set for the expansion of any vector field on S^3.

Tensor harmonics

The transverse traceless tensor harmonics $(G_{ij})_{lm}^n(\chi,\theta,\phi)$ are tensor eigenfunctions of the Laplacian operator on S^3 which are transverse and traceless. That is, they satisfy the eigenvalue equation

$$G_{ij}^{(n)}{}_{|k}{}^{|k} = -(n^2-3)G_{ij}^{(n)}, \quad n=3,4,5,\dots \tag{A10}$$

and the transverse and traceless conditions

$$G_{ij}^{(n)|i}=0, \quad G_i^{(n)i}=0 . \tag{A11}$$

The most general solution to (A11) and (A12) is a sum of solutions

$$G_{ij}^{(n)}(\chi,\theta,\phi)=\sum_{l=2}^{n-1}\sum_{m=-l}^{l} C_{lm}^n (G_{ij})_{lm}^n(\chi,\theta,\phi) , \tag{A12}$$

where C_{lm}^n are a set of arbitrary constants. As in the vector case they may be classified as odd or even. Explicit expressions for $(G_{ij}^o)_{lm}^n$ and $(G_{ij}^e)_{lm}^n$ are given in Ref. 20.

Using the transverse vector harmonics $(S_i^o)_{lm}^n$ and $(S_i^e)_{lm}^n$, we may construct traceless tensor harmonics $(S_{ij}^o)_{lm}^n$ and $(S_{ij}^e)_{lm}^n$ defined, both for odd and even, by (n,l,m suppressed)

$$S_{ij}=S_{i|j}+S_{j|i} \tag{A13}$$

and thus $S_i^i=0$ since S_i is transverse. In addition, the S_{ij} may be shown to satisfy

$$S_{ij}^{|j}=-(n^2-4)S_i , \tag{A14}$$

$$S_{ij}^{|ij}=0 , \tag{A15}$$

$$S_{ij|k}^{|k}=-(n^2-6)S_{ij} . \tag{A16}$$

Using the scalar harmonics Q_{lm}^n, we may construct two tensors $(Q_{ij})_{lm}^n$ and $(P_{ij})_{lm}^n$ defined by (n,l,m suppressed)

$$Q_{ij}=\tfrac{1}{3}\Omega_{ij}Q, \quad n=1,2,3 \tag{A17}$$

and

$$P_{ij}=\frac{1}{(n^2-1)}Q_{|ij}+\tfrac{1}{3}\Omega_{ij}Q, \quad n=2,3,4 . \tag{A18}$$

The P_{ij} are traceless, $P_i^i=0$, and in addition, may be shown to satisfy

$$P_{ij}^{|j}=-\tfrac{2}{3}(n^2-4)P_i , \tag{A19}$$

$$P_{ij|k}^{|k}=-(n^2-7)P_{ij} , \tag{A20}$$

$$P_{ij}^{|ij}=\tfrac{2}{3}(n^2-4)Q . \tag{A21}$$

The six tensor harmonics Q_{ij}, P_{ij}, S_{ij}^o, S_{ij}^e, G_{ij}^o, and G_{ij}^e constitute a complete orthogonal set for the expansion of any symmetric second-rank tensor field on S^3.

Orthogonality and normalization

The normalization of the scalar, vector, and tensor harmonics is fixed by the orthogonality relations. We denote

the integration measure on S^3 by $d\mu$. Thus

$$d\mu=d^3x(\det\Omega_{ij})^{1/2}=\sin^2\chi \sin\theta \, d\chi \, d\theta \, d\phi . \tag{A22}$$

The Q_{lm}^n are normalized so that

$$\int d\mu Q_{lm}^n Q_{l'm'}^{n'}=\delta^{nn'}\delta_{ll'}\delta_{mm'} . \tag{A23}$$

This implies

$$\int d\mu (P_i)_{lm}^n (P^i)_{l'm'}^{n'}=\frac{1}{(n^2-1)}\delta^{nn'}\delta_{ll'}\delta_{mm'} \tag{A24}$$

and

$$\int d\mu (P_{ij})_{lm}^n (P^{ij})_{l'm'}^{n'}=\frac{2(n^2-4)}{3(n^2-1)}\delta^{nn'}\delta_{ll'}\delta_{mm'} . \tag{A25}$$

The $(S_i)_{lm}^n$, both odd and even, are normalized so that

$$\int d\mu (S_i)_{lm}^n (S^i)_{l'm'}^{n'}=\delta^{nn'}\delta_{ll'}\delta_{mm'} . \tag{A26}$$

This implies

$$\int d\mu (S_{ij})_{lm}^n (S^{ij})_{l'm'}^{n'}=2(n^2-4)\delta^{nn'}\delta_{ll'}\delta_{mm'} . \tag{A27}$$

Finally, the $(G_{ij})_{lm}^n$, both odd and even, are normalized so that

$$\int d\mu (G_{ij})_{lm}^n (G^{ij})_{l'm'}^{n'}=\delta^{nn'}\delta_{ll'}\delta_{mm'} . \tag{A28}$$

The information given in this appendix about the spherical harmonics is all that is needed to perform the derivations presented in the main text. Further details may be found in Refs. 19 and 20.

APPENDIX B: ACTION AND FIELD EQUATIONS

The action (5.8) is

$$I=I_0(\alpha,\phi,N_0)+\sum_n I_n , \tag{B1}$$

where I_0 is the action of the unperturbed model (4.2):

$$I_0=-\tfrac{1}{2}\int dt \, N_0 e^{3\alpha}\left[\frac{\dot{\alpha}^2}{N_0^2}-e^{-2\alpha}-\frac{\dot{\phi}^2}{N_0^2}+m^2\phi^2\right] . \tag{B2}$$

I_n is quadratic in the perturbations and may be written

$$I_n=\int dt (L_g^n+L_m^n) , \tag{B3}$$

where

$$L_g^n=\tfrac{1}{2}e^\alpha N_0\left[\tfrac{1}{3}(n^2-\tfrac{5}{2})a_n^2+\frac{(n^2-7)}{3}\frac{(n^2-4)}{(n^2-1)}b_n^2-2(n^2-4)c_n^2-(n^2+1)d_n^2+\tfrac{2}{3}(n^2-4)a_n b_n\right.$$

$$+g_n\left[\tfrac{2}{3}(n^2-4)b_n+\tfrac{2}{3}(n^2+\tfrac{1}{2})a_n\right]+\frac{1}{N_0^2}\left[-\frac{1}{3(n^2-1)}k_n^2+(n^2-4)j_n^2\right]\Bigg]$$

$$+\frac{1}{2}\frac{e^{3\alpha}}{N_0}\left\{-\dot{a}_n^2+\frac{(n^2-4)}{(n^2-1)}\dot{b}_n^2+(n^2-4)\dot{c}_n^2+\dot{d}_n^2\right\}$$

$$+\dot\alpha\left[-2a_n\dot a_n+8\frac{(n^2-4)}{(n^2-1)}b_n\dot b_n+8(n^2-4)c_n\dot c_n+8d_n\dot d_n\right]$$

$$+\dot\alpha^2\left[-\tfrac{3}{2}a_n{}^2+6\frac{(n^2-4)}{(n^2-1)}b_n{}^2+6(n^2-4)c_n{}^2+6d_n{}^2\right]+g_n[2\dot\alpha\,\dot a_n+\dot\alpha^2(3a_n-g_n)]$$

$$+e^{-\alpha}\left[k_n\left[-\tfrac{2}{3}\dot a_n-\tfrac{2}{3}\frac{(n^2-4)}{(n^2-1)}\dot b_n+\tfrac{2}{3}\dot\alpha g_n\right]-2(n^2-4)\dot c_n j_n\right]\Bigg\}\tag{B4}$$

and

$$L_m^n=\tfrac{1}{2}N_0e^{3\alpha}\Bigg[\frac{1}{N_0{}^2}(\dot f_n{}^2+6a_n\dot f_n\dot\phi)-m^2(f_n{}^2+6a_nf_n\phi)-e^{-2\alpha}(n^2-1)f_n{}^2$$

$$+\frac{3}{2}\left[\frac{\dot\phi^2}{N_0{}^2}-m^2\phi^2\right]\left[a_n{}^2-\frac{4(n^2-4)}{(n^2-1)}b_n{}^2-4(n^2-4)c_n{}^2-4d_n{}^2\right]+\frac{\dot\phi^2}{N_0{}^2}g_n{}^2$$

$$-g_n\left[2m^2f_n\phi+3m^2a_n\phi^2+2\frac{\dot f_n\dot\phi}{N_0{}^2}+3\frac{a_n\dot\phi^2}{N_0{}^2}\right]-2\frac{e^{-\alpha}}{N_0{}^2}k_nf_n\dot\phi\Bigg].\tag{B5}$$

The full expressions for π_α and π_ϕ are

$$\pi_\alpha=\frac{e^{3\alpha}}{N_0}\Bigg[-\dot\alpha+\sum_n\left[-a_n\dot a_n+\frac{4(n^2-4)}{(n^2-1)}b_n\dot b_n+4(n^2-4)c_n\dot c_n+4d_n\dot d_n\right]$$

$$+\dot\alpha\sum_n\left[-\tfrac{3}{2}a_n{}^2+\frac{6(n^2-4)}{(n^2-1)}b_n{}^2+6(n^2-4)c_n{}^2+6d_n{}^2\right]+\sum_n g_n[\dot a_n+\dot\alpha(3a_n-g_n)+\tfrac{1}{3}e^{-\alpha}k_n]\Bigg],\tag{B6}$$

$$\pi_\phi=\frac{e^{3\alpha}}{N_0}\Bigg\{\dot\phi+\sum_n\left[3a_n\dot f_n+\dot\phi\left[\tfrac{3}{2}a_n{}^2-\frac{4(n^2-4)}{(n^2-1)}b_n{}^2-4(n^2-4)c_n{}^2-4d_n{}^2\right]\right]$$

$$+\sum_n[\dot\phi g_n{}^2-g_n(\dot f_n+3a_n\dot\phi)-e^{-\alpha}k_nf_n]\Bigg\}.\tag{B7}$$

The classical field equations may be obtained from the action (B1) by varying with respect to each of the fields in turn. Variation with respect to α and ϕ gives two field equations, similar to those obtained in Sec. IV, but modified by terms quadratic in the perturbations:

$$N_0\frac{d}{dt}\left[\frac{1}{N_0}\frac{d\phi}{dt}\right]+3\frac{d\alpha}{dt}\frac{d\phi}{dt}+N_0{}^2m^2\phi=\text{quadratic terms},\tag{B8}$$

$$N_0\frac{d}{dt}\left[\frac{\dot\alpha}{N_0}\right]+3\dot\phi^2-N_0{}^2e^{-2\alpha}-\tfrac{3}{2}(-\dot\alpha^2+\dot\phi^2-N_0{}^2e^{-2\alpha}+N_0{}^2m^2\phi^2)=\text{quadratic terms}.\tag{B9}$$

Variation with respect to the perturbations a_n, b_n, c_n, d_n, and f_n leads to five field equations:

$$N_0\frac{d}{dt}\left[e^{3\alpha}\frac{\dot a_n}{N_0}\right]+\tfrac{1}{3}(n^2-4)N_0{}^2e^\alpha(a_n+b_n)+3e^{3\alpha}(\dot\phi\dot f_n-N_0{}^2m^2\phi f_n)=N_0{}^2[3e^{3\alpha}m^2\phi^2-\tfrac{1}{3}(n^2+2)e^\alpha]g_n$$

$$+e^{3\alpha}\dot\alpha\dot g_n-\tfrac{1}{3}N_0\frac{d}{dt}\left[e^{2\alpha}\frac{k_n}{N_0}\right],\tag{B10}$$

$$N_0\frac{d}{dt}\left[e^{3\alpha}\frac{\dot b_n}{N_0}\right]-\tfrac{1}{3}(n^2-1)N_0{}^2e^\alpha(a_n+b_n)=\tfrac{1}{3}(n^2-1)N_0{}^2e^\alpha g_n+\tfrac{1}{3}N_0\frac{d}{dt}\left[e^{2\alpha}\frac{k_n}{N_0}\right],\tag{B11}$$

$$\frac{d}{dt}\left[e^{3\alpha}\frac{\dot c_n}{N_0}\right]=\frac{d}{dt}\left[e^{2\alpha}\frac{j_n}{N_0}\right],\tag{B12}$$

$$N_0 \frac{d}{dt} \left[e^{3\alpha} \frac{\dot{d}_n}{N_0} \right] + (n^2 - 1) N_0^2 e^{\alpha} d_n = 0 , \tag{B13}$$

$$N_0 \frac{d}{dt} \left[e^{3\alpha} \frac{\dot{f}_n}{N_0} \right] + 3 e^{3\alpha} \dot{\phi} \dot{a}_n + N_0^2 [m^2 e^{3\alpha} + (n^2 - 1)e^{\alpha}] f_n = e^{3\alpha} (-2N_0^2 m^2 \phi g_n + \dot{\phi} \dot{g}_n - e^{-\alpha} \phi k_n) . \tag{B14}$$

In obtaining (B10)—(B14), the field equations (B8) and (B9) have been used and terms cubic in the perturbations have been dropped.

Variation with respect to the Lagrange multipliers k_n, j_n, g_n, and N_0 leads to a set of constraints. Variation with respect to k_n and j_n leads to the momentum constraints:

$$\dot{a}_n + \frac{(n^2 - 4)}{(n^2 - 1)} \dot{b}_n + 3 f_n \dot{\phi} = \dot{\alpha} g_n - \frac{e^{-\alpha}}{(n^2 - 1)} k_n , \tag{B15}$$

$$\dot{c}_n = e^{-\alpha} j_n . \tag{B16}$$

Variation with respect to g_n gives the linear Hamiltonian constraint:

$$3a_n(-\dot{\alpha}^2 + \dot{\phi}^2) + 2(\dot{\phi} \dot{f}_n - \dot{\alpha} \dot{a}_n) + N_0^2 m^2 (2f_n \phi + 3a_n \phi^2) - \tfrac{2}{3} N_0^2 e^{-2\alpha} [(n^2 - 4)b_n + (n^2 + \tfrac{1}{2})a_n]$$
$$= \tfrac{2}{3} \dot{\alpha} e^{-\alpha} k_n + 2g_n (-\dot{\alpha}^2 + \dot{\phi}^2) . \tag{B17}$$

Finally, variation with respect to N_0 yields the Hamiltonian constraint, which we write as

$$\tfrac{1}{2} e^{3\alpha} \left[-\frac{\dot{\alpha}^2}{N_0^2} + \frac{\dot{\phi}^2}{N_0^2} - e^{-2\alpha} + m^2 \phi^2 \right] = \text{quadratic terms} . \tag{B18}$$

[1] A. H. Guth, Phys. Rev. D 23, 347 (1981).

[2] A. D. Linde, Phys. Lett. 108B, 389 (1982).

[3] S. W. Hawking and I. G. Moss, Phys. Lett. 110B, 35 (1982).

[4] A. Albrecht and P. J. Steinhardt, Phys. Rev. Lett. 48, 120 (1982).

[5] S. W. Hawking, Phys. Lett. 115B, 295 (1982).

[6] A. H. Guth and S. Y. Pi, Phys. Rev. Lett. 49, 1110 (1982).

[7] J. M. Bardeen, P. J. Steinhardt, and M. S. Turner, Phys. Rev. D 28, 679 (1983).

[8] S. W. Hawking, Phys. Lett. B 150B, 339 (1985).

[9] V. A. Rubakov, M. V. Sazhin, and A. V. Veryaskin, Phys. Lett. 115B, 189 (1982).

[10] S. W. Hawking, Pontif. Accad. Sci. Varia 48, 563 (1982).

[11] J. B. Hartle and S. W. Hawking, Phys. Rev. D 28, 2960 (1983).

[12] S. W. Hawking, in Relativity, Groups and Topology II, Les Houches 1983, Session XL, edited by B. S. DeWitt and R. Stora (North-Holland, Amsterdam, 1984).

[13] S. W. Hawking, Nucl. Phys. B239, 257 (1984).

[14] S. W. Hawking and Z. C. Wu, Phys. Lett. 151B, 15 (1985).

[15] S. W. Hawking and D. N. Page, DAMTP report, 1984 (unpublished).

[16] S. W. Hawking and J. C. Luttrell, Nucl. Phys. B247, 250 (1984).

[17] L. P. Grishchuk, Zh. Eksp. Teor. Fiz. 67, 825 (1974) [Sov. Phys. JETP 40, 409 (1975)]; Ann. N.Y. Acad. Sci. 302, 439 (1977).

[18] A. A. Starobinsky, Pis'ma Zh. Eksp. Teor. Fiz. 30, 719 (1979) [JETP Lett. 30, 682 (1979)].

[19] E. M. Lifschitz and I. M. Khalatnikov, Adv. Phys. 12, 185 (1963).

[20] U. H. Gerlach and U. K. Sengupta, Phys. Rev. D 18, 1773 (1978).

PHYSICAL REVIEW D   VOLUME 32, NUMBER 10   15 NOVEMBER 1985

Arrow of time in cosmology

S. W. Hawking

University of Cambridge, Department of Applied Mathematics and Theoretical Physics, Silver Street,
Cambridge CB3 9EW, England
(Received 29 April 1985)

The usual proof of the *CPT* theorem does not apply to theories which include the gravitational field. Nevertheless, it is shown that *CPT* invariance still holds in these cases provided that, as has recently been proposed, the quantum state of the Universe is defined by a path integral over metrics that are compact without boundary. The observed asymmetry or arrow of time defined by the direction of time in which entropy increases is shown to be related to the cosmological arrow of time defined by the direction of time in which the Universe is expanding. It arises because in the proposed quantum state the Universe would have been smooth and homogeneous when it was small but irregular and inhomogeneous when it was large. The thermodynamic arrow would reverse during a contracting phase of the Universe or inside black holes. Possible observational tests of this prediction are discussed.

I. INTRODUCTION

Physics is time symmetric. More accurately, it can be shown[1] that any quantum field theory that has (a) Lorentz invariance, (b) positive energy, and (c) local causality, i.e., $\phi(x)$ and $\phi(y)$ commute (or anticommute) if x and y are spacelike separated, is invariant under *CPT* where *C* means interchange particles with antiparticles, *P* means replace left hand by right hand, and *T* means reverse the direction of motion of all particles. In most situations, the effect of any *C* or *P* noninvariance can be neglected, so that the interactions ought to be invariant under *T* alone.

In fact, if one takes the gravitational field into account, the Universe that we live in does not satisfy any of the three conditions listed above. The Universe is not Lorentz invariant because spacetime is not flat, or even asymptotically flat. The energy density is not positive definite because gravitational potential energy is negative. In a certain sense the total energy of the Universe is zero because the positive energy of the matter is exactly compensated by the negative gravitational potential energy. Finally, the concept of local causality ceases to be well defined if the spacetime metric itself is quantized because one cannot tell if x and y are spacelike separated. Nevertheless, I shall show in Sec. III of this paper that the universe is invariant under *CPT* if, as has been recently proposed,[2–4] it is in the quantum state defined by a path integral over compact four-metrics without boundary. This is a nontrivial result because an arbitrary quantum state for the Universe is not, in general, invariant under *CPT*.

The Universe that we live in certainly does not appear time symmetric, as anyone who has watched a movie being shown backward can testify: one sees events that are never witnessed in ordinary life, like pieces of a cup gathering themselves together off the floor and jumping back onto a table. One can distinguish a number of different "arrows of time" that express the time asymmetry of the Universe. (1) The thermodynamic arrow: the direction of

time in which entropy increases. (2) The electrodynamic arrow: the fact that one uses retarded solutions of the field equations rather than advanced ones. (3) The psychological arrow: the fact that we remember events in the past but not in the future. (4) The cosmological arrow: the direction in time in which the universe is expanding.

I shall take the point of view that the first arrow implies the second and third. In the case of the psychological arrow this follows because human beings (or computers, which are easier to talk about) are governed by the thermodynamic arrow, like everything else in the Universe. In the case of electrodynamics, one can express the vector potential $A_\mu(x)$ as a sum of a contribution from sources in the past of x plus a surface integral at past infinity. One can also express $A_\mu(x)$ as a sum of a contribution from sources in the future of x plus a surface integral at future infinity. The boundary conditions that give rise to the thermodynamic arrow imply that there is no incoming radiation in the past. Thus the surface integral in the past is zero and the electromagnetic field can be expressed as an integral over sources in the past. On the other hand, the boundary conditions that give rise to the thermodynamic arrow do not prevent the possibility of outgoing radiation in the future. This means that the surface integral in the future is strongly correlated with the contribution from sources in the future. It therefore cannot be neglected.

The accepted explanation for the thermodynamic arrow of time is that for some reason the Universe started out in a state of high order or low entropy. Such states occupy only a very small fraction of the volume of phase space accessible to the Universe. As the Universe evolves in time it will tend to move around phase space ergodically. At a later time therefore there is a high probability that the Universe will be found in a state of disorder or higher entropy because such states occupy most of phase space. Consider, for example, a system consisting of a number N of gas molecules in a rectangular box which is divided into two by a partition with a small hole in it. Suppose that at some initial time, say 10 o'clock, all the molecules are in the left-hand side of the box. Such configurations occupy only one part in 2^N of the available $6N$-dimensional phase space. As time goes on, the system will move around phase space on a constant-energy surface. At a later time there will be a high probability of finding the system in a more disordered state with molecules in both halves of the box. Thus entropy will increase with time. Of course, if one waits long enough, one will eventually see all the molecules returning to one half of the box. However, for macroscopic values of N, the time taken is likely to be much longer than the age of the Universe.

Suppose, on the other hand, that the Universe satisfied a *final* condition that was in a state of high order. In that case it would be likely to be in a more disordered state at earlier times and entropy would decrease with time. However, as remarked above, the psychological arrow is determined by the thermodynamic arrow. Thus, if the thermodynamic arrow were reversed, the psychological arrow would be reversed as well: we would define time to run in the other direction and we would still say that entropy increased with time. However, the cosmological arrow provides an independent definition of the direction of time with which we can compare the thermodynamic, psychological, and electrodynamic arrows. In the early 1960s Hogarth[5] and Hoyle and Narlikar[6] tried to connect the electrodynamic and cosmological arrows using the Wheeler-Feynman[7] direct-particle-interaction formulation of electrodynamics. At a summer school held[8] at Cornell in 1963 their work was criticized by a Mr. X (generally assumed to be Richard Feynman) on the grounds that they had implicitly assumed the thermodynamic arrow. They also got the "wrong" answer in that they predicted retarded potentials in a steady-state universe but advanced ones in an evolutionary universe without continual creation of matter. It is now generally accepted that we live in an evolutionary universe.

Another proposal to explain the thermodynamic arrow of time has been put forward by Penrose.[9] It is based on the prediction of classical general relativity[10] that there will be spacetime singularities both in the past, at the big bang, and in the future at the big crunch, if the whole universe recollapses, or in black holes if only local regions collapse. Penrose's proposal is that the Weyl tensor should be zero at singularities in the past. This would mean that the Universe would start off in a smooth and uniform state of high order. However, the Weyl tensor would not, in general, be zero at singularities in the future which could be irregular and disordered.

There are several objections which can be raised to Penrose's proposal. First, it is rather *ad hoc*. Why should the Weyl tensor be zero on past singularities but not on future ones? In effect, one is putting in the thermodynamic arrow by hand. Second, it is based on the prediction of singularities in classical general relativity. However, it is generally believed that the gravitational field has to be quantized in order to be consistent with other field theories which are quantized. It is not clear whether singularities occur in quantum gravity or how to

impose Penrose's boundary condition at them, if they do. Finally, Penrose's proposal does not explain why the cosmological and thermodynamic arrows should agree. With Penrose's boundary condition the thermodynamic arrow would agree with the cosmological arrow during the expanding phase of the Universe but it would disagree if the Universe were to start recollapsing.

The *CPT* invariance of the quantum state of the Universe defined by a path integral over compact metrics implies that if there is a certain probability of the Universe expanding, there must be an equal probability of it contracting. In order for the thermodynamic and cosmological arrows to agree in both the expanding and contracting phases, one requires boundary conditions which imply that the Universe is in a smooth state of high order when it is small but that it may be in an inhomogeneous disordered state when it is large. In Sec. IV it will be shown that the results of Ref. 11 imply that this is indeed the case for the quantum state defined by a path integral over compact metrics. This means that during the expansion phase the Universe starts out in a smooth state of high order but that, as it expands, it becomes more inhomogeneous and disordered. Thus the thermodynamic and cosmological arrows agree. However, when the Universe starts to recollapse, it has to get back to a smooth state when it is small. This means that disorder will decrease with time during the contracting phase and the thermodynamic arrow will be reversed. It will thus still agree with the cosmological arrow.

It should be emphasized that this reversal of the thermodynamic arrow of time is not caused by the gravitational fields or quantum effects at the point of maximum expansion of the Universe. Rather it is a result of the boundary condition that the Universe should be in a state of high order when it is small and it would occur in any theory which had this boundary condition as has been pointed out by a number of authors.[12,13] The only way that quantum gravity comes into the question of the arrow of time is that it provides a natural justification for the boundary condition.

One might ask what would happen to an observer (or computer) who survived from the expanding phase to the contracting one. One might think that one was free to enclose the observer or computer in a container that was so well insulated that he would be unaffected by the reversal of the thermodynamic arrow outside. If he were then to open a little window in his spaceship, he would see time going backward outside. The answer to this apparent paradox is that the observer's thermodynamic arrow, and hence his psychological arrow, would reverse at around the time of maximum expansion of the Universe, not because of effects that propagated into the spacecraft through the walls, but because of the boundary condition that the spacecraft be in a state of low entropy at late times when the Universe is small again. The contents of the memory of the observer or computer would increase during the expansion phase as the observer recorded observations but it would decrease during the contracting phase because the psychological arrow would be reversed and the observer would remember events in his future rather than his past.

The prediction that the thermodynamic arrow would reverse if the Universe started to recontract may not have much practical importance because the Universe is not going to recollapse for a long time, if it ever does. However, we are fairly confident that localized regions of the Universe will collapse to form black holes. If one was in such a region, it would seem just like the whole Universe was collapsing around one. One might therefore expect that the region would become smooth and ordered, just like the whole Universe would if it recollapsed. Thus one would predict that the thermodynamic arrow of time should be reversed inside black holes. One would expect this reversal to occur only after one has fallen through the event horizon, so one would not be able to tell anyone outside about it. This and other consequences of the point of view adopted in this paper will be considered further in Sec. V. Section II will be a brief review of the canonical formulation of quantum gravity. In Sec. III it will be shown that the quantum state of the Universe defined by a path integral over compact metrics is invariant under CPT. Despite this invariance it will be shown in Sec. IV that the results of Ref. 11 imply that there is a thermodynamic arrow because the inhomogeneities in the Universe are small when the Universe is small but that they grow as the Universe expands.

II. CANONICAL QUANTUM GRAVITY

In the canonical approach the quantum state of the Universe is represented by a wave function $\Psi(h_{ij},\phi_0)$ which is a function of the three-metric h_{ij} and the matter field configuration ϕ_0 on a three-surface S. The interpretation of the wave function is that $|\Psi(h_{ij},\phi_0)|^2$ is the (unnormalized) probability of finding a three-surface S with three-metric h_{ij} and matter field configuration ϕ_0. The wave function is not an explicit function of time because there is no invariant definition of time in a curved space which is not asymptotically flat. In fact, the position in time of the surface S is determined implicitly by the three-metric h_{ij}. This means that $\Psi(h_{ij},\phi_0)$ obeys the zero-energy Schrödinger equation:

$$H\Psi(h_{ij},\phi_0)=0 . \tag{2.1}$$

This equation can be decomposed into two parts: the momentum constraint and the Wheeler-DeWitt equation. The momentum constraint is

$$\left[\frac{\delta\Psi}{\delta h_{ij}}\right]_{|i}=8\pi T^{0j}\Psi . \tag{2.2}$$

It implies that the wave function is the same on three-metrics h_{ij} and matter field configurations ϕ_0 that are related by a coordinate transformation. The Wheeler-DeWitt equation is

$$\left[-G_{ijkl}\frac{\delta^2}{\delta h_{ij}\delta h_{kl}}+h^{1/2}(-{}^3R+16\pi T_{00})\right]\Psi=0 , \tag{2.3}$$

where

$$G_{ijkl}=\tfrac{1}{2}h^{-1/2}(h_{ik}h_{jl}+h_{il}h_{jk}-h_{ij}h_{kl}) .$$

It can be regarded as a second-order wave equation for Ψ on the infinite-dimensional space called superspace which is the space of all three-metrics h_{ij} and matter field configurations ϕ_0.

Any solution of Eqs. (2.2) and (2.3) represents a possible quantum state of the Universe. However, it seems reasonable to suppose that the Universe is not just in some arbitrary state but that its state is picked out or preferred in some way. As explained in Ref. 4, the most natural choice of quantum state is that for which the wave function is given by a path integral over compact metrics:

$$\Psi(h_{ij},\phi_0)=\int_C d[g_{\mu\nu}]d[\phi]\exp(-\widetilde{I}[g_{\mu\nu},\phi]) , \tag{2.4}$$

where $\widetilde{I}$ is the Euclidean action and the path integral is taken over four-metrics $g_{\mu\nu}$ and matter field configurations ϕ on compact four-manifolds which are bounded by the three-surface S with the induced three-metric h_{ij} and matter field configuration ϕ_0. The contour of integration in the space of all four-metrics has to be deformed from Euclidean (i.e., positive definite) metrics to complex metrics in order to make the path integral converge.[14,15] The proposal that the quantum state is given by (2.4) seems to give predictions that are in agreement with observation.[4,11,16]

III. THE CPT THEOREM

The precise statement of CPT invariance in flat spacetime is that the vacuum expectation values of bosonic quantum field operators $\phi(x)$ satisfy

$$\langle\phi(x_1)\phi(x_2)\cdots\phi(x_n)\rangle$$
$$=[\langle\phi^\dagger(-x_1)\phi^\dagger(-x_2)\cdots\phi^\dagger(-x_n)\rangle]^* . \tag{3.1}$$

In the case of fermion fields there is a factor of $(-1)^{F+J}$ where F is the fermion number and J is the number of undotted spinor indices. In the case of asymptotically flat spacetime one can formulate and prove CPT invariance in a similar way in terms of the vacuum expectation values of field operators at infinity.[17] However, although asymptotic flatness may be a reasonable approximation for local systems, one does not expect it to apply to the whole Universe. One therefore does not have any flat or asymptotically flat region in which one can define the TP operation $x\rightarrow-x$. All that one has is a wave function $\Psi(h_{ij},\phi_0)$ which is not an explicit function of time. However, one can introduce a concept of time by replacing the dependence of Ψ on $h^{1/2}$, the square root of the determinant of the three-metric h_{ij}, by its conjugate momentum, the trace of the second fundamental form of S. One defines the Laplace transform

$$\Phi(\widetilde{h}_{ij},K_E,\phi_0)=\int_0^\infty d[h^{1/2}]\exp\left[-\frac{m_p{}^2K_E}{18\pi}\int K_E h^{1/2}d^3x\right]\Psi(h_{ij},\phi_0) , \tag{3.2}$$

where $\tilde{h}_{ij}$ is the three-metric defined up to a conformal factor and K_E is the trace of the Euclidean second fundamental form. The Laplace transform Φ is holomorphic for $\mathrm{Re}(K_E) > 0$. This means that one can analytically continue Φ in K_E to Lorentzian values $K_L = iK_E$ of the trace of the second fundamental form. Then $|\Phi(\tilde{h}_{ij}, K_L, \phi_0)|^2$ is proportional to the probability of finding a three-surface S with the conformal three-metric $\tilde{h}_{ij}$, the rate of expansion K_L and the matter field configuration ϕ_0.

Consider first the case in which one has only fields like the gravitational field and real scalar fields which are invariant under C and P. The Euclidean action $\tilde{I}$ is real for Euclidean (i.e., positive definite) four-metrics $g_{\mu\nu}$ and real scalar fields ϕ. The contour of integration in the path integral (2.4) has to be deformed from Euclidean to complex metrics in order to make the integral converge. However, there will be an equal contribution from metrics with a complex action $\tilde{I}$ and from metrics with the complex conjugate action $(\tilde{I})^*$. Thus the wave function $\Psi(h_{ij}, \phi_0)$ will be real. This implies that

$$\Phi(\tilde{h}_{ij}, K_E, \phi_0) = \Phi^*(\tilde{h}_{ij}, K_E^*, \phi_0) \tag{3.3}$$

for complex K_E. In particular, this implies

$$\Phi(\tilde{h}_{ij}, K_L, \phi_0) = \Phi^*(\tilde{h}_{ij}, -K_L, \phi_0) \tag{3.4}$$

for real K_L. Equation (3.4) is the statement of T invariance for the quantum state of the Universe. It implies that the probability of finding a contracting three-surface is the same as that of finding an expanding one, i.e., if the wave function represents an expanding phase of the Universe, then it will also represent a contracting one.

Consider now a situation in which one has charged fields, for example, a complex scalar field ϕ. The wave function Ψ will now be a functional of the three-metric h_{ij} and the complex field configuration ϕ_0 on S. In the Euclidean path integral (2.4) for Ψ one has to integrate over independent field configurations ϕ and $\tilde{\phi}$ on the Euclidean background $g_{\mu\nu}$ where $\phi = \phi_0$ and $\tilde{\phi} = \phi_0^*$ on S. The Euclidean action $\tilde{I}[g_{\mu\nu}, \phi, \tilde{\phi}]$ is no longer necessarily real but

$$\tilde{I}[g_{\mu\nu}, \phi, \tilde{\phi}] = \tilde{I}^*[g_{\mu\nu}, \tilde{\phi}, \phi] \tag{3.5}$$

This implies

$$\Phi(\tilde{h}_{ij}, K_L, \phi) = \Phi^*(\tilde{h}_{ij}, -K_L, \phi^*) \tag{3.6}$$

Equation (3.6) is a statement of the invariance of the quantum state of the universe under CT.

Finally one can consider fields, such as chiral fermions, which are not invariant under P. To deal with fermions one should introduce a triad of covectors e_i^a on S and should regard the wave function Ψ as a functional of the e_i^a and the fermion field ψ_0 on S. The path integral representation of the wave function is then

$$\Psi(e_i^a, \psi_0) = \int_C d[e_\mu^a] d[\psi] d[\tilde{\psi}] \exp(-\tilde{I}[e_\mu^a, \psi, \tilde{\psi}]), \tag{3.7}$$

where on S, $\psi = \psi_0$ and $\tilde{\psi} = \bar{\psi}_0$. The oriented triad e_i^a on S defines a directed unit normal e_μ^0 to S. The path integral (3.7) is taken over all compact four-geometries which are bounded by S and for which e_μ^0 points inward.

The Euclidean action will obey

$$\tilde{I}[e_\mu^a, \psi, \tilde{\psi}] = \tilde{I}^*[-e_\mu^a, \psi^C, \tilde{\psi}^C], \tag{3.8}$$

where $\psi^C = C\psi^*$ is the charge conjugate field and C is the charge conjugation matrix. This implies

$$\Psi(e_i^a, \psi) = \Psi^*(-e_i^a, \psi^C). \tag{3.9}$$

One can regard (3.9) as the expression of the CPT invariance of the quantum state of the Universe because changing the sign of the triad e_i^a not only reverses the spatial directions, and so carries out the operation P, but it also reverses the direction of the orientated normal to S, e_μ^0. Alternatively, one can consider the Laplace transform Φ

$$\Phi(\tilde{e}_i^a, K_L, \psi) = \Phi^*(-\tilde{e}_i^a, -K_L, \psi^C), \tag{3.10}$$

where $\tilde{e}_i^a$ is the triad in S defined up to a positive multiplicative factor.

It is clear that this proof of the CPT invariance of the quantum state defined by a path integral over compact metrics would apply equally well if there were higher derivative terms in the gravitational action. In the case of an action containing quadratic terms in the curvature, the wave function Ψ could be taken to be a function of the three-metric h_{ij}, the second fundamental form K^{ij}, and the matter field configuration ϕ_0. For fields that are invariant under C and P, the wave function $\Psi(h_{ij}, K_E^{ij}, \phi_0)$ would be real for real Euclidean values of the second fundamental form K_E^{ij}. This implies that

$$\Psi(h_{ij}, K_L^{ij}, \phi_0) = \Psi^*(h_{ij}, -K_L^{ij}, \phi_0). \tag{3.11}$$

One can regard (3.11) as an expression of the T invariance of the quantum state. The extension to fields that are not invariant under C and P is straightforward. One can also apply similar arguments to the corresponding quantum state in Kaluza-Klein theories.

IV. THE INCREASE OF DISORDER

In Ref. 11 it was argued that the wave function $\Psi(h_{ij}, \phi_0)$ can be approximated by a sum of terms of the form

$$\Psi_0(\alpha, \phi) \prod_n \Psi_n(\alpha, \phi, a_n, b_n, c_n, d_n, f_n). \tag{4.1}$$

The wave function Ψ_0 describes a homogeneous isotropic closed Universe of radius e^α containing a homogeneous massive scalar field ϕ. The quantities $a_n, b_n, \ldots, f_n$ are the coefficients of harmonics of order n which describe perturbations from homogeneity and isotropy.

One can substitute (4.1) into the Wheeler-DeWitt equation and keep terms to all orders in the "background" quantities α and ϕ but only to second order in the "perturbations" $a_n, b_n, \ldots, f_n$. One obtains a second-order wave equation for Ψ_0 on the two-dimensional "minisuperspace" parametrized by the coordinates α and ϕ. The path integral (2.4) for the wave function implies that $\Psi_0 \to 1$ as $\alpha \to -\infty$. One can integrate the wave equation with this boundary condition.[18] One finds that Ψ_0 starts to oscillate rapidly. This allows one to apply the WKB approximation

$$\Psi_0 = \text{Re}(C e^{iS}) . \tag{4.2}$$

The trajectories of ∇S in the (α, ϕ) plane correspond to solutions of the classical field equations for a homogeneous isotropic Universe with a homogeneous massive scalar field. The trajectories corresponding to Ψ_0 start out at large values of $|\phi|$. They have a period of exponential expansion in which $|\phi|$ decreases followed by a period of matter dominated expansion in which ϕ oscillates around zero with decreasing amplitude. They reach a point of maximum expansion and then recontract in a time symmetric manner.

The perturbation wave functions Ψ_n can be further decomposed as follows:

$$\Psi_n = {}^S\Psi_n(\alpha,\phi,a_n,b_n,f_n) \, {}^V\Psi_n(\alpha,\phi,c_n) \, {}^T\Psi_n(\alpha,\phi,d_n) . \tag{4.3}$$

The wave function ${}^T\Psi_n$ describes gravitational wave perturbations parametrized by the coefficients d_n of the transverse traceless harmonics on the three-sphere. The wave function ${}^V\Psi_n$ describes the effect of gauge transformations which correspond to coordinate transformations on the three-sphere parametrized by the coefficients c_n of the vector harmonics. The wave function ${}^S\Psi_n$ parametrized by the coefficients a_n, b_n, and f_n of the scalar harmonics describe two gauge degrees of freedom and one physical degree of freedom of density perturbations. In situations in which the WKB approximation can be applied to the background wave function Ψ_0, the perturbation wave functions obey decoupled Schrödinger equations of the form

$$i\frac{\partial {}^T\Psi_n}{\partial t} = {}^T H \, {}^T\Psi_n , \tag{4.4}$$

where t is the time parameter of the solution of the classical field equations that corresponds to Ψ_0 via the WKB approximation.

One can evaluate the perturbation wave functions directly from the path integral expression (2.4) for the wave function. Consider, for example, the gravitational wave perturbations. One can regard them as quantum fields parametrized by d_n' propagating on a homogeneous isotropic background metric of the form

$$ds^2 = -N(t)^2 dt^2 + e^{2\alpha'(t)} d\Omega_3^2 , \tag{4.5}$$

where $d\Omega_3^2$ is the metric on the unit three-sphere, if the lapse function N is real everywhere, the metric (4.5) has a Lorentzian signature and cannot be compact and nonsingular. However, I shall consider complex background fields $(N(t),\alpha'(t),\phi'(t))$ such that at some value $t=t_0$, N is negative imaginary. The metric then has a Euclidean signature at $t=t_0$ and will be regular and compact if $\alpha' = -\infty$, $d\alpha'/dt = iN e^{-\alpha'}$, and $d_n' = 0$. The argument of N will vary continuously with t. When N becomes real, the metric will become Lorentzian. One can express the perturbation wave functions as path integrals on these backgrounds, e.g.,

$$ {}^T\Psi_n(\alpha,\phi,d_n) = \int d[d_n'] \exp(-\tilde{I}[\alpha',\phi',d_n']) , \tag{4.6}$$

where the path integral is taken over all gravitational wave perturbations d_n' on all regular compact background fields described by $\alpha'(t)$ and $\phi'(t)$.

The path integral over d_n' in a given background field is Gaussian and therefore can be evaluated as

$$(\det\Delta)^{-1/2}\exp(-\tilde{I}^{\text{cl}}[d_n]) , \tag{4.7}$$

where Δ is a differential operator and

$$\frac{e^{3\alpha}}{2iN}\left\{ d_n' \frac{d}{dt} d_n' + 4\frac{d\alpha'}{dt} d_n'^2 \right\} \tag{4.8}$$

is the action of a solution of the classical field equations for a perturbation d_n' on the given background with $d_n' = 0$ at $t=t_0$ and $d_n' = d_n$ at the location $t=t_1$ of the three-surface S.

One expects the dominant contribution to the path integral (4.6) to come from backgrounds which are close to solutions of the classical background equations. These solutions will be Euclidean (N imaginary) at $t=t_0$ and they will become Lorentzian in those regions of the (α,ϕ) plane in which Ψ_0 oscillates and the WKB approximation can be applied. In such a background the classical field equation for d_n' is

$$\left[-\frac{d}{dt}\left\{ \frac{e^{3\alpha'}}{iN}\frac{d}{dt} \right\} + iN e^{\alpha'}(n^2-1) \right] d_n' = 0 . \tag{4.9}$$

In the region of the (α',ϕ') plane in which the WKB approximation can be applied and N is real, one can regard Eq. (4.9) as a harmonic oscillator equation for the variable $x = \exp(3/2\alpha')d_n'$ with the time-dependent frequency $\nu = \exp(-\alpha')(n^2-1)^{1/2}$. If α' were independent of t, the solution of (4.9) that obeys the above boundary conditions is

$$d_n' = d_n \frac{\sin\nu\tau}{\sin\nu\tau_1} , \tag{4.10}$$

where $\tau = \int_{t_0}^t N \, dt$.

Of course α' will vary with t but (4.10) will still be a good approximation provided that the adiabatic approximation holds, i.e., $|\dot\alpha'/N|$, the rate of change of α', is small compared to the frequency ν. In the Euclidean region near $t=t_0$, this will be true because $|\dot\alpha'/N| < e^{-\dot\alpha'}$. In the Lorentzian region it will be true for perturbation modes whose wavelength ν^{-1} is small compared to the horizon distance $N/\dot\alpha'$. For such modes

$$\frac{d}{dt}d_n' = N\nu d_n' \cot\nu\tau . \tag{4.11}$$

For t_1 in the region in which the WKB approximation can be applied and for $n \gg 1$, the imaginary part of $\nu\tau_1$, which arises from the Euclidean region near $t=t_0$, will be less than $-i$. This means that the real part of the Euclidean action (4.8) will be $\frac{1}{2}\nu e^{3\alpha}d_n^2 = \frac{1}{2}\nu x^2$. The imaginary part of the Euclidean action will be small. It will give rise to a phase factor in ${}^T\Psi_n$ which can be removed by a canonical transformation of variables. Thus the perturbation wave function will have the ground-state form

$$^{T}\Psi_n(d_n) = B \exp[-\tfrac{1}{2}\nu\exp(3\alpha)d_n{}^2]$$

$$= B\,e^{-\nu x^2/2}\,. \tag{4.12}$$

The vector perturbation wave function $^{V}\Psi_n(c_n)$ describes a gauge degree of freedom and does not have any physical significance. The scalar perturbation, which is a function $^{S}\Psi_n(a_n,b_n,f_n)$ describes two gauge degrees of freedom and one physical degree of freedom. A similar analysis and use of the adiabatic approximation shows that this physical degree of freedom is in its ground state when the wavelength of the perturbation is less than the horizon size during the period of exponential expansion. Thus at early times in the exponential expansion, i.e., when the Universe is small, the physical perturbation modes of the Universe have their minimum excitation. The Universe is in a state that is as ordered and homogeneous as it can be consistent with the uncertainty principle. This ordered state is not only an initial state for the expansion phase of the Universe but it is also a final state for the contracting phase because the WKB trajectories for Ψ_0 return to the same region of the (α,ϕ) plane and the perturbation wave functions depend only on the position in this plane.

On the other hand, the perturbation modes are not in their ground state when the Universe is large because in this case the adiabatic approximation breaks down when the wavelength of the perturbation becomes greater than the horizon size during the period of exponential expansion. Detailed calculations[11] show that when the scalar perturbation modes reenter the horizon during the matter-dominated era, they are in a highly excited state and give rise to a scale-free spectrum of density fluctuations $\delta\rho/\rho$. These density inhomogeneities provide the initial conditions necessary for the formation of galaxies and other structures in the Universe. The perturbation wave functions are still in a very special state because their phase factors have to be such that when they are evolved according to the Schrödinger equation, they will return to their ground-state form when the Universe recontracts. However, this special nature of the perturbation wave functions would not be noticed by an observer who makes the usual coarse-grained measurements. All he would notice was that during the expansion the Universe had evolved from a homogeneous, ordered state to an inhomogeneous, disordered state. Thus he would say that the thermodynamic arrow pointed in the direction of time in which the Universe was expanding. On the other hand, an observer in the contracting phase would feel that the Universe was evolving from a state of disorder to one of order. He would therefore ascribe the opposite direction to the thermodynamic arrow and would also find that it agreed with the cosmological arrow.

The connection between the thermodynamic and cosmological arrows should hold in models that are more general than the one considered in Ref. 11 because it depends only on the fact that the adiabatic approximation should hold for small perturbations on "small" three-geometries but not for perturbations on "large" three-geometries. Thus one might expect that it would also hold in models that allowed for the formation of black holes as a result of the gravitational collapse of density

fluctuations produced during the expansion. This would mean that the thermodynamic arrow would reverse inside a black hole. This is currently under investigation.

V. CONSEQUENCES

Are there any observable consequences of the prediction that the thermodynamic arrow should reverse in a recontracting phase of the Universe or inside a black hole? Of course, one could wait until the Universe recollapsed or one could jump into a black hole. However, the probability distribution of the density parameter $\Omega=\rho/\rho_{crit}$ seems to be concentrated at $\Omega=1$ (Ref. 16). Thus one would have to wait a very long time for the collapse of the Universe. On the other hand, if one jumped into a black hole, one would not be able to tell anyone outside. Furthermore, if the thermodynamic arrow did reverse, one would not remember it because it would now be in one's future rather than in the past.

In principle it is possible to determine from the present positions and velocities of clusters of galaxies that they developed from an initial configuration with very low peculiar velocities. In a similar way it should therefore be possible to calculate whether they will evolve to a state with low peculiar velocities at some time in the future. The difficulty is that on the basis of the inflationary model, one would expect the value of Ω for the presently observed Universe to be equal to one to one part in 10^4. Thus one would expect the Universe to expand by a further factor of at least 10^4 before it began to recontract. In this extra expansion other clusters of galaxies which we have not yet observed would appear over the horizon and their gravitational fields could have a significant effect on the behavior of clusters near us. Thus it would seem very difficult to make an experimental test of the prediction that the thermodynamic arrow would reverse if the Universe began to recontract.

A better bet would seem to be to study the inflow of matter into a black hole. At least in principle this is a situation that we ought to be able to observe with some accuracy. However, on the basis of classical general relativity, one might expect the boundary of the region of high spacetime curvature not to be spacelike, as it is in the Schwarzschild solution, but to be null, like the Cauchy horizon in the Reissner-Nordström or Kerr solutions. If this were the case, the behavior of the matter and metric on the brink of the quantum era would depend on the entire future history of infall into the black hole. Merely to observe the infall for a limited period of time would be insufficient to determine whether or not the thermodynamic arrow of time reversed near the region of high curvature. Clearly more work has to be done on the classical and quantum aspects of gravitational collapse.

One might think that the *CPT* theorem implied that all the baryons in the Universe would have to decay into leptons before the Universe began to recollapse and that the leptons would be reassembled into antibaryons in the collapsing phase. If this were the case, one could disprove the proposed "no boundary" condition for the Universe if one could show that the observed value of Ω was such that the Universe should begin to recollapse before all the baryons had decayed. However, what the *CPT* theorem implies is just that the probability of finding an expanding three-surface with a matter configuration of baryons is the same as that of finding a contracting three-surface with a matter configuration of antibaryons. This requirement is no restriction at all because the two three-surfaces can merely be the same three-surface viewed with different orientation of time: reversing the orientation of time and space interchanges the labels, baryons, and antibaryons. Thus the *CPT* invariance of the quantum state of the Universe does not imply any limit on the lifetime of the proton. In any case, we certainly do not observe baryons changing into antibaryons as they fall into a black hole.

To sum up, the proposal that spacetime is compact without boundary implies that the quantum state of the Universe is invariant under *CPT*. Despite this, one would observe an increase in (coarse-grained) entropy during an expansion phase of the Universe. However, it seems difficult to test the prediction that entropy should decrease during a contracting phase of the Universe or inside a black hole.

Note added in proof. Since this paper was submitted for publication a paper by Don Page has appeared [following paper, Phys. Rev. D 32, 2496 (1985)]. In it he questions my conclusion that the thermodynamic arrow of time would reverse in a contracting phase of the universe or in a black hole. My conclusion was based on the fact that the wave function Ψ went exactly into 1 as one goes to $\alpha = -\infty$ on a null geodesic in the α, ϕ plane. This would imply that Ψ was not oscillating at large negative α and therefore that all the classical Lorentzian contracting solutions would have to bounce at a small radius. At the bounce one could apply an analysis similar to that in Ref. 11 to show that all the inhomogeneous modes were in their ground state. This would mean that the inhomogeneity would decrease in the collapsing phase and therefore that the thermodynamic arrow of time would be reversed.

Page has pointed out however that even at large negative α, there might be a small oscillating component in the wave function. This would arise from complex stationary points in the path integral over compact metrics that were near to the Lorentzian metric which started with an inflationary expansion, reached a maximum radius and then recollapsed to zero radius without boundary. Although the amplitude of this oscillating component would be small, its frequency would be very high. It would therefore correspond to an appreciable probability flux of classical solutions in the WKB approximation.[16] One would not expect the inhomogeneous perturbations about such solutions to be in their ground state when the solution recollapsed because the adiabatic approximation used in Ref. 11 would break down. There is thus no reason for the thermodynamic arrow of time to reverse in these solutions. Similarly one would not expect it to reverse inside black holes.

I think that Page may well be right in his suggestion. In that case the two main results of this paper that are correct are, first, that the wave function is invariant under *CPT*, though this does not imply that the individual classical solutions that correspond to the wave function via the WKB approximation are invariant under *CPT*, second, that the classical solutions, which start out with an inflationary period, will have a well-defined thermodynamic arrow of time.

---

[1]R. F. Streater and A. S. Wightman, *PCT, Spin, Statistics and All That* (Benjamin, New York, 1964).

[2]S. W. Hawking, in *Astrophysical Cosmology: Proceedings of the Study Week on Cosmology and Fundamental Physics*, edited by H. A. Brück, G. V. Coyne, and M. S. Longair (Pontificiae Academiae Scientiarum Scripta Varia, Vatican City, 1982), pp. 563—574.

[3]J. B. Hartle and S. W. Hawking, Phys. Rev. D 28, 2960 (1983).

[4]S. W. Hawking, Nucl. Phys. B239, 257 (1984).

[5]J. E. Hogarth, Proc. R. Soc. London A267, 365 (1962).

[6]F. Hoyle and J. V. Narlikar, Proc. R. Soc. London A277, 1 (1964).

[7]J. A. Wheeler and R. P. Feynman, Rev. Mod. Phys. 17, 157 (1945); 21, 425 (1949).

[8]*The Nature of Time*, edited by T. Gold and D. L. Schumacher (Cornell University Press, Ithaca, 1967).

[9]R. Penrose, in *General Relativity: An Einstein Centenary Survey,* edited by S. W. Hawking and W. Israel (Cambridge University Press, England, 1979).

[10]S. W. Hawking and G. F. R. Ellis, *The Large Scale Structure of Spacetime* (Cambridge University Press, England, 1973).

[11]J. J. Halliwell and S. W. Hawking, Phys. Rev. D 31, 1777 (1985).

[12]P. C. W. Davies, *The Physics of Time Asymmetry* (Surrey University Press/California University Press, Berkeley, 1974), Sec. 7.4.

[13]T. Gold, in *La Structure et l'Evolution de l'Universe*, 11th International Solvay Congress (Edition Stoops, Brussels, 1958); Am. J. Phys. 30, 403 (1962); in *Recent Developments in General Relativity* (Pergamon-MacMillan, New York, 1962); H. Bondi, Observatory 82, 133 (1962); D. L. Schumacher, Proc. Cambridge Philos. Soc. 60, 575 (1964); M. Gell-Mann, comments in Proceedings of the Temple University Panel on Elementary Particles and Relativistic Astrophysics (unpublished); Y. Ne'eman, Int. J. Theor. Phys. 3, 1 (1970); P. T. Landsberg, Stud. Gen. 23, 1108 (1970); W. J. Cocke, Phys. Rev. 160, 1165 (1967); H. Schmidt, J. Math. Phys. 7, 495 (1966).

[14]G. W. Gibbons, S. W. Hawking, and M. J. Perry, Nucl. Phys. B138, 141 (1978).

[15]S. W. Hawking, in *Relativity, Groups and Topology*, Les Houches, 1983, edited by B. S. DeWitt and R. Stora (North-Holland, Amsterdam, 1984).

[16]S. W. Hawking and D. N. Page, report 1985 (unpublished).

[17]S. W. Hawking, Commun. Math. Phys. 87, 395 (1982).

[18]S. W. Hawking and Z. C. Wu, Phys. Lett. 151B, 15 (1985).

THE NO-BOUNDARY PROPOSAL AND THE ARROW OF TIME

S. W. HAWKING

Department of Applied Mathematics and Theoretical Physics
University of Cambridge, U.K.

When I began research nearly 30 years ago, my supervisor, Dennis Sciama, set me to work on the arrow of time in cosmology. I remember going to the university library in Cambridge to look for a book called *The Direction of Time* by the German philosopher, Reichenbach [Reichenbach, 1956]. However, I found the book had been taken out by the author, J. B. Priestly, who was writing a play about time, called *Time and the Conways*. Thinking that this book would answer all my questions, I filled in a form to force Priestly to return the book to the library, so I could consult it. However, when I eventually got hold of the book I was very disappointed. It was rather obscure, and the logic seemed to be circular. It laid great stress on causation, in distinguishing the forward direction of time from the backward direction. But in physics, we believe there are laws that determine the evolution of the universe uniquely. Suppose state A evolved into state B. Then one could say that A caused B. But one could equally well look at it in the other direction of time, and say that B caused A. So causality does not define a direction of time.

My supervisor suggested I look at a paper by a Canadian, called Hogarth [Hogarth, 1962]. This applied to cosmology, a direct action formulation of electrodynamics. It claimed to derive a connection between the expansion of the universe and the electromagnetic arrow of time. That is, whether one got retarded or advanced solutions of Maxwell's equations. The paper said that one would obtain retarded solutions in a steady state universe, but advanced solutions in a big bang universe. This was seized on by Hoyle and Narlikar [Hoyle and Narlikar, 1964] as further evidence, if any were needed, that the steady state theory was correct. However, now that no one except Hoyle believes that the universe is in a steady state, one must conclude that the basic premise of the paper was incorrect.

Shortly after this, there was a meeting on the direction of time at Cornell in 1964 [Gold, 1967]. Among the participants there was a Mr. X, who felt the proceedings were so worthless that he didn't want his name associated with them. It was an open secret that Mr. X was Feynman.

Mr. X said that the electromagnetic arrow of time didn't come from an action at a distance formulation of electrodynamics, but from ordinary statistical mechanics. Guided by his comments, I came to the following understanding of the arrow of time. The important point is that the trajectories of a system should have the boundary condition that they are in a small region of phase space at a certain time. In general, the evolution equations of physics will then imply that at other times the trajectories will be spread out over a much larger region of phase space. Suppose the boundary condition of being in a small region is an initial condition (see Figure 1). Then this will mean that the system will begin in an ordered state,

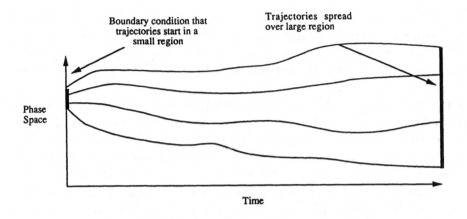

Fig. 1. Evolution of a system with an initial boundary condition.

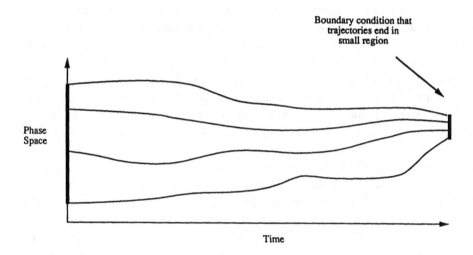

Fig. 2. Evolution of a system with a final boundary condition.

and will evolve to a more disordered state. Entropy will increase with time and the second law of thermodynamics will be satisfied.

On the other hand, suppose the boundary condition of being in a small region of phase space was a final condition instead of an initial condition (see Figure 2). Then at early times the trajectories would be spread out over a large region, and they would narrow down to a small region as time increased. Thus disorder and entropy would decrease with time rather than increase. However, any intelligent beings who observed this behavior would also be living in a universe in which entropy decreased with time. We don't know exactly how the human brain works in detail but we can describe the operation of a computer. One can consider all possible trajectories of a computer interacting with its surroundings. If one imposes a final boundary condition on these trajectories, one can show that the correlation between the computer memory and the surroundings is greater at early times than at late times. In other words, the computer remembers the future but not the past. Another way of seeing this is to note that when a computer records something in memory, the total entropy increase. Thus computers remember things in the direction of time in which entropy increases. In a universe in which entropy is decreasing in time, computer memories will work backward. They will remember the future and forget the past.

Although we don't really understand the workings of the brain, it seems reasonable to assume that we remember in the same direction of time that computers do. If it were the opposite direction, one could make a fortune with a computer that remembered who won tomorrow's horse races. This means that the psychological arrow of time, our subjective sense of time, is the same as the thermodynamic arrow of time, the direction in which entropy increases. Thus, in a universe in which entropy was decreasing with time, any intelligent beings would also have a subjective sense of time that was backward. So the second law of thermodynamics is really a tautology. Entropy increases with time because we define the direction of time to be that in which entropy increases.

There are, however, two non-trivial questions one can ask about the arrow of time. The first is, why should there be a boundary condition at one end of time but not the other? It might seem more natural to have a boundary condition at both ends of time, or at neither. As I will discuss, the former possibility would mean that the arrow of time would reverse, while in the latter case there would be no well defined arrow of time. The second question is, given that there is a boundary condition at one end of time, and hence a well defined arrow of time, why should this arrow point in the direction of time in which the universe is expanding? Is there a deep connection or is it just an accident?

I realized that the problem of the arrow of time should be formulated in the manner I have described. But at that time in 1964, I could think of no good reason why there should be a boundary condition at one end of time. I also needed something more definite and less airy-fairy than the arrow of time for my PhD. I therefore switched to singularities and black holes. They were a lot easier. But I retained an interest in the problem of the direction of time. This surfaced again in 1983, when Jim Hartle and I formulated the no-boundary proposal for the universe [Hartle and Hawking, 1983]. This was the suggestion that the quantum state of the universe was determined by a path integral over positive definite metrics on closed spacetime manifolds. In other words, the boundary condition of the universe was that it had no boundary.

The no-boundary condition determined the quantum state of the universe, and thus what happened in it. It should therefore determine whether there was an arrow of time, and which way it pointed. In the paper that Hartle and I wrote, we applied the no-boundary condition to models with a cosmological constant and a conformally invariant scalar field. Neither of these gave a universe like we live in. However, a minisuperspace model with a minimally coupled scalar field gave an inflationary period that could be arbitrarily long [Hawking, 1984]. This would be followed by radiation and matter dominated phases, like in the chaotic inflationary model. Thus it seemed that the no-boundary condition would account for the observed expansion of the universe. But would it explain the observed arrow of time? In other words, would departures from a homogeneous and isotropic expansion be small when the universe is small, and grow larger as the universe got bigger? Or would the no-boundary condition predict the opposite behavior? Would the departures be small when the universe was large and large when the universe was small? In this latter case, disorder would decrease as the universe expanded. This would mean that the thermodynamic arrow pointed in the opposite way to the cosmological arrow. In other words, people living in such a universe would say that the universe was contracting, rather than expanding.

To answer the question, of what the no-boundary proposal predicted for the arrow of time, one needed to understand how perturbations of a Friedmann model would behave. Jonathan Halliwell and I studied this problem. We expanded perturbations of a minisuperspace model in spherical harmonics, and expanded the Hamiltonian to second order [Halliwell and Hawking, 1984]. This gave us a Wheeler–Dewitt equation,

$$[m_p^{-2} G_{ijkl} \frac{\partial^2}{\partial h_{ij} \partial h_{kl}} - m_p^2 h^{\frac{1}{2}}\,^{(3)}R] \Psi(h_{ij}) = 0 \,,$$

$$h_{ij} = a^2 (\Omega_{ij} + \epsilon_{ij}) \,,$$

for the wave function of the universe. We solved this as a background minisuperspace wave function times wave functions for the perturbation modes. These perturbation mode wave functions obeyed Schroedinger equations, which we could solve approximately. To obtain the boundary conditions for these Schroedinger equations, we used a semiclassical approximation to the no-boundary condition.

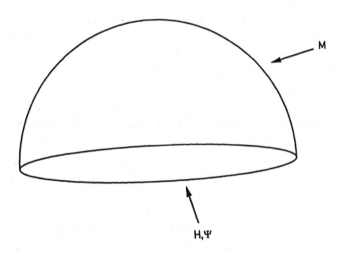

Fig. 3. The no-boundary condition.

Consider a three geometry and scalar field that are a small perturbation of a three sphere and a constant field (see Figure 3). The wave function at this point in superspace will be given by a path integral over all Euclidean four geometries and scalar fields that have only that boundary. One would expect the dominant contribution to this path integral to come from a saddle point. That is, a complex solution of the field equations which has the given geometry and field on one boundary, and which has no other boundary. The wave function for the perturbation mode will then be

$$\Psi(h) = \int d[g]e^{-I}.$$

In this way, Halliwell and I calculated the spectrum of perturbations predicted by the no-boundary condition. The exact shape of this spectrum doesn't matter for the arrow of time. What is important is that, when the radius of the universe is small and the saddle point is a complex solution that expands monotonically, the amplitudes of the perturbations are small. This means that the trajectories corresponding to different probable histories of the universe, are in a small region of phase space when the universe is small. As the universe gets larger, the amplitudes of some of these perturbations will go up. Because the evolution of the universe is governed by a Hamiltonian, the volume of phase space remains unchanged. Thus while the perturbations are linear, the region of phase space that the trajectories are in will change shape only by some matrix of determinant one. In other words, an initially spherical region will evolve to an ellipsoidal region of the same volume. Eventually however, some of the perturbations can grow so large that they become nonlinear. The volume of phase space is still left unchanged by the evolution, but in general the initially spherical region will be deformed into long thin filaments. These can spread out and occupy a large region of phase space. Thus one gets an arrow of time. The universe is nearly homogeneous and isotropic when it is small. But it is more irregular when it is large. In other words, disorder increases as the universe expands. So the thermodynamic and cosmological arrows of time agree, and people living in the universe will say it is expanding rather than contracting.

In 1985 I wrote a paper in which I pointed out that these results about perturbations would explain both why there was a thermodynamic arrow, and why it should agree with the cosmological arrow [Hawking, 1985]. But I made what I now realize was a great mistake. I thought that the no-boundary condition would imply that the perturbations would be small whenever the radius of the universe was small. That is, the perturbations would be small not only in the early stages of the expansion, but also in the late stages of a universe that collapsed again. This would mean that the trajectories of the system would be that subset that lies in a small region of phase space, at both the beginning and the end of time. But they would spread out over a much larger region at times in between. This would mean that disorder would increase during the expansion, but decrease again during the contraction (see Figure 4). So the thermodynamic arrow would point forward in the expansion phase, and backward in the contracting phase. In other words, the thermodynamic and cosmological arrows would agree in both expanding and contracting phases. Near the time of maximum expansion, the entropy of the universe would be a maximum. This would mean that an intelligent being who continued from the expanding to the contracting phase would not observe the arrow of time pointing backward. Instead, his subjective sense of time would be in the opposite direction in the contracting phase. So he would not remember that he had come from the expanding phase because that would be in his subjective future.

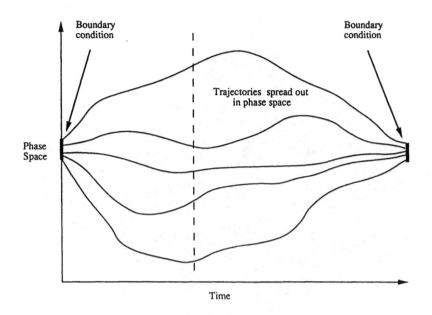

Fig. 4. Evolution of a system with initial and final boundary conditions.

If the thermodynamic arrow of time were to reverse in a contracting phase of the universe, one might also expect it to reverse in gravitational collapse to form a black hole. This would raise the possibility of an experimental test of the no-boundary condition. If the reversal took place only inside the horizon it would not be much use because someone that observed it could not tell the rest of us. But one might hope that there would be slight effects that could be detected outside the horizon.

The idea that the arrow of time would reverse in the contracting phase had a satisfying ring to it. But shortly after having my papers accepted by the Physical Review, discussions with Raymond Laflamme and Don Page convinced me that the prediction of reversal was wrong. I added a note to the proofs saying that entropy would continue to increase during the contraction, but I fell ill with pneumonia before I could write a paper to explain it properly. So I want to take this opportunity to show how I went wrong, and what the correct result is.

One reason I made my mistake was that I was misled by computer solutions of the Wheeler–Dewitt equation for a minisuperspace model of the universe [Hawking and Wu, 1985]. In these solutions, the wave function didn't oscillate in a so-called "forbidden region" at very small radius. I now realize that these computer solutions had the wrong boundary conditions (see Figure 5). But at the time, I interpreted them as indicating that the Lorentzian four geometries that corresponded to the WKB approximation didn't collapse to zero radius. Instead, I thought they would bounce and expand again (see Figure 6). My feelings were strengthened when I found that there was a class of classical solutions that oscillated. The computer

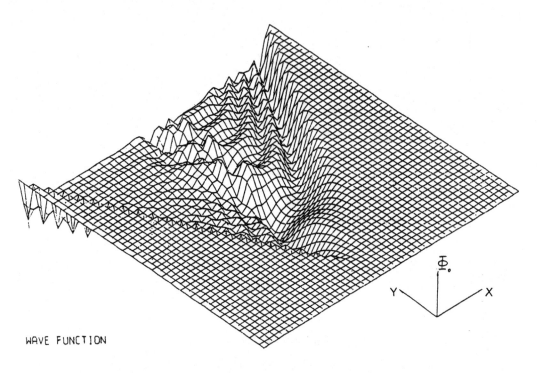

WAVE FUNCTION

Fig. 5. The wave function for a homogeneous, isotropic universe with a scalar field. The wave function does not oscillate near the lines $y = |x|$.

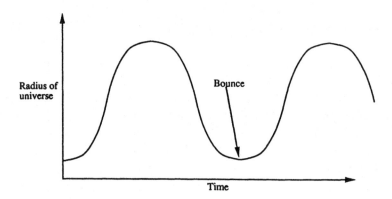

Fig. 6. A quasi-periodic solution for a Friedmann universe filled with a massive scalar field.

calculations of the wave function seemed to correspond to a superposition of these solutions. The oscillating solutions were quasi-periodic. So it seemed natural to suppose that the boundary conditions on the perturbations should be that they were small whenever the radius was small. This would have led to an arrow of time that pointed forward in the expanding phase, and backward in the contracting phase, as I have explained.

I set my research student, Raymond Laflamme, to work on the arrow of time in more general situations than a homogeneous and isotropic Friedmann background. He soon found a major objection to my ideas. Only a few solutions, like the spherically symmetric Friedmann models, can bounce when they collapse. Thus the wave function for something like a black hole could not be concentrated on nonsingular solutions. This made me realize that there could be a difference between the start of the expansion, and the end of the contraction. The dominant contributions to the wave functions for either, would come from saddle points that corresponded to complex solutions of the field equations. These solutions have been studied in detail by my student, Glenn Lyons [Lyons, 1992]. When the radius of the universe is small, there are two kinds of solutions (see Figure 7). One would be an almost Euclidean complex solution that started like the north pole of a sphere and expanded monotonically up to the given radius. This would correspond to the start of the expansion. But the end of the contraction would correspond to a solution that started in a similar way, but then had a long, almost Lorentzian period of expansion followed by contraction to the given radius. The wave function for perturbations about the first kind of solution would be heavily damped, unless the perturbations were small and in the linear regime. But the wave function for perturbations about the solution that expanded and contracted could be large for large perturbation amplitudes. This would mean that the perturbations would be small at one end of time, but could be large and nonlinear at the other end. So disorder and irregularity would increase during the expansion, and would continue to increase during the contraction. There would be no reversal of the arrow of time at the point of maximum expansion.

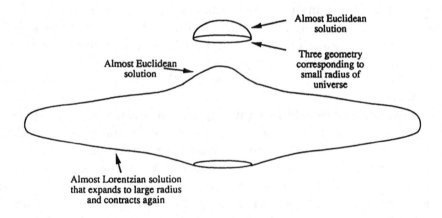

Fig. 7. Two possible saddle points in the path integral for the wave function of a given radius.

Glenn Lyons, Raymond Laflamme and I have studied how the arrow of time manifests itself in the various perturbation modes. It makes sense to talk about the arrow of time only for modes that are shorter than the horizon scale at the time concerned. Modes that are longer than the horizon just appear as a homogeneous background. There are two kinds of behavior for perturbation modes within the horizon. They can oscillate or they can have power law growth or decay. Modes that oscillate are the tensor modes that correspond to gravitational waves, and scalar modes that correspond to density perturbations of wavelength less than the Jeans length. On the other hand, density perturbations longer than the Jeans length have power law growth and decay.

Perturbation modes that oscillate will have an amplitude that varies adiabatically as an inverse power of the radius of the universe:

$$Aa^p = \text{constant},$$

where A is the amplitude of the oscillating perturbation, a is the radius of the universe and p is some positive number. This means they will be essentially time symmetric about the time of maximum expansion. In other words, the amplitude of the perturbation will be the same at a given radius during the expansion, as at the same radius during the contracting phase. So if they are small when they come within the horizon during expansion, which is what the no-boundary condition predicts, they will remain small at all times. They will not become nonlinear, and they will not show an arrow of time. By contrast, density perturbations on scales longer than the Jeans length will grow in amplitude in general

$$A = Ba^p + Ca^{-q},$$

where p and q are positive. They will be small when they come within the horizon during the expansion. But they will grow during the expansion, and continue to grow during the contraction. Eventually, they will become nonlinear. At this stage, the trajectories will spread out over a large region of phase space.

So the no-boundary condition predicts that the universe is in a smooth and ordered state at one end of time. But irregularities increase while the universe expands and contracts again. These irregularities lead to the formation of stars and galaxies, and hence to the development of intelligent life. This life will have a subjective sense of time, or psychological arrow, that points in the direction of increasing disorder.

The one remaining question is why this psychological arrow should agree with the cosmological arrow. In other words, why do we say the universe is expanding rather than contracting? The answer to this comes from inflation, combined with the weak anthropic principle. If the universe had started to contract a few billion years ago, we would indeed observe it to be contracting. But inflation implies that the universe should be so near the critical density that it will not stop expanding for much longer than the present age. By that time, all the stars will have burnt out. The universe will be a cold dark place, and any life will have died out long before. Thus the fact that we are around to observe the universe, means that we must be in the expanding, rather than the contracting phase. This is the explanation why the psychological arrow agrees with the cosmological arrow.

So far I have been talking about the arrow of time on a macroscopic, fluid dynamical scale. But the inflationary model depends on the existence of an arrow of time on a much smaller, microscopic scale. During the inflationary phase, practically the entire energy content of the universe is in the single homogeneous mode of a scalar field. The amplitude of this mode, changes only slowly with time, and its energy–momentum tensor causes the universe to expand in an accelerating, exponential way. At the end of the inflationary period, the amplitude of the homogeneous mode begins to oscillate. The idea is that these coherent homogeneous oscillations of the scalar field cause the creation of short wavelength particles of other fields, with a roughly thermal spectrum. The universe expands thereafter, like the hot big bang model.

This inflationary scenario implicitly assumes the existence of a thermodynamic arrow of time that points in the direction of the expansion. It wouldn't work if the arrow of time had been in the opposite direction. Normally, people brush the assumption of an arrow of time under the carpet. But in this case, one can show that this microscopic arrow also seems to follow from the no-boundary condition. One can introduce extra matter fields, coupled to the scalar field. If one expands them in spherical harmonics, one obtains a set of Schroedinger equations with oscillating coefficients. The no-boundary condition tells you that the matter fields start in their ground state. One then finds that the matter fields become excited when the scalar field begins to oscillate. Presumably, the back reaction will damp the oscillations of the scalar field, and the universe will go over to a radiation dominated phase. Thus, the no-boundary proposal seems to explain the arrow of time on microscopic as well as on macroscopic scales.

I have told you how I came to the wrong conclusion, and what I now think is the correct result about what the no-boundary condition predicts for the arrow of time. This was my greatest mistake, or at least my greatest mistake in science. I once thought there ought to be a journal of recantations, in which scientists could admit their mistakes. But it might not have many contributors.

REFERENCES

Gold, T. (1967) *The Nature of Time*, Cornell University Press, New York.

Halliwell, J. J. and Hawking, S. W. (1984) The Origin of Structure in the Universe, *Phys. Rev.* **D31**, 8.

Hartle, J. B. and Hawking, S. W. (1983) Wave Function of the Universe, *Phys. Rev.* **D28**, 2960–2975.

Hawking, S. W. (1984) The Quantum State of the Universe, *Nucl. Phys.* **B239**, 257.

Hawking, S. W. (1985) The Arrow of Time in Cosmology, *Phys. Rev.* **D32**, 2489.

Hawking, S. W. and Wu, Z. C. (1985) Numerical Calculations of Minisuperspace Cosmological Models, *Phys. Lett.* **B151**, 15.

Hogarth, J. E. (1962) Cosmological Considerations of the Absorber Theory of Radiation, *Proc. of the Royal Soc.* **A267**, 365.

Hoyle, F. and Narlikar, J. V. (1964) Time Symmetric Electrodynamics and the Arrow of Time in Cosmology, *Proc. of the Royal Soc.* **A273**, 1.

Lyons, G. W. (1992) Complex Solutions for the Scalar Field Model of the Universe, submitted to *Phys. Rev.* **D**.

Page, D. N. (1985) Will Entropy Decrease if the Universe Collapses? *Phys. Rev.* **D32**, 2496–2499.

Reichenbach, H. (1956) *The Direction of Time*, University of California Press, Berkeley.

THE COSMOLOGICAL CONSTANT IS PROBABLY ZERO

S.W. HAWKING

Department of Applied Mathematics and Theoretical Physics,
Silver Street, Cambridge, CB3 9EW, England

Received 12 August 1983
Revised manuscript received 24 October 1983

It is suggested that the apparent cosmological constant is not necessarily zero but that zero is by far the most probable value. One requires some mechanism like a three-index antisymmetric tensor field or topological fluctuations of the metric which can give rise to an effective cosmological constant of arbitrary magnitude. The action of solutions of the euclidean field equations is most negative, and the probability is therefore highest, when this effective cosmological constant is very small.

The cosmological constant is probably the quantity in physics that is most accurately measured to be zero: observations of departures from the Hubble Law for distant galaxies place an upper limit of the order of

$$|\Lambda|/m_p^2 < 10^{-120} , \tag{1}$$

where m_p is the Planck mass. On the other hand, one might expect that the zero point energies of quantum fluctuations would produce an effective or induced Λm_p^{-2} of order one if the quantum fluctuations were cut off at the Planck mass. Even if this were renormalized exactly to zero, one would still get a change in the effective Λ of order $\mu^4 m_p^{-2}$ whenever a symmetry in the theory was spontaneously broken, where μ is the energy at which the symmetry was broken. There are a large number of symmetries which seem to be broken in the present epoch of the universe, including chiral symmetry, electroweak symmetry and possibly, supersymmetry. Each of these would give a contribution to Λ that would exceed the upper limit (1) by at least forty orders of magnitude.

It is very difficult to believe that the bare value of Λ is fine tuned so that after all the symmetry breakings, the effective Λ satisfies the inequality (1). What one would like to find is some mechanism by which the effective value of Λ could relax to zero. Although there have been a number of attempts to find such a

mechanism (see e.g. refs. [1,2]), I think it is fair to say that no satisfactory scheme has been suggested. In this paper, I want to propose instead a very simple idea: the cosmological constant can have any value but it is much more probable for it to have a value very near zero. A preliminary version of this argument was given in ref. [3].

My proposal requires that a variable effective cosmological constant be generated in some manner and that the path integral includes all, or some range, of values of this effective cosmological constant. One possibility would be to include the value of the cosmological constant in the variables that are integrated over in the path integral. A more attractive way would be to introduce a three-index antisymmetric tensor field $A_{\mu\nu\rho}$. This would have gauge transformations of the form

$$A_{\mu\nu\rho} \to A_{\mu\nu\rho} + \nabla_{[\mu} C_{\nu\rho]} , \tag{2}$$

The action of the field is F^2 where F is the field strength formed from A:

$$F_{\mu\nu\rho\sigma} = \nabla_{[\mu} A_{\nu\rho\sigma]} . \tag{3}$$

Such a field has no dynamics: the field equations imply that F is a constant multiple of the four-index antisymmetric tensor $\epsilon_{\mu\nu\rho\sigma}$. However, the F^2 term in the action behaves like an effective cosmological

403

Volume 134B, number 6 PHYSICS LETTERS 26 January 1984

constant [4]. Its value is not determined by field equations. Three-index antisymmetric tensor fields arise naturally in the dimensional reduction of $N = 1$ supergravity in eleven dimensions to $N = 8$ supergravity in four dimensions. Other mechanisms that would give an effective cosmological constant of arbitrary magnitude include topological fluctuations of the metric [5] and a scalar field ϕ with a potential term $V(\phi)$ but no kinetic term. In this last case, the gravitational field equations could be satisfied only if ϕ was constant. The potential $V(\phi)$ then acts as an effective cosmological constant.

In the path integral formulation of quantum theory, the amplitude to go from a field configuration $\phi_1(x)$ on the surface $t = t_1$ to a configuration $\phi_2(x)$ on $t = t_2$ is

$$\langle \phi_2, t_2 | \phi_1, t_1 \rangle = \int d[\phi] \exp(iI[\phi]) , \qquad (4)$$

where $d[\phi]$ is a measure on the space of all field configurations $\phi(x, t)$, $I[\phi]$ is the action of the field configuration and the integral is over all field configurations which agree with ϕ_1 and ϕ_2 at $t = t_1$ and $t = t_2$ respectively. The integral (4) oscillates and does not converge. One can improve the situation by making a rotation to euclidean space by defining a new coordinate $\tau = it$. The transition amplitude then becomes

$$\langle \phi_2, \tau_2 | \phi_1, \tau_1 \rangle = \int d[\phi] \exp(-\tilde{I}[\phi]) , \qquad (5)$$

where $\tilde{I} = -iI$ is the euclidean action which is bounded below for well behaved field theories in flat space. One can interpret $\exp(-\tilde{I}[\phi])$ as being proportional to the probability of the euclidean field configuration $\phi(x, \tau)$. One calculates amplitudes like (5) in euclidean space and then analytically continues them in $\tau_2 - \tau_1$ back to real time separations.

One can adopt a similar euclidean approach in the case of gravity [6,7]. There is a difficulty because the euclidean gravitational action is not bounded below. This can be overcome by dividing the space of all positive definite metrics up into equivalence classes under conformal transformations. In each equivalence class one integrates over the conformal factor on a contour which is parallel to the imaginary axis [8,3]. The dominant contribution to the path integral comes from metrics which are near to solutions of the field equations. Of particular interest are solutions in which the dynamical matter fields, i.e. the matter fields apart

from $A_{\mu\nu\rho}$ or ϕ are near their ground state values over a large region. This would be a reasonable approximation to the universe at the present time. The ground state of the matter fields plus the contribution of the $A_{\mu\nu\rho}$ or ϕ fields will generate an effective cosmological constant Λ_e. If the effective value Λ_e is positive, the solutions are necessarily compact and their four-volume is bounded by that of the solution of greatest symmetry, the four-sphere of radius $(3\Lambda_e^{-1})^{1/2}$. The euclidean action $\tilde{I}$ will be negative and will be bounded below by

$$-3\pi m_p^2/\Lambda_e . \qquad (6)$$

If Λ_e is negative, the solutions can be either compact or non-compact [5]. If they are compact, the action $\tilde{I}$ will be finite and positive. If they are non-compact, $\tilde{I}$ will be infinite and positive.

The probability of a given field configuration will be proportional to

$$\exp(-\tilde{I}) . \qquad (7)$$

If Λ_e is negative, $\tilde{I}$ will be positive and the probability will be exponentially small. If Λ_e is positive, the probability will be of the order of

$$\exp(3\pi m_p^2/\Lambda_e) . \qquad (8)$$

Clearly, the most probable configurations will be those with very small values of Λ_e. This does not imply that the effective cosmological constant will be small everywhere in these configurations. In regions in which the dynamical fields differ from the ground state values there can be an apparent cosmological constant as in the inflationary model of the universe.

[1] F. Wilczek, in: The very early universe, eds. G.W. Gibbons, S.W. Hawking and S.T.C. Siklos (Cambridge U.P., Cambridge, 1983).
[2] A.D. Dolgov, in: The very early universe, eds. G.W. Gibbons, S.W. Hawking and S.T.C. Siklos (Cambridge U.P., Cambridge, 1983).
[3] S.W. Hawking, The cosmological constant. Phil. Trans. Roy. Soc. A., to be published.
[4] A. Aurilia, H. Nicolai and P.K. Townsend. Nucl. Phys. B176 (1980) 509.
[5] S.W. Hawking, Nucl. Phys. B144 (1978) 349.
[6] S.W. Hawking, The path integral approach to quantum gravity, in: General relativity: an Einstein centenary survey, eds. S.W. Hawking and W. Isreal (Cambridge U.P., Cambridge, 1979).
[7] S.W. Hawking, Euclidean quantum gravity, in: Recent developments in gravitation, Cargese Lectures, eds. M. Levy and S. Deser (1978).
[8] G.W. Gibbons, S.W. Hawking and M.J. Perry, Nucl. Phys. B138 (1978) 141.

404

PHYSICAL REVIEW D   VOLUME 37, NUMBER 4   15 FEBRUARY 1988

Wormholes in spacetime

S. W. Hawking

Department of Applied Mathematics and Theoretical Physics, University of Cambridge, Silver Street, Cambridge CB3 9EW, England
(Received 28 October 1987)

Any reasonable theory of quantum gravity will allow closed universes to branch off from our nearly flat region of spacetime. I describe the possible quantum states of these closed universes. They correspond to wormholes which connect two asymptotically Euclidean regions, or two parts of the same asymptotically Euclidean region. I calculate the influence of these wormholes on ordinary quantum fields at low energies in the asymptotic region. This can be represented by adding effective interactions in flat spacetime which create or annihilate closed universes containing certain numbers of particles. The effective interactions are small except for closed universes containing scalar particles in the spatially homogeneous mode. If these scalar interactions are not reduced by sypersymmetry, it may be that any scalar particles we observe would have to be bound states of particles of higher spin, such as the pion. An observer in the asymptotically flat region would not be able to measure the quantum state of closed universes that branched off. He would therefore have to sum over all possibilities for the closed universes. This would mean that the final state would appear to be a mixed quantum state, rather than a pure quantum state.

I. INTRODUCTION

In a reasonable theory of quantum gravity the topology of spacetime must be able to be different from that of flat space. Otherwise, the theory would not be able to describe closed universes or black holes. Presumably, the theory should allow all possible spacetime topologies. In particular, it should allow closed universes to branch off, or join onto, our asymptotic flat region of spacetime. Of course, such behavior is not possible with a real, nonsingular, Lorentzian metric. However, we now all know that quantum gravity has to be formulated in the Euclidean domain. There, it is no problem: it is just a question of plumbing. Indeed, it is probably necessary to include all possible topologies for spacetime to get unitarity.

Topology change is not something that we normally experience, at least, on a macroscopic scale. However, one can interpret the formation and subsequent evaporation of a black hole as an example: the particles that fell into the hole can be thought of as going off into a little closed universe of their own. An observer in the asymptotically flat region could not measure the state of the closed universe. He would therefore have to sum over all possible quantum states for the closed universe. This would mean that the part of the quantum state that was in the asymptotically flat region would appear to be in a mixed state, rather than a pure quantum state. Thus, one would lose quantum coherence.[1,2]

If it is possible for a closed universe the size of a black hole to branch off, it is also presumably possible for little Planck-size closed universes to branch off and join on. The purpose of this paper is to show how one can describe this process in terms of an effective field theory in flat spacetime. I introduce effective interactions which create, or destroy, closed universes containing certain numbers of particles. I shall show that these effective in-

teractions are small, except for scalar particles. There is a serious problem with the very large effective interactions of scalar fields with closed universes. It may be that these interactions can be reduced by supersymmetry. If not, I think we will have to conclude that any scalar particles that we observe are bound states of fermions, like the pion. Maybe this is why we have not observed Higgs particles.

I base my treatment on general relativity, even though general relativity is probably only a low-energy approximation to some more fundamental quantum theory of gravity, such as superstrings. For closed universes of the Planck size, any higher-order corrections induced from string theory will change the action by a factor ~ 1. So the effective field theory based on general relativity should give answers of the right order of magnitude.

In Sec. II, I describe how closed universes or wormholes can join one asymptotically Euclidean region to another, or to another part of the same region. Solutions of the Wheeler-DeWitt equation that correspond to such wormholes are obtained in Sec. III. These solutions can also be interpreted as corresponding to Friedmann universes. It is an amusing thought that our Universe could be just a rather large wormhole in an asymptotically flat space.

In Sec. IV, I calculate the vertex for the creation or annihilation of a wormhole containing a certain number of particles. Section V contains a discussion of the initial quantum state in the closed-universe Fock space. There are two main possibilities: either there are no closed universes present initially, or there is a coherent state which is an eigenstate of the creation plus annihilation operators for each species of closed universe. There will be loss of quantum coherence in the first case, but not the second. This is described in Sec. VI. The interactions between wormholes and particles of different spin in asymptotically flat space are discussed in Sec.

VII. Finally, in Sec. VIII, I conclude that wormholes will have to be taken into account in any quantum theory of gravity, including superstrings.

This paper supercedes earlier work of mine[3-5] on the loss of quantum coherence. These papers were incorrect in associating loss of coherence with simply connected spaces with nontrivial topology, rather than with wormholes.

II. WORMHOLES

What I am aiming to do is to calculate the effect of closed universes that branch off on the behavior of ordinary, nongravitational particles in asymptotically flat space at energies low compared to the Planck mass. The effect will come from Euclidean metrics which represent a closed universe branching off from asymptotically flat space. One would expect that the effect would be greater, the larger the closed universe. Thus one might expect the dominant contribution would come from metrics with the least Euclidean action for a given size of closed universe. In the $R = 0$ conformal gauge, these are conformally flat metrics:

$$ds^2 = \Omega^2 dx^2 ,$$

$$\Omega = 1 + \frac{b^2}{(x - x_0)^2} .$$

At first sight, this looks like a metric with a singularity at the point x_0. However, the blowing up of the conformal factor near x_0 means that the space opens out into another asymptotically flat region, joined to the first asymptotically flat region by a wormhole of coordinate radius b and proper radius $2b$. The other asymptotic region can be a separate asymptotically flat region of the Universe, or it can be another part of the first asymptotic region. In the latter case, the conformal factor will be modified slightly by the interaction between the two ends of the wormhole, or handle to spacetime.[6] However, the change will be small when the separation of the two ends is large compared to $2b$, the size of the wormhole. Typically, b will be of the order of the Planck length, so it will be a good approximation to neglect the interactions between wormholes. This conformally flat metric is just one example of a wormhole. There are, of course, nonconformally flat closed universes that can join onto asymptotically flat space. Their effects will be similar, but will involve gravitons in the asymptotically flat space. Since it is difficult to observe gravitons, I shall concentrate on conformally flat closed universes.

I shall consider a set of matter fields ϕ in the closed universe. Spin-1 gauge fields are conformally invariant. In the case of matter fields of spin $\frac{1}{2}$ and 0, the effect of any mass will be small for wormholes of the Planck size. I shall therefore take the matter fields ϕ to be conformally invariant. The effect of mass could be included as a perturbation.

In order to find the effect of the closed universe or wormhole on the matter fields ϕ in the asymptotically flat spaces, one should calculate the Green's functions

$$\langle \phi(y_1)\phi(y_2) \cdots \phi(y_r)\phi(z_1)\phi(z_2) \cdots \phi(z_s) \rangle ,$$

where $y_1, \ldots, y_r$ and $z_1, \ldots, z_s$ are points in the two asymptotic regions (which may be the same region). This can be done by performing a path integration over all matter fields ϕ and all metrics $g_{\mu\nu}$ that have one or two asymptotically flat regions and a handle or wormhole connecting them. Let S be a three-sphere, which is a cross section of the closed universe or wormhole. One can then factorize the path integral into a part

$$\langle 0 \mid \phi(y_1) \cdots \phi(y_r) \mid \psi \rangle ,$$

which depends on the fields on one side of S, and a part

$$\langle \psi \mid \phi(z_1) \cdots \phi(z_s) \mid 0 \rangle ,$$

which depends on the fields on the other side of S. Strictly speaking, one can factorize in this way only when the regions at the two ends of the wormhole are separate asymptotic regions. However, even when they are the same region, one can neglect the interaction between the ends and factorize the path integral if the ends are widely separated.

In the above $\mid 0 \rangle$ represented the usual particle scattering vacuum state defined by a path integral over asymptotically Euclidean metrics and matter fields that vanish at infinity. $\mid \psi \rangle$ represented the quantum state of the closed universe or wormhole on the surface S. This can be described by a wave function Ψ which depends on the induced metric h_{ij} and the values ϕ_0 of the matter fields on S. The wave function obeys the Wheeler-DeWitt equation

$$\left[-m_P^{-2} G_{ijkl} \frac{\delta^2}{\delta h_{ij} \delta h_{kl}} - m_P^2 h^{1/2}\, {}^3R + \tfrac{1}{2} h^{1/2} T^{nn}\left(\phi_0, -i\frac{\delta}{\delta\phi_0}\right) \right] \Psi[h_{ij}, \phi_0] = 0 ,$$

where

$$G_{ijkl} = \tfrac{1}{2} h^{1/2}(h_{ik}h_{jl} + h_{il}h_{jk} - h_{ij}h_{kl}) .$$

The wave function also obeys the momentum constraint

$$\left[-2im_P^2 \left[\frac{\delta}{\delta h_{ij}} \right]_{|j} + T^{ni}\left(\phi_0, -i\frac{\delta}{\delta\phi_0}\right) \right] \Psi[h_{ij}, \phi_0] = 0 .$$

III. WORMHOLE EXCITED STATES

The solutions of the Wheeler-DeWitt equation that correspond to wormholes, that is, closed universes connecting two asymptotically Euclidean regions, form a Hilbert space $\mathcal{H}_w$ with the inner product

$$\langle \psi_1 | \psi_2 \rangle = \int d[h_{ij}] d[\phi_0] \Psi_1^* \Psi_2 \; .$$

Let $| \psi_i \rangle$ be a basis for $\mathcal{H}_w$. Then one can write the Green's function in the factorized form

$$\langle \phi(y_1) \cdots \phi(y_r) \phi(z_1) \cdots \phi(z_s) \rangle = \sum \langle 0 | \phi(y_1) \cdots \phi(y_r') | \psi_i \rangle \langle \psi_i | \phi(z_1) \cdots \phi(z_s) | 0 \rangle \; .$$

What are these wormhole excited states $| \psi_i \rangle$? To find them one would have to solve the full Wheeler-DeWitt and momentum constraint equations. This is too difficult, but one can get an idea of their nature from mode expansions.[7] One can write the three-metric h_{ij} on the surface S as

$$h_{ij} = \sigma^2 a^2 (\Omega_{ij} + \epsilon_{ij}) \; .$$

Here $\sigma^2 = 2/3\pi m_P^2$ is a normalization factor, Ω_{ij} is the metric on the unit three-sphere, and ϵ_{ij} is a perturbation, which can be expanded in harmonics on the three-sphere:

$$\epsilon_{ij} = \sum_{n,l,m} [6^{1/2} a_{nlm} \tfrac{1}{3} \Omega_{ij} Q_{lm}^n + 6^{1/2} b_{nlm} (P_{ij})_{lm}^n + 2^{1/2} c_{nlm}^0 (S_{ij}^0)_{lm}^n + 2^{1/2} c_{nlm}^e (S_{ij}^e)_{lm}^n + 2 d_{nlm}^0 (G_{ij}^0)_{lm}^n + 2 d_{nlm}^e (G_{ij}^e)_{lm}^n] \; .$$

The $Q(x^i)$ are the standard scalar harmonics on the three-sphere. The $P_{ij}(x^i)$ are given by (suppressing all but i,j indices)

$$P_{ij} = \frac{1}{n^2 - 1} Q_{|ij} + \tfrac{1}{3} \Omega_{ij} Q \; .$$

They are traceless, $P_i^i = 0$. The S^{ij} are defined by

$$S_{ij} = S_{i|j} + S_{j|i} \; ,$$

where S_i are the transverse vector harmonics, $S_i^{|i} = 0$. The G_{ij} are the transverse traceless tensor harmonics $G_i^i = G_{ij}^{|j} = 0$. Further details about harmonics and their normalization can be found in Ref. 7.

Consider a conformally invariant scalar field ϕ. One can describe it in terms of hyperspherical harmonics on the surface S:

$$\phi_0 = \sigma^{-1} a^{-1} \sum f_n Q_n \; .$$

The wave function Ψ is then a function of coefficients a_n, b_n, c_n, d_n, and f_n and the scale factor a.

One can expand the Wheeler-DeWitt operator to all orders in a and to second order in the other coefficients. In this approximation, the different modes do not interact with each other, but only with the scale factor a. However, the conformal scalar coefficients f_n do not even interact with a. One can therefore write the wave function as a sum of products of the form

$$\Psi = \Psi_0(a, a_i, b_i, c_i, d_i) \prod \psi_n(f_n) \; .$$

The part of the Wheeler-DeWitt operator that acts on ψ_n is

$$-\frac{d^2}{df_n^2} + (n^2 + 1) f_n^2 \; .$$

It is therefore natural to take them to be harmonic-oscillator wave functions

$$\psi_{nm} = \left[\frac{\beta^2}{\pi 2^{2m} (m!)^2} \right]^{1/4} e^{-\beta^2 f_n^2 / 2} H_m(\beta f_n) \; ,$$

where $\beta^4 = (n^2 + 1)$ and H_m are Hermite polynomials. The wave functions ψ_{nm} can then be interpreted as corresponding to the closed universe containing m scalar particles in the nth harmonic mode.

The treatment for spin-$\tfrac{1}{2}$ and -1 fields is similar. The appropriate data for the fields on S can be expanded in harmonics on the three-sphere. The main difference is that the lowest harmonic is not the $n = 0$ homogeneous mode, as in the scalar case, but has $n = \tfrac{1}{2}$ or 1. Again, the coefficients of the harmonics appear in the Wheeler-DeWitt equation to second order only as fermionic[8] or bosonic harmonic oscillators, with a frequency independent of a. One can therefore take the wave functions to be fermion or boson harmonic-oscillator wave functions in the coefficients of the harmonics. They can then be interpreted as corresponding to definite numbers of particles in each mode.

In the gravitational part of the wave function, Ψ_0, the coefficients a_n, b_n, and c_n reflect gauge degrees of freedom. They can be made zero by a diffeomorphism of S and suitable lapse and shift functions. The coefficients d_n correspond to gravitational wave excitations of the closed universe. However, gravitons are very difficult to observe. I shall therefore take these modes to be in their ground state.

The scale factor a appears in the Wheeler-DeWitt equation as the operator

$$\frac{\partial^2}{\partial a^2} - a^2 \; .$$

I shall assume that the zero-point energies of each mode are either subtracted or canceled by fermions in a supersymmetric theory. The total wave function Ψ will then satisfy the Wheeler-DeWitt equation if the gravitational part Ψ_0 is a harmonic-oscillator wave function in a with unit frequency and level equal to the sum E of the energies of the matter-field harmonic oscillators.

The wave function Ψ_0 will oscillate for $a < r_0$ $= (2E)^{1/2}$. In this region one can use the WKB approximation[7,9,10] to relate it to a Lorentzian solution of the classical field equations. This solution will be a $k = +1$ Friedmann model filled with conformally invariant matter. The maximum radius of the Friedmann model will be $a = r_0$. For $a > r_0$, the wave function will be exponential. Thus, in this region it will correspond to a Euclidean metric. This will be the wormhole metric described in Sec. II, with $b = 1/2\sigma r_0$. These excited state solutions were first found in Ref. 11, but their significance as wormholes was not realized. Notice that the wave function is exponentially damped at large a, whereas the cosmological wave functions described in Refs. 7, 9, and 10 tend to grow exponentially at large a. The difference here is that one is looking at the closed universe from an asymptotically Euclidean region, instead of from a compact Euclidean space, as in the cosmological case. This changes the sign of the trace K surface term in the gravitational action.

IV. THE WORMHOLE VERTEX

One now wants to calculate the matrix element of the product of the values of ϕ at the points $y_1, y_2, \ldots, y_r$, between the ordinary, flat-space vacuum $\langle 0 |$ and the closed-universe state $| \psi \rangle$. This is given by the path integral

$$\langle 0 | \phi(y_1) \cdots \phi(y_r) | \psi \rangle = \int d[h_{ij}] d[\phi_0] \Psi[h_{ij}, \phi_0] \int d[g_{\mu\nu}] d[\phi] \phi(y_1) \cdots \phi(y_r) e^{-I[g,\phi]} .$$

The gravitational field is required to be asymptotically flat at infinity, and to have a three-sphere S with induced metric h_{ij} as its inner boundary. The scalar field ϕ is required to be zero at infinity, and to have the value ϕ_0 on S.

In general, the positions of the points y_i cannot be specified in a gauge-invariant manner. However, I shall be concerned only with the effects of the wormholes on low-energy particle physics. In this case the separation of the points y_i can be taken to be large compared to the Planck length, and they can be taken to lie in flat Euclidean space. Their positions can then be specified up to an overall translation and rotation of Euclidean space.

Consider first a wormhole state $| \psi \rangle$ in which only the $n = 0$ homogeneous scalar mode is excited above its ground state. The integral over the wave function Ψ of the wormhole can then be replaced in the above by

$$\int da \, df_0 \, \psi_E(a) \psi_{0m}(f_0) .$$

The path integral will then be over asymptotically Euclidean metrics whose inner boundary is a three-sphere S of radius a and scalar fields with the constant value f_0 on S. The saddle point for the path integral will be flat Euclidean space outside a three-sphere of radius a centered on a point x_0 and the scalar field

$$\phi = \frac{a\sigma f_0}{(x - x_0)^2}$$

(the energy-momentum tensor of this scalar field is zero). The action of this saddle point will be $(a^2 + f_0^2)/2$. The determinant Δ of the small fluctuations about the saddle point will be independent of f_0. Its precise form will not be important.

The integral over the coefficient f_0 of the $n = 0$ scalar harmonic will contain a factor of

$$\int df_0 f_0^r e^{-f_0^2} H_m(f_0) .$$

This will be zero when m, the number of particles in the mode $n = 0$, is greater than r, the number of points y_i in the correlation function. This is what one would expect, because each particle in the closed universe must be created or annihilated at a point y_i in the asymptotically flat region. If $r > m$, particles may be created at one point y_i and annihilated at another point y_j without going into the closed universe. However, such matrix elements are just products of flat-space propagators with matrix elements with $r = m$. It is sufficient therefore to consider only the case with $r = m$.

The integral over the radius a will contain a factor

$$\int da \, a^m e^{-a^2} H_E(a) \Delta(a) ,$$

where $E = m$ is the level number of the radial harmonic oscillator. For small m, the dominant contribution will come from $a \sim 1$, that is, wormholes of the Planck size. The value $C(m)$ of this integral will be ~ 1.

The matrix element will then be

$$D(m) \prod \frac{\sigma}{(y_i - x_0)^2} ,$$

where $D(m)$ is another factor ~ 1. One now has to integrate over the position x_0 of the wormhole, with a measure of the form $m_P^4 dx_0^4$, and over an orthogonal matrix O which specifies its orientation with respect to the points y_i. The $n = 0$ mode is invariant under O, so this second integral will have no effect, but the integral over x_0 will ensure the energy and momentum are conserved in the asymptotically flat region. This is what one would expect, because the Wheeler-DeWitt and momentum constraint equations imply that a closed universe has no energy or momentum.

The matrix element will be the same as if one was in

flat space with an effective interaction of the form

$$F(m)m_P^{4-m}\phi^m(c_{0m}+c_{0m}^\dagger) \, ,$$

where $F(m)$ is another coefficient ~ 1 and c_{0m} and $c_{0m}^\dagger$ are the annihilation and creation operators for a closed universe containing m scalar particles in the $n=0$ homogeneous mode.

In a similar way, one can calculate the matrix elements of products of ϕ between the vacuum and a closed-universe state containing m_0 particles in the $n=0$ mode, m_1 particles in the $n=1$ mode, and so on. The energy-momentum tensor of scalar fields with higher harmonic angular dependence will not be zero. This will mean that the saddle-point metric in the path integral for the matrix element will not be flat space, but will be curved near the surface S. In fact, for large particle numbers, the saddle-point metric will be the conformally flat wormhole metrics described in Sec. II. However, the saddle-point scalar fields will have a Q_n angular dependence and a $\sigma^{n+1}/(x-x_0)^{n+2}$ radial dependence in the asymptotic flat region. This radial decrease is so fast that closed universes with higher excited harmonics will not give significant matrix elements, except for that containing two particles in the $n=1$ modes. By the constraint equations or, equivalently, by averaging over the orientation O of the wormhole, the matrix element will be zero unless the two particles are in a state that is invariant under O. The matrix element for such a universe will be the same as that produced by an effective interaction of the form

$$\nabla\phi\nabla\phi(c_{12}+c_{12}^\dagger)$$

with a coefficient ~ 1.

In a similar way one can calculate the matrix elements for universes containing particles of spin $\frac{1}{2}$ or higher. Again, the constraint equations or averaging over O mean that the matrix element is nonzero only for closed-universe states that are invariant under O. This means that the corresponding effective interactions will be Lorentz invariant. In particular, they will contain even numbers of spinor fields. Thus, fermion number will be conserved mod 2: the closed universes are bosons.

The matrix elements for universes containing spin-$\frac{1}{2}$ particles will be equivalent to effective interactions of the form

$$m_P^{4-3m/2}\psi^m d_m + \text{c.c.} \, ,$$

where ψ^m denotes some Lorentz-invariant combination of m spinor fields ψ or their adjoints $\bar\psi$, and d_m is the annihilation operator for a closed universe containing m spin-$\frac{1}{2}$ particles in $n=\frac{1}{2}$ modes. One can neglect the effect of closed universes with spin-$\frac{1}{2}$ particles in higher modes.

In the case of spin-1 gauge particles, the effective interaction would be of the form

$$m_P^{4-2m}[(F_{\mu\nu})^m(g_m+g_m^\dagger)] \, ,$$

where g_m is the annihilation operator for a closed

universe containing m spin-1 particles in $n=1$ modes. As before, the higher modes can be neglected.

V. THE WORMHOLE INITIAL STATE

What I have done is introduce a new Fock space $\mathcal{F}_w$ for closed universes, which is based on the one wormhole Hilbert space $\mathcal{H}_w$. The creation and annihilation operators $c_{nm}^\dagger$, c_{nm}, etc., act on $\mathcal{F}_w$ and obey the commutation relations for bosons. The full Hilbert space of the theory, as far as asymptotically flat space is concerned, is isomorphic to $\mathcal{F}_p\otimes\mathcal{F}_w$, where $\mathcal{F}_p$ is the usual flat-space particle Fock space.

The distinction between annihilation and creation operators is a subtle one because the closed universe does not live in the same time as the asymptotically flat region. If both ends of the wormhole are in the same asymptotic region, one can say that a closed universe is created at one point and is annihilated at another. However, if a closed universe branches off from our asymptotically flat region, and does not join back on, one would be free to say either (1) it was present in the initial state and was annihilated at the junction point x_0, (2) it was not present initially, but was created at x_0 and is present in the final state, or (3) as Sidney Coleman (private communication) has suggested, one might have a coherent state of closed universes in both the initial and final states, in such a way that they were both eigenstates of the annihilation plus creation operators $c_{nm}+c_{nm}^\dagger$, etc., with some eigenvalue q.

In this last case, the closed-universe sector of the state would remain unchanged and there would be no loss of quantum coherence. However, the initial state would contain an infinite number of closed universes. Such eigenstates would not form a basis for the Fock space of closed universes.

Instead, I shall argue that one should adopt the second possibility: there are no closed universes in the initial state, but closed universes can be created and appear in the final state. If one takes a path-integral approach, the most natural quantum state for the Universe is the so-called "ground" state, or, "no boundary" state.[8] This is the state defined by a path integral over all compact metrics without boundary. Calculations based on minisuperspace models[7-11] indicate that this choice of state leads to a universe like we observe, with large regions that appear nearly flat. One can then formulate particle scattering questions in the following way: one asks for the conditional probability that one observes certain particles on a nearly flat surface S_2 given that the region is nearly asymptotically Euclidean and is in the quantum state defined by conditions on the surfaces S_1 and S_3 to either side of S_2, and at great distance from it in the positive and negative Euclidean-time directions, respectively. One then analytically continues the position of S_2 to late real time. It then measures the final state in the nearly flat region. One continues the positions of both S_1 and S_3 to early real time. One gives the time coordinate of S_1 a small positive imaginary part, and the time coordinate of S_3 a small negative imaginary part. The initial state is then defined by data on the surfaces S_1 and S_3.

If one adopts the formulation of particle scattering in terms of conditional probabilities, one would impose the conditions on the surfaces S_1 and S_3 in the nearly flat region. However, one would not impose conditions on any closed universes that branched off or joined on between S_1 and S_3, because one could not observe them. Thus, the initial or conditional state would not contain any closed universes. A closed universe that branched off between S_1 and S_2 (or between S_2 and S_3) would be regarded as having been created. If it joined up again between S_1 and S_2 (S_2 and S_3, respectively), it would be regarded as having been annihilated again. Otherwise, it would be regarded as part of the final state. An observer in the nearly flat region would be able to measure only the part of the final state on S_2 and not the state of the closed universe. He would therefore have to sum over all possibilities for the closed universes. This summation would mean that the part of the final state that he could observe would appear to be in a mixed state rather than in a pure quantum state.

VI. THE LOSS OF QUANTUM COHERENCE

Let $|\alpha_i\rangle$ be a basis for the flat-space Fock space $\mathcal{F}_p$ and $|\beta\rangle_j$ be a basis for the wormhole Fock space $\mathcal{F}_w$. In case (2) above, in which there are no wormholes initially, the initial, or conditional, state can be written as the state

$$\lambda^i |\alpha_i\rangle |O\rangle_w ,$$

where $|O\rangle_w$ is the zero closed-universe state in $\mathcal{F}_w$. The final state can be written as

$$\mu^{ij} |\alpha_i\rangle |\beta_j\rangle .$$

However, an observer in the nearly flat region can measure only the states $|\alpha_i\rangle$ on S_2, and not the closed-universe states $|\beta_j\rangle$. He would therefore have to sum over all possible states for the closed universes. This would give a mixed state in the $\mathcal{F}_p$ Fock space with density matrix

$$\rho_k^i = \mu^{ij} \bar{\mu}_{kj} .$$

The matrix ρ^{ik} will be Hermitian and positive semidefinite, if the final state is normalized in $\mathcal{H}$:

$$\mathrm{tr}\rho = \mu^{ij} \bar{\mu}_{ij} = 1 .$$

These are the properties required for it to be interpreted as the density matrix of a mixed quantum state. A measure of the loss quantum coherence is

$$1 - \mathrm{tr}(\rho^2) = 1 - \mu^{ij} \mu^{kl} \bar{\mu}_{il} \bar{\mu}_{kj} .$$

This will be zero if the final state is a pure quantum state. Another measure is the entropy which can be defined as

$$-\mathrm{tr}(\rho \ln\rho) .$$

This again will be zero for a pure quantum state.

If case (3) above is realized, the initial closed-universe state is not the no-wormhole state $|O\rangle_w$, but a coherent state $|q\rangle_w$ such that

$$(c_{nm} + c_{nm}^\dagger) |q\rangle_w = q_{nm} |q\rangle_w .$$

The effective interactions would leave the closed-universe sector in the same coherent state. Thus the final state would be the product of some state in $\mathcal{F}_p$ with the coherent state $|q\rangle_w$. There would be no loss of quantum coherence, but one would have effective ϕ^m and other interactions whose coefficients would depend on the eigenvalues q_{nm}, etc. It would seem that these could have any value.

VII. WORMHOLE EFFECTIVE INTERACTIONS

There will be no significant interaction between wormholes, unless they are within a Planck length of each other. Thus, the creation and annihilation operators for wormholes are practically independent of the positions in the asymptotically flat region. This means that the effective propagator of a wormhole excited state is $\delta^4(p)$. Using the propagator one can calculate Feynman diagrams that include wormholes, in the usual manner.

The interactions of wormholes with m scalar particles in the $n=0$ mode are alarmingly large. The $m=1$ case would be a disaster; it would give the scalar field a propagator that was independent of position because a scalar particle could go into a wormhole whose other end was at a great distance in the asymptotically flat region. Suppose, however, that the scalar field were coupled to a Yang-Mills field. One would have to average over all orientations of the gauge group for the closed universe. This would make the matrix element zero, except for closed-universe states that were Yang-Mills singlets. In particular, the matrix element would be zero for $m=1$. A special case is the gauge group Z_2. Such fields are known as twisted scalars. They can reverse sign on going round a closed loop. They will have zero matrix elements for m odd because one will have to sum over both signs.

Consider now the matrix element for the scalar field, and its complex conjugate, between the vacuum and a closed universe containing a scalar particle and antiparticle in the $n=0$ mode. This will be nonzero, because a particle-antiparticle state contains a Yang-Mills singlet. It would give an effective interaction of the form

$$m_P^2 \, \mathrm{tr}(\phi\bar{\phi})(c_{011} + c_{011}^\dagger) ,$$

where c_{011} is the annihilation operator for a closed universe with one scalar particle and one antiparticle in the $n=0$ mode. This again would be a disaster; with two of these vertices one could make a closed loop consisting of a closed universe [propagator, $\delta^4(p)$] and a scalar particle (propagator, $1/p^2$). This closed loop would be infrared divergent. One could cut off the divergence by giving the scalar particle a mass, but the effective mass would be the Planck mass. One might be able to remove this mass by renormalization, but the creation of closed universes would mean that a scalar particle would lose quantum coherence within a Planck length. The

$m = 4$ matrix element will give a large ϕ^4 effective vertex.

There seems to be four possibilities in connection with wormholes containing only scalar particles in the $n = 0$ mode.

(1) They may be reduced or canceled in a supersymmetric theory.

(2) The scalar field may be absorbed as a conformal factor in the metric. This could happen, however, only for one scalar field that was a Yang-Mills singlet.

(3) It may be that any scalar particle that we observe is a bound state of particles of higher spin, such as the pion.

(4) The universe may be in a coherent state $|q\rangle_w$ as described above. However, one would then have the problem of why the eigenvalues q should be small or zero. This is similar to the problem of why the θ angle should be so small, but there are now an infinite number of eigenvalues.

In the case of particles of spin $\frac{1}{2}$, the exclusion principle limits the occupation numbers of each mode to zero or 1. Averaging over the orientation O of the wormhole will mean that the lowest-order interaction will be for a wormhole containing one fermion and one antifermion. This would give an effective interaction of the form

$$m_P \psi \bar{\psi} (d_{11} + d_{11}^\dagger) \, ,$$

where d_{11} is the annihilation operator for a closed universe containing a fermion and an antifermion in $n = \frac{1}{2}$ modes. This would give the fermion a mass of the order of the Planck mass. However, if the fermion were chiral, this interaction would cancel out under averaging over orientation and gauge groups. This is because there is no two-chiral-fermion state that is a singlet under both groups. This suggests that supersymmetry might ensure the cancellation of the dangerous interactions with wormholes containing scalar particle in the $n = 0$ mode. Conformally flat wormholes, such as those considered in this paper, should not break supersymmetry.

For chiral fermions, the lowest-order effective interaction will be of the four-Fermi form

$$m_P^{-2} \, \mathrm{tr}(\psi_1 \gamma^\mu \bar{\psi}_1 \psi_2 \gamma_\mu \bar{\psi}_2)(d_{1111} + d_{1111}^\dagger) \, ,$$

where d_{1111} is the annihilation operator for a wormhole containing a fermion and an antifermion each of species 1 and 2. This would lead to baryon decay, but with a lifetime $\sim 10^{50}$ yr. There will also be Yukawa-type effective interactions produced by closed universes containing one scalar particle, one fermion, and one antifermion.

VIII. CONCLUSION

It would be tempting to dismiss the idea of wormholes by saying that they are based on general relativity, and we now all know that string theory is the ultimate theory of quantum gravity. However, string theory, or any other theory of quantum gravity, must reduce to general relativity on scales large compared to the Planck length. Even at the Planck length, the differences from general relativity should be only ~ 1. In particular, the ultimate theory of quantum gravity should reproduce classical black holes and black-hole evaporation. It is difficult to see how one could describe the formation and evaporation of a black hole except as the branching off of a closed universe. I would therefore claim that any reasonable theory of quantum gravity, whether it is supergravity, or superstrings, should allow little closed universes to branch off from our nearly flat region of spacetime.

The effect of these closed universes on ordinary particle physics can be described by effective interactions which create or destroy closed universes. The effective interactions are small, apart from those involving scalar fields. The scalar field interactions may cancel because of supersymmetry. Or, any scalar particles that we observe may be bound states of particles of higher spin. Near a wormhole of the Planck size, such a bound state would behave like the higher-spin particles of which it was made. A third possibility is that the universe is in a coherent $|q\rangle_w$ state. I do not like this possibility because it does not seem to agree with the "no boundary" proposal for the quantum state of the Universe. There also would not seem to be any way to specify the eigenvalues q. Yet the values of the eigenvalues for large particle numbers cannot be zero if these interactions are to reproduce the results of semiclassical calculations on the formation and evaporation of macroscopic black holes.

The effects of little closed universes on ordinary particle physics may be small, apart, possibly, for scalar particles. Nevertheless, it raises an important matter of principle. Because there is no way in which we could measure the quantum state of closed universes that branch off from our nearly flat region, one has to sum over all possible states for such universes. This means that the part of the final state that we can measure will appear to be in a mixed quantum state, rather than a pure state. I think even Gross[12] will agree with that.

[1] R. M. Wald, Commun. Math. Phys. **45**, 9 (1975).

[2] S. W. Hawking, Phys. Rev. D **14**, 2460 (1976).

[3] S. W. Hawking, D. N. Page, and C. N. Pope, Nucl. Phys. **B170**, 283 (1980).

[4] S. W. Hawking, in *Quantum Gravity 2: A Second Oxford Symposium*, edited by C. J. Isham, R. Penrose, and D. W. Sciama (Clarendon, Oxford, 1981).

[5] S. W. Hawking, Commun. Math. Phys. **87**, 395 (1982).

[6] C. W. Misner, Ann. Phys. (N.Y.) **24**, 102 (1963).

[7] J. J. Halliwell and S. W. Hawking, Phys. Rev. D **31**, 1777

(1985).

[8] P. D. D'Eath and J. J. Halliwell, Phys. Rev. D **35**, 1100 (1987).

[9] S. W. Hawking, Nucl. Phys. **B239**, 257 (1984).

[10] S. W. Hawking and D. N. Page, Nucl. Phys. **B264**, 185 (1986).

[11] J. B. Hartle and S. W. Hawking, Phys. Rev. D **28**, 2960 (1983).

[12] D. J. Gross, Nucl. Phys **B236**, 349 (1984).

Nuclear Physics B335 (1990) 155–165
North-Holland

DO WORMHOLES FIX THE CONSTANTS OF NATURE?

S.W. HAWKING

*Department of Applied Mathematics and Theoretical Physics, University of Cambridge,
Silver Street, Cambridge CB3 9EW, UK*

Received 1 August 1989

This paper examines the claim that the wormhole effects that cause the cosmological constant to be zero, also fix the values of all the other effective coupling constants. It is shown that the assumption that wormholes can be replaced by effective interactions is valid in perturbation theory, but it leads to a path integral that does not converge. Even if one ignores this difficulty, the probability measure on the space of effective coupling constants diverges. This does not affect the conclusion that the cosmological constant should be zero. However, to find the probability distribution for other coupling constants, one has to introduce a cutoff in the probability distribution. The results depend very much on the cutoff used. For one choice of cutoff at least, the coupling constants do not have unique values, but have a gaussian probability distribution.

1. Introduction

The aim of this paper is to discuss whether wormholes introduce an extra degree of uncertainty into physics, over and above that normally associated with quantum mechanics [1, 2]. Or whether, as Coleman [3] and Preskill [4] have suggested, the uncertainty is removed by the same mechanism that makes the cosmological constant zero.

Wormholes [5–7] are four-dimensional positive-definite (or euclidean) metrics that consist of narrow throats joining large, nearly flat regions of space-time. One of the original motivations for studying them was to provide a complete quantum treatment of gravitational collapse and black-hole evaporation. If one accepts the "no boundary" proposal [8] for the quantum state of the universe, the class of positive-definite metrics in the path integral, can not have any singularities or edges. There thus has to be somewhere for the particles that fell into the hole, and the antiparticles to the emitted particles, to go to. (In general, these two sets of particles will be different, and so they can not just annihilate with each other.) A wormhole leading off to another region of space-time, would seem to be the most reasonable possibility [5]. If this is indeed the case, one would not be able to measure the part of the quantum state that went down the wormhole. Thus there would be loss of quantum coherence, and the final quantum state in our region of the universe would

285

be a mixed state, rather than a pure quantum state. This would represent an extra degree of uncertainty that was introduced into physics by quantum gravity, over and above the uncertainty normally associated with quantum theory. The entropy of the density matrix of the final state would be a measure of this extra degree of uncertainty.

If macroscopic wormholes occur in the formation and evaporation of black holes, one would expect that there would also be a whole spectrum of wormholes down to the Planck size, and maybe beyond. One might expect that such very small wormholes would be branching off from our region of space-time all the time. So how is it that quantum coherence seems to be conserved in normal situations? The answer [9, 10] seems to be that for microscopic wormholes, the extra degree of uncertainty can be absorbed into an uncertainty about the values of physical coupling constants. The argument goes as follows:

Step 1. Because Planck-size wormholes are much smaller than the scales on which we can observe, one would not see wormholes as such. Instead, they would appear as point interactions, in which a number of particles appeared or disappeared from our region of the universe. Energy, momentum, and gauge charges would be conserved in these interactions, so they could be represented, at least in perturbation theory, by the addition of gauge invariant effective interaction terms $\theta_i(\phi)$ to the lagrangian, where ϕ are the low-energy effective fields in the large regions [5, 6]. It is implicitly assumed that there is a discrete spectrum of wormhole states labelled by the index i. This will be discussed in another paper [11].

Step 2. The strengths of the effective interactions will depend on the amplitudes for the wormholes to join on. This in turn will depend on what is at the other end of the wormholes. In the dilute wormhole approximation, each wormhole is assumed to connect two large regions, and the amplitudes are assumed to depend only on the vertex functions θ_i at each end. Thus the effect of wormholes smaller than the scale on which we can observe, can be represented by a bi-local effective addition to the action [10]:

$$-\tfrac{1}{2}\sum \Delta^{ij}\int \mathrm{d}^4x\sqrt{g(x)}\,\theta_i(x)\int \mathrm{d}^4y\sqrt{g(y)}\,\theta_j(y).$$

The position independent matrix Δ^{ij} can be set to the unit matrix by a choice of the basis of wormhole state and normalization of the vertex functions θ_i. The question of the sign of the bi-local action will be discussed later.

Step 3. The bi-local action can be transformed into a sum of local additions to the action by using the identity [10]

$$\exp\left[\tfrac{1}{2}\int \mathrm{d}^4x\sqrt{g(x)}\,\theta(x)\int \mathrm{d}^4y\sqrt{g(y)}\,\theta(y)\right]$$

$$= (\pi)^{-1/2}\int \mathrm{d}\alpha\exp\left[-\tfrac{1}{2}\alpha^2\right]\exp\left[-\alpha\int \mathrm{d}^4x\sqrt{g(x)}\,\theta(x)\right].$$

This means that the path integral

$$Z = \int d[\phi] \exp\left[-\int d^4x \sqrt{g}\, L\right] \exp\left[-\tfrac{1}{2}\sum \int d^4x \sqrt{g(x)}\, \theta_i(x) \int d^4y \sqrt{g(y)}\, \theta_i(y)\right]$$

becomes

$$Z = \int d\alpha_i\, P(\alpha_i) Z(\alpha_i),$$

where

$$P(\alpha_i) = \exp\left[-\tfrac{1}{2}\sum \alpha_j \alpha_j\right],$$

$$Z(\alpha_i) = \int d[\phi] \exp\left[-\int d^4x \sqrt{g}\left(L + \sum \alpha_j \theta_j\right)\right].$$

This can be interpreted as dividing the quantum state of the universe into noninteracting super selection sectors labelled by the parameters α_i. In each sector, the effective lagrangian is the ordinary lagrangian L, plus an α dependent term, $\sum \alpha_i \theta_i$. The different sectors are weighted by the probability distribution $P(\alpha)$. Thus the effective interactions θ_i do not have unique values of their couplings. Rather, there is a spread of possible couplings α_i. This smearing of the physical coupling constants is the reflection for Planck-scale wormholes of the extra degree of uncertainty introduced by black-hole evaporation. It means that even if the underlying theory is superstrings, the effective theory of quantum gravity will appear to be unrenormalizable, with an infinite number of coupling constants that can not be predicted, but have to be fixed by observation [2].

Coleman [3] however has suggested that the probability distributions for the coupling constants are entirely concentrated at certain definite values, that could, in principle, be calculated. The argument is based on a proposal for explaining the vanishing of the cosmological constant [12], and goes as follows:

Step 4. The probability distribution $P(\alpha)$ for the α parameters should be modified by the factor $Z(\alpha)$ which is given by the path integral over all low energy fields ϕ with the effective interactions $\sum \alpha_i \theta_i$.

Step 5. The path integral for $Z(\alpha)$ does not converge, because the Einstein–Hilbert action is not bounded below. However, one might hope that an estimate for $Z(\alpha)$ could be obtained from the saddle point in the path integral, that is, from solutions of the euclidean field equations. If one takes the gravitational action to be

$$\int d^4x \sqrt{g}\left(\Lambda(\alpha) - \frac{1}{16\pi G(\alpha)} R + O(R^2)\right),$$

the saddle point will be a sphere of radius $\sqrt{3/8\pi G \Lambda}$ and action $-3/8G^2(\alpha)\Lambda(\alpha)$.

If one just took a single sphere, $Z(\alpha)$ would be $\exp(3/8G^2\Lambda)$. However, Coleman argues that there can be many such spheres connected by wormholes. Thus

$$Z(\alpha) = \exp\left(\exp\left(\frac{3}{8G^2\Lambda}\right)\right).$$

Either the single or the double exponentials blow up so rapidly, as Λ approaches zero from above, that the probability distribution will be concentrated entirely at those α for which $\Lambda = 0$ [3,12].

Step 6. The argument to fix the other effective couplings takes at least two alternative forms:

(i) Coleman's original proposal [3] was that the effective action for a single sphere should be expanded in a power series in Λ. The leading term will be $-3/8G\Lambda$, but there will be higher-order corrections arising from the higher powers of the curvature in the effective action:

$$\Gamma = -\frac{3}{8G\Lambda} + f(\hat{\alpha}) + \Lambda g(\hat{\alpha}) + \ldots,$$

where $\hat{\alpha}$ are the directions in the α parameter space orthogonal to the direction in which $\Lambda(\alpha)$ varies. The higher-order corrections to Γ would not make much difference if $Z(\alpha) = e^{-\Gamma}$. But if

$$Z(\alpha) = \exp(\exp(-\Gamma)),$$

then

$$\frac{\delta Z(\alpha)}{Z(\alpha)} = e^{-\Gamma}\delta\Gamma.$$

The factor, $e^{-\Gamma}$, will be very large for Λ small and positive. Thus a small correction to Γ will have a big effect on the probability. This would cause the probability distribution to be concentrated entirely at the minimum of the coefficient, $f(\hat{\alpha})$, in the power series expansion of Γ (always assuming that f has a minimum). Similarly, one would expect the probability distribution to be concentrated entirely at the minimum of the minimum of the higher coefficients in the power series expansion. This would lead to an infinite number of conditions on the α parameters. It is hoped that these would cause the probability distribution to be concentrated entirely at a single value of the effective couplings, α.

(ii) An alternative mechanism for fixing the effective couplings has been suggested by Preskill [4]. If the dominant term in Γ is $-3/G^2\Lambda$, one might expect that the probability distribution would be concentrated entirely at $G(\alpha) = 0$, as well as at $\Lambda(\alpha) = 0$. However, we know that $G(\alpha) \neq 0$, because we observe gravity. So there must be some minimum value of $G(\alpha)$. One would expect that the probability distribution would be concentrated entirely at this minimum value, and one would hope that the minimum would occur at a single value of the effective couplings, α.

This paper will examine the validity of the above steps. Steps 1 and 2 are usually assumed without any supporting calculations. However, an explicit calculation is given in sect. 2, for the case of a scalar field. This confirms that wormholes can indeed be replaced by a bi-local action, at least for the calculation of low-energy Green functions in perturbation theory. The sign of the bi-local action is that required for the use of the identity in step 3. However, the sign also means that the path integral does not converge, even in the case of a scalar field on a background geometry. Thus the procedure of using the effective actions to calculate a background geometry for each set of α parameters, is suspect. However, if one is prepared to accept it, one would indeed expect that Γ would diverge on a hypersurface in α space, on which $\Lambda = 0$. Thus the cosmological constant will be zero, without any uncertainty. However, to calculate the probability distributions of the other effective coupling constants, one has to introduce a cutoff for the divergent probability measure. Different cutoffs will give different answers. Indeed, a natural cutoff will just give the probability distribution $P(\alpha)$ for all effective couplings except the cosmological constant. Thus one can not conclude that the effective couplings will be given unique values by wormholes.

2. The bi-local action

In this section, it will be shown that scalar field Green functions on a class of wormhole backgrounds can be calculated approximately from a bi-local addition to the scalar field action in flat space. In particular, the sign of the bi-local action will be obtained. The wormhole backgrounds will be taken to be hyperspherically symmetric, like all the specific examples considered so far. This means that they are conformally flat. For definiteness, the conformal factor will be taken to be

$$ds^2 = \left(1 + \frac{b^2}{(x - x_0)^2}\right)^2 dx^2.$$

This is the wormhole solution for a conformally scalar field [13], or a Yang–Mills field [14]. In the case of a minimally coupled scalar [7], the conformal factor will have the same asymptotic form at infinity, and near x_0, the infinity in the other asymptotically euclidean region. The conformal factors will differ slightly in the region of the throat, but this will just make the bi-local action slightly different.

The metric given above appears to be singular at the point x_0. However, one can see that this is really infinity in another asymptotically euclidean region, by introducing new coordinates that are asymptotically euclidean in the other region

$$y^\nu = \frac{O_\mu^\nu (x^\mu - x_0^\mu)}{(x - x_0)^2} + y_0^\nu,$$

where O_μ^ν is an orthogonal matrix. In order to study low-energy physics in the asymptotically euclidean regions, one needs to know the Green functions for points $x_1, x_2, \ldots$ and $y_1, y_2, \ldots$ in the two regions, far from the throat. Consider the Green function for a point x in one asymptotic region, and a point y in the other. Since the wormhole metric given above has $R = 0$, the conformally and minimally coupled scalar fields will have the same Green functions. One can therefore calculate the Green function using conformal invariance as

$$G(x, y) = \Omega(\tilde{x})^{-1} \frac{1}{(\tilde{x} - x)^2} \Omega(x)^{-1},$$

where $\tilde{x}$ is the image of the point y under the transformation above. For x and y far from the wormhole ends, x_0 and y_0,

$$\Omega(x) \approx 1, \qquad \Omega(\tilde{x}) \approx \frac{b^2}{(y - y_0)^2}, \qquad \tilde{x} \approx x_0.$$

Thus

$$G(x, y) \approx \frac{b^2}{(x - x_0)^2 (y - y_0)^2}.$$

This is what one would have obtained from a bi-local interaction of the form

$$-\tfrac{1}{2} \int d^4 x_0 \, b\phi(x_0) \int d^4 y_0 \, b\phi(y_0).$$

Note that the bi-local action has a negative sign. This is because the Green functions are positive.

Now consider two points x_1, x_2 and y_1, y_2 in each asymptotic region. The four-point function will contain a term, $G(x_1, y_1)G(x_2, y_2)$, which will be given approximately by the bi-local action

$$-\tfrac{1}{2} \int d^4 x_0 \, b^2 \phi^2(x_0) \int d^4 y_0 \, b^2 \phi^2(y_0).$$

In general, Green functions involving n-points in each asymptotic region will be given by bi-local actions with vertex functions $\theta(x)$ of the form, $b^n \phi^n(x)$. If one takes gravitational interactions into account, one would expect that the bi-local action would be multiplied by a factor e^{-I_n}, where $I_n \propto n/G$ is the action for a wormhole containing n scalar particles.

One can also consider higher-order corrections to the Green functions on a wormhole background which arise because the image, $\tilde{x}$, of the point, y, is not exactly at x_0. These will be reproduced by bi-local actions involving vertex functions containing derivatives of the scalar field. Only those vertex functions that are scalar combinations of derivatives will survive averaging over the orthogonal matrix O, which specifies the rotation of one asymptotically euclidean region with respect to the other. Thus the vertex terms and the effective action will be Lorentz invariant. It seems that any scalar polynomial in the scalar field and its covariant derivatives can occur as a vertex function.

Earlier this year, B. Grinstein and J. Maharana [15] performed a similar calculation.

3. Convergence of the path integral

The bi-local action has a negative sign, so it appears in the path integral as a positive exponential. This is what is required in order to introduce the α parameters using the identity in step 3. If the bi-local action had the opposite sign, the integral over the α parameters would be $\int d\alpha e^{+\alpha^2/2}$, which would not converge. On the other hand, because the bi-local action is negative, the path integral will not converge. This is true even in the case of the path integral over a scalar field on a non-dynamic wormhole background. There will be vertex functions of the form ϕ^n for each n. In the case of even n, the integral $\int d^4y\, \phi^n(y)$ will be positive. This means that $-\int d^4x\, \phi^n(x)$, the other part of the bi-local action, will give ϕ an effective potential that is unbounded below. Thus the path integral over ϕ, with the bi-local effective action, will not converge. This does not mean that scalar field theory on non-dynamical wormhole backgrounds is not well defined. What it *does* mean is that a bi-local action gives a reasonable approximation to the effect of wormholes on low-energy Green functions, in perturbation theory. But one should not take the bi-local action too literally. One can see this if one considers introducing the α parameters. One will then get a scalar potential which is a polynomial in ϕ, with α-dependent coefficients. For certain values of the α, there will be metastable states, and decay of the false vacuum. But these obviously have no physical reality. The moral therefore is that one can use a bi-local action to represent the effect of wormholes in perturbation theory. But one should be wary of using the bi-local action to calculate non-perturbative effects, like vacuum states.

This is even more true of the effective gravitational interactions of wormholes. It is not clear whether there is a direct contribution of wormholes to the cosmological constant, i.e. whether any of the vertex functions contain a constant term. This would show up only in the pure trace contribution to linearized gravitational Green functions in the presence of a wormhole. So far, these have not been calculated. However, even if there is no direct wormhole contribution to the cosmological constant, there will be indirect contributions arising from loops involving other effective interactions. These will be cut off on the scale of the wormholes, that is, on the scale on which the wormholes no longer appear to join on at a single point. In a similar manner, there does not seem to be a wormhole that makes a direct contribution to the Einstein lagrangian, R, and hence to Newton's constant. By analogy with the case of wormholes with electromagnetic and fermion fields, one would expect that such a wormhole would have to contain just a single graviton. However, its effect would average to zero under rotations of the wormhole, described by the matrix O. However, there will again be indirect contributions to $1/G$ from loops involving other effective interactions.

There are convergence problems with gravitational path integrals, even in the absence of wormholes. The Einstein–Hilbert action $-\int d^4x \sqrt{g}\,(R/16\pi G - \Lambda)$ is not bounded below, because conformal transformations of the metric can make the action arbitrarily negative. Still, one might hope that the dominant contribution to the path integral would come from metrics that were saddle points of the action, that is they were solutions of the euclidean field equations. The spherical metric given in sect. 1 has the lowest action of any solution of the euclidean field equations with a given value of Λ. One might therefore expect that

$$\Gamma = -\frac{3}{8G^2\Lambda}\,.$$

The problem of the convergence of the path integral is much worse however, if one replaces wormholes with a bi-local action. If there were a direct wormhole contribution to the effective cosmological constant, the path integral would contain a factor e^{CV^2}, where V is the volume of space-time. If the constant C were negative, the integral over α would not converge. But if C were positive, the path integral would diverge. Even rotating the contour of the conformal factor to the imaginary axis would not help, because in four dimensions it would leave the volume real and positive. One might still hope that the saddle point of the effective action would give an estimate of the path integral. However, the bi-local action would give rise to an effective cosmological constant of value $-2CV$. Unless this were balanced by a very large positive cosmological constant of non-wormhole origin, the action of any compact solution of the euclidean field equations would be positive. So it would be suppressed, rather than enhanced, as in the case of the sphere. Even if there were a large positive non-wormhole cosmological constant, it would not give a solution of infinite volume, with zero effective cosmological constant. One might still use the α identity, replace the bi-local action with a weighted sum over path integrals with an α-dependent cosmological constant. But if gravitational path integrals can be made sense of only by taking the saddle point, one should presumably also take the saddle point in the integral over α. In the case of a single exponential, this would give

$$\alpha + \frac{3\sqrt{C}}{8G^2\left(\Lambda_0 + \alpha\sqrt{C}\right)^2} = 0,$$

and in the case of a double exponential

$$-\frac{2}{\alpha} + \frac{3\sqrt{C}}{8G^2\left(\Lambda_0 + \alpha\sqrt{C}\right)^2} = 0,$$

where Λ_0 is the non-wormhole contribution to the cosmological constant. In either case, the effective cosmological constant at the saddle point will be of the order of Λ_0.

4. The divergence of the probability measure

Suppose, as one often does, one ignores problems about the convergence of the path integral. Then, as described in sect. 1, there will be a probability measure on the space of the α parameters

$$\mu(\alpha) = P(\alpha)Z(\alpha),$$

where $P(\alpha) = \exp[\Sigma - \frac{1}{2}\alpha_i\alpha_i]$ and $Z(\alpha) = \exp[-\Gamma(\alpha)]$ or $\exp[\exp[-\Gamma(\alpha)]]$. If

$$\Gamma \approx -\frac{3}{8G^2(\alpha)\Lambda(\alpha)}$$

and $G^2\Lambda$ vanishes on some surface K in α space, the measure $\mu(\alpha)$ will diverge. That is to say, the total measure of α space will be infinite.

The total measure of the part of α space for which $G^2\Lambda > \epsilon > 0$ may well be finite. In this case, one could say that

$$G^2\Lambda = 0,$$

with probability one. Since we observe that $G \neq 0$, one could deduce that

$$\Lambda = 0.$$

However, with such a badly divergent probability measure, this is about the only conclusion one could draw. To go further, and to try to argue as in sect. 1, that the probability measure is concentrated entirely at a certain point in α space, one has to introduce some cutoff in the probability measure. One then takes the limit as the cutoff is removed. The trouble is, different ways of cutting off the probability measure will give different results. And it hard to see why one cut-off procedure should be preferred to another.

One can cut off the probability measure by introducing a function F on α, which is zero on the surface K where $1/\Gamma = 0$, and which is positive for small negative $1/\Gamma$. One then cuts the region $0 \leqslant F < \epsilon$ out of α space. One would expect the probability measure on the rest of α space to be finite, and therefore to give a well-defined probability distribution for the effective coupling constants. If $Z(\alpha)$ is given by a double exponential, the probability distribution will be highly concentrated near the minimum of Γ on the surface, $F = \epsilon$. Thus, in the limit ϵ tends to zero, the probability would be concentrated entirely at a single point of α space. But the point will depend on the choice of the function F, and different choices will give different results. For example, Coleman's procedure [3] is equivalent to choosing $F = \Lambda$. On the other hand, Preskill [4] has suggested using a cutoff on the volume of space-time. This would be equivalent to using

$$F = G^2 \Lambda^2 .$$

But if you minimise $G^2\Lambda$ for fixed $G^2\Lambda^2$, you would drive G to zero and Λ to a non-zero value, if G can be zero anywhere in α space. This is not what one wants. One therefore has to suppose that G is bounded away from zero, at least in the region of α space in which the bi-local action is a reasonable approximation for wormholes.

It seems therefore that one can get different results by different methods of cutting off the divergence in the probability measure. There does not seem to be a unique preferred cutoff. A possible candidate would be to use Γ or $Z(\alpha)$ themselves to define the cutoff; for example, $F = -1/\Gamma$. This would lead to $\Lambda = 0$, but the other effective couplings would be distributed with the probability distribution $P(\alpha)$. In this case, wormholes would have introduced an extra degree of uncertainty into physics. This uncertainty would reflect the fact that we can observe only our large region of the universe, and not the major part of space-time, which is down a wormhole, beyond our ken.

References

[1] S.W. Hawking, Commun. Math. Phys. 87 (1982) 395
[2] S.W. Hawking and R. Laflamme, Phys. Lett. B209 (1988) 39
[3] S. Coleman, Nucl. Phys. B310 (1988) 643

[4] J. Preskill, Nucl. Phys. B323 (1989) 141

[5] S.W. Hawking, Phys. Lett. B195 (1987) 337

[6] S.W. Hawking, Phys. Rev. D37 (1988) 904

[7] S.B. Giddings and A. Strominger, Nucl. Phys. B306 (1988) 890

[8] J.B. Hartle and S.W. Hawking, Phys. Rev. D28 (1983) 2960

[9] S. Coleman, Nucl. Phys. B307 (1988) 864

[10] I. Klebanov, L. Susskind and T. Banks, Nucl. Phys. B317 (1989) 665

[11] S.W. Hawking and D. Page, in preparation

[12] S.W. Hawking, Phys. Lett. B134 (1984) 402

[13] J.J. Halliwell and R. Laflamme, Santa Barbara ITP preprint NSF-ITP-89-41 (1989)

[14] A. Hosoya and W. Ogura, Phys. Lett. B225 (1989) 117

[15] B. Grinstein and J. Maharana, Fermilab preprint FERMILAB-PUB-89/121-T (1989)

Commun. Math. Phys. 148, 345–352 (1992)

Selection Rules for Topology Change *

G. W. Gibbons and S. W. Hawking

D.A.M.T.P., Silver Street, Cambridge CB3 9EW, UK

Received September 7, 1991; in revised form October 25, 1991

Abstract. It is shown that there are restrictions on the possible changes of topology of space sections of the universe if this topology change takes place in a compact region which has a Lorentzian metric and spinor structure. In particular, it is impossible to create a single wormhole or attach a single handle to a spacetime but it is kinematically possible to create such wormholes in pairs. Another way of saying this is that there is a $\mathbb{Z}_2$ invariant for a closed oriented 3-manifold Σ which determines whether Σ can be the spacelike boundary of a compact manifold M which admits a Lorentzian metric and a spinor structure. We evaluate this invariant in terms of the homology groups of Σ and find that it is the mod 2 Kervaire semi-characteristic.

Introduction

There has been great interest recently in the possibility that the topology of space may change in a semi-classical theory of quantum gravity in which one assumes the existence of an everywhere non-singular Lorentzian metric $g_{\alpha\beta}^L$ of signature $-+++$. In particular, Thorne, Frolov, Novikov and others have speculated that an advanced civilization might at some time in our future be able to change the topology of space sections of the universe so that they developed a wormhole or handle [1–3]. If one were to be able to control such a topology change, it would have to occur in a compact region of spacetime without singularities at which the equations broke down and without extra unpredictable information entering the spacetime from infinity. Thus if we assume, for convenience, that space is compact now, then the suggestion amounts to saying that the 4-dimensional spacetime manifold M, which we assume to be smooth and connected, is compact with boundary $\partial M = \Sigma$ consisting of 2 connected components, one of which has topology S^3 and the other of which has topology $S^1 \times S^2$, and both are spacelike

---

* e-mail addresses: GWG1@phx.cam.ac.uk, SWH1@phx.cam.ac.uk

with respect to the Lorentzian metric $g_{\alpha\beta}^L$. If $(M, g_{\alpha\beta}^L)$ is assumed time-oriented, which we will justify later, then the S^3 component should be the past boundary of M and the $S^1 \times S^2$ component should be the future boundary of M. Spacetimes of this type have previously been thought to be of no physical interest because a theorem of Geroch [4] states that they must contain closed timelike curves. In the last few years, however, people have begun to consider seriously whether such causality violating spacetimes might be permitted by the laws of physics. One of the main results of this paper is that even if causality violations are allowed, there is an even greater obstacle to considering such a spacetime as physically reasonable – it does not admit an $SL(2, \mathbb{C})$ spinor structure and therefore it is simply not possible on purely kinematical grounds to contemplate a civilization, no matter how advanced constructing a wormhole of this type, provided one assumes that the existence of two-component Weyl fermions is an essential ingredient of any successful theory of nature. We will discuss later the extent to which one might circumvent this result by appealing to more exotic possibilities such as $Spin^c$ structures.

It appears, however, that there is no difficulty in imagining an advanced civilization constructing a pair of wormholes, i.e. that the final boundary is the connected sum of 2 copies of $S^1 \times S^2, S^1 \times S^2 \# S^1 \times S^2$. Thus one may interpret our results as providing a new topological conservation law for wormholes, they must be conserved modulo 2. More generally we are able to associate with any closed orientable 3-manifold Σ a topological invariant, call it u (for universe) such that $u = 0$ if Σ

(1) bounds a smooth connected compact Lorentz 4-manifold M which admits an $SL(2, \mathbb{C})$ spinor structure;
(2) is spacelike with respect to the Lorentz metric $g_{\alpha\beta}^L$,

and $u = 1$ otherwise.

We shall show that this invariant is additive modulo 2 under disjoint union of 3-manifolds,

$$u(\Sigma_1 \cup \Sigma_2) = u(\Sigma_1) + (\Sigma_2) \bmod 2.$$

Under the connected sum it satisfies

$$u(\Sigma_1 \# \Sigma_2) = u(\Sigma_1) + u(\Sigma_2) + 1 \bmod 2.$$

The connected sum, $X \# Y$ of two manifolds X, Y of the same dimension n is obtained by removing an n-ball B^n and from X and Y and gluing the two manifolds together across the common S^{n-1} boundary component so created. We shall also show that $u(S^3) = 1$, and $u(S^1 \times S^2) = 0$. The result that one cannot create a single wormhole then follows immediately from the formula for disjoint unions while the fact that one can create pairs of wormholes follows from the formula for connected sums. Another consequence of these formulae is that for the disjoint union of k S^3's, $u = k$ modulo 2. In particular, this prohibits the "creation from nothing" of a single S^3 universe with a Lorentz metric and spinor structure.

Our invariant u may be expressed in terms of rather more familiar topological invariants of 3-manifolds. In fact,

$$u = \dim_{\mathbb{Z}_2}(H_0(\Sigma; \mathbb{Z}_2) \oplus H_1(\Sigma; \mathbb{Z}_2)) \bmod 2,$$

where $H_0(\Sigma; \mathbb{Z}_2)$ is the zeroth and $H_1(\Sigma; \mathbb{Z}_2)$ the first homology group of Σ with $\mathbb{Z}_2$ coefficients. Thus $\dim_{\mathbb{Z}_2} H_0(\Sigma; \mathbb{Z}_2) \bmod 2$ counts the number of connected components modulo 2. The right-hand side of this expression for u is sometimes referred to as the mod 2 Kervaire semi-characteristic.

So far we have considered the case where the space sections of the universe are closed. We can extend these results to cases where the space sections of the universe may be non-compact but the topology change takes place in a compact region bounded by a timelike tube. Such spacetimes may be obtained from the ones we have considered by removing a tubular neighbourhood of a timelike curve.

It seems that a selection rule of this type derived in this paper occurs only if one insists on an everywhere non-singular Lorentzian metric. If one gives up the Lorentzian metric and passes to a Riemannian metric or if one adopts a "first order formalism" in which one treats the vierbein field as the primary variable and allows the legs of the vierbein to become linearly dependent at some points in spacetime then our selection rule would not necessarily apply. However, in the context of asking what an advanced civilization is capable of neither of these possibilities seems reasonable. At the quantum level, however, both are rather natural and in view of the existence of a number of examples there seems to be little reason to doubt that the topology of space can fluctuate at the quantum level. For the purposes of the present paper we will adhere to the assumption of an everywhere non-singular Lorentz metric.

Spin-Cobordism and Lorentz-Cobordism

Every closed oriented 3-manifold admits a $Spin(3) \equiv SU(2)$ spin structure. If the 3-manifold is not simply connected the spin structure is not unique. The set of spin structures is in $1-1$-correspondence with elements of $H^1(\Sigma; \mathbb{Z}_2)$, the first cohomology group of the 3-manifold Σ with $\mathbb{Z}_2$ coefficients. Given a closed oriented 3-manifold Σ one can always find a spin-cobordism, that is there always exists a compact orientable 4-manifold M with boundary $\partial M = \Sigma$ and such that M admits a $Spin(4) \equiv SU(2) \times SU(2)$ spin structure which when restricted to the boundary Σ coincides with any given spin structure on Σ [5].

A closed 3-manifold Σ is said to admit a Lorentz-cobordism if one can find a compact 4-manifold M whose boundary $\partial M = \Sigma$ together with an everywhere non-singular Lorentzian metric with respect to which the boundary Σ is spacelike. A necessary and sufficient condition for a Lorentz-cobordism is that the manifold M should admit a line field $\mathbf{V}$, i.e. a pair $(\mathbf{V}, -\mathbf{V})$ at each point, where $\mathbf{V}$ is a non-zero vector which is transverse to the boundary ∂M. To show this one uses the fact that any compact manifold admits a Riemannian metric $g_{\alpha\beta}^R$. If one has a line field $\mathbf{V}$, one can define a Lorentzian metric $g_{\alpha\beta}^L$ by

$$g^{L\alpha\beta} = g^{R\alpha\beta} - 2V^\alpha V^\beta / (g_{\alpha\beta}^R V^\alpha V^\beta).$$

Alternatively, given a Lorentzian metric $g_{\alpha\beta}^L$ one can diagonalize it with respect to the Riemannian metric $g_{\alpha\beta}^R$. One can choose $\mathbf{V}$ to be the eigenvector with negative eigenvalue. The Lorentzian metric $g_{\alpha\beta}^L$ will be time-orientable if and only if one can choose a consistent sign for $\mathbf{V}$. For physical reasons we shall generally assume time-orientability. If $M, g_{\alpha\beta}^L$ is not time-orientable, it will have a double cover that is, with twice as many boundary components.

If one has a time-orientable Lorentz-cobordism, the various connected components of the boundary lie either in the past or in the future. Thus one might think that one should specify in the boundary data for a Lorentz-cobordism a specification of which connected components lie in the future and which lie in the past. However, it is not difficult to show that given a time-oriented Lorentz-cobordism for which a particular component lies in, say the future, one can construct another time-oriented Lorentz-cobordism for which that component lies in the past and the remaining components are as they were in the first Lorentz-cobordism. The construction is as follows. Let Σ be the component in question. Consider the Riemannian product metric on $\Sigma \times I$, where I is the closed interval $-1 \le t \le 1$. Now by virtue of being a closed orientable 3-manifold Σ admits an everywhere non-vanishing vector field $\mathbf{U}$ which may be normalized to have unit length with respect to the metric on Σ. To give $\Sigma \times I$ a time-orientable Lorentz metric we choose as our everywhere non-vanishing unit timelike vector field $\mathbf{V}$:

$$\mathbf{V} = a(t)\frac{\partial}{\partial t} + b(t)\mathbf{U},$$

where $a^2 + b^2 = 1$ and $a(t)$ passes smoothly and monotonically from -1 at $t = +1$ to $+1$ at $t = 1$. Thus $\mathbf{V}$ is outward directed on both boundary components. One can now attach a copy of $\Sigma \times I$ with this metric, or its time reversed version, to the given Lorentz-cobordism so reversing the direction of time at the boundary desired component. Of course, one will have to arrange that the metrics match smoothly but this is always possible. Considered in its own right the spacetime we have just used could serve as a model for the "creation from nothing" of a pair of twin universes. In general, it will not be geodesically complete and it contains closed timelike curves inside the Cauchy Horizons which occur at the two values of t for which $a^2 = b^2$. However, it is a perfectly valid Lorentz-cobordism.

If a Lorentzian spacetime admits an $SL(2, \mathbb{C})$ spinor structure it must be both orientable and time-orientable and in addition admit a $Spin(4)$ structure [9, 10]. For example, since any closed orientable 3-manifold is a spin manifold, the time reversing product metric we constructed above admits an $SL(2, \mathbb{C})$ structure. By contrast the next example, which could be said to represent the creation of a single, i.e. connected, universe from nothing, does not admit an $SL(2, \mathbb{C})$ spinor structure because it is not time-orientable. Let Σ be a closed connected orientable Riemannian 3-manifold admitting a free involution Γ which is an isometry of the 3-metric on Σ. A Lorentz-cobordism for Σ is obtained by taking $\Sigma \times I$ as before but now with the product Lorentzian metric, i.e. with $a = 1$ and $b = 0$. One now identifies points under the free $\mathbb{Z}_2$ action which is the composition of the involution Γ acting on Σ and reversal of the time coordinate t on the interval I, $-1 \le t \le 1$. Because its double cover has no closed time like curves, the identified space has none either. Of course, it may be that two points x^α and x'^α lying on a timelike curve γ in $\Sigma \times I$ are images of one another under the involution Γ. On the identified space $(\Sigma \times I)/\Gamma$ the timelike curve γ will thus intersect itself. However, the two tangent vectors at the identified point lie in different halves of the light cone at that point. Thus a particle moving along such a curve may set out into the future and subsequently return from the future or *vice versa*. This is not what is meant by a closed timelike curve because if such a curve has a discontinuity in its tangent vector at some point the two tangent vectors must lie in the same half of the light cone at that point.

The special case when Σ is the standard round 3-sphere and the involution Γ is the antipodal map gives a Lorentz-cobordism for a single S^3 universe. If one modifies the product metric by multiplying the metric on Σ by a square of scale factor which is a non-vanishing even function of time one obtains a Friedman-Lemaitre-Robertson-Walker metric. Identifying points in the way described above is referred to as the "elliptic interpretation". A particular case arises when one considers de-Sitter spacetime. If one regards this as a quadric in 5-dimensional Minkowski spacetime the identification is of antipodal points on the quadric. In this case there are no timelike or lightlike curves joining antipodal points, however, there remains a number of difficulties with this interpretation from the point of view of physics [11], not the least of which is the absence of a spinor structure. In fact, as we shall see below, this problem is quite general: there is no spin-Lorentz-cobordism for a single S^3 universe.

A necessary and sufficient condition for the existence of a line field transverse to the boundary ∂M of a compact manifold M is, by a theorem of Hopf, the vanishing of the Euler characteristic $\chi(M)$. Given an oriented cobordism M of Σ, one can obtain another cobordism by taking the connected sum of M and a compact four manifold without boundary. Under connected sums of 4-manifolds the Euler characteristic obeys the equation

$$\chi(M_1 \# M_2) = \chi(M-1) + \chi(M_2) - 2.$$

Thus we can increase the Euler characteristic by two by taking the connected sum with $S^2 \times S^2$ and decrease it by two by taking the connected sum with $S^1 \times S^3$. Therefore, if we start with a spin-cobordism for which the Euler characteristic is even we may, by taking connected sums, obtain an orientable spin-cobordism with zero Euler characteristic and hence a spin-Lorentz-cobordism. On the other hand, if the initial spin-cobordism had odd Euler characteristic we would be obliged to take connected sums with closed 4-manifolds with odd Euler characteristic in order to obtain a Lorentz-cobordism. Examples of such manifolds are $\mathbb{RP}^4$ which has Euler characteristic 1 and $\mathbb{CP}^2$ which has Euler characteristic 3. However, the former is not orientable while the latter, though orientable, is not a spin-manifold. In fact, quite generally, it is easy to see that any four-dimensional closed spin manifold must have even Euler characteristic and thus it is not possible, by taking connected sums, to find a spin-Lorentz-cobordism if the initial spin-cobordism had odd Euler characteristic. To see that a closed spin 4-manifold has even Euler characteristic recall from Hodge theory that on a closed orientable 4-manifold one has, using Poincaré duality:

$$\chi = 2 - 2b_1 + b_2^+ + b_2^-,$$

where b_1 is the first Betti number and b_2^+ and b_2^- are the dimensions of the spaces of harmonic 2-forms which are self-dual or anti-self-dual, respectively. On the other hand, from the Atiyah-Singer theorem the index of the Dirac operator with respect to some, and hence all, Riemannian metrics on a closed 4-manifold is given by

$$\text{index}(\text{Dirac}) = (b_2^+ - b_2^-)/8.$$

The index of the Dirac operator is always an integer, in fact on a closed 4-manifold it is always an even integer. It follows therefore that for a spin 4-manifold χ must be even. The arguments we have just given suggest, but do not prove, that the Euler characteristic of any spin-cobordism for a closed 3-manifold Σ is a property only of Σ. This is in fact true, as we shall show in the next section. It then follows from our discussion above that we may identify our invariant $u(\Sigma)$ with the Euler characteristic mod 2 of any spin cobordism for Σ.

Even without the results of the next section it is easy to evaluate our invariant $u(\Sigma)$ for a number of 3-manifolds of interest using comparatively elementary arguments. Suppose there were a spin-Lorentz-cobordism M for S^3. Then one could glue M across the S^3 to a four-ball, B^4. The Euler characteristic of the resulting closed manifold would be the Euler characteristic of M, which is zero, plus the Euler characteristic of the four-ball, which is one. It is clear that the unique spin structure induced on the boundary would extend to the interior of the 4-ball and so one obtains a contradiction. The same contradiction would result if we took the disjoint union of an odd number of S^3's. If we take the disjoint union of an even number of S^3's it is easy to construct spin-Lorentz-cobordisms. Thus although there exists a spin-Lorentz-cobordism with two S^3's in the past and two in the future, our results show that one cannot slice this spin-Lorentz-cobordism by a spacelike hypersurface diffeomorphic to S^3 which disconnects the spacetime. If this were possible we would have obtained a spin-Lorentz-cobordism for three S^3's which is impossible. In the language of particle physics: there is a 4-fold vertex but no 3-fold vertex.

If we regard $S^1 \times S^2$ as the boundary of $S^1 \times B^3$, where B^3 is the 3-ball we may fill it in with $S^1 \times B^3$. There are two possible spin structures to consider but in both cases they extend to the interior and one obtains a spin-cobordism with vanishing Euler characteristic. Starting with the flat product Riemannian metric on $S^1 \times B^3$ it is easy to find an everywhere non-vanishing unit vector field $\mathbf{V}$ which is outward pointing on the boundary: one simply takes a linear combination of the radial vector field on the 3-ball and the standard rotational vector field on the circle S^1 with radius-dependent coefficients such that the coefficient of the radial vector field vanishes at the origin of the 3-ball and the coefficient of the circular vector field vanishes on the boundary of the 3-ball. As with our product example above the resulting spacetime will, in general, be incomplete and have closed timelike curves but it is a valid spin-Lorentz-cobordism.

These results are sufficient to justify the claim in the introduction that wormholes must be created in pairs according to the Lorentzian point of view. One can also establish easily enough, using suitable connected sums of spin-Lorentz-cobordisms, that our invariant $u(\Sigma)$ is well defined and has the stated behaviour under disjoint union and connected sum of 3-manifolds as long as one fixes a spin structure on the boundary. However, our invariant is independent of the choice of spin structure on the boundary, as we have seen in the examples given above. In order not to have to keep track of the spin structure on the boundary it is advantageous to proceed in a slightly different fashion by using some $\mathbb{Z}_2$-cohomology theory. This we shall do in the next section.

The Euler Characteristic and the Kervaire Semi-Characteristic

The calculations which follow owe a great deal to conversations with Michael Atiyah, Nigel Hitchin, and Graeme Segal for which we are grateful. We begin by recalling the following exact sequence of homomorphisms of cohomology groups for an orientable cobordism M of a closed orientable 3-manifold Σ, the coefficient group being $\mathbb{Z}_2$:

$$0 \to H^0(M) \to H^0(\Sigma) \to H^1(M, \Sigma) \to H^1(M) \to H^1(\Sigma) \to H^2(M, \Sigma) \to H^2(M) \to \dots .$$

Now if we define W to be the image of $H^2(M, \Sigma)$ in $H^2(M)$ under the last homomorphism, and we use Lefshetz-Poincaré duality between relative cohomology and absolute homology groups together with the fact that the compact manifold M is connected we obtain the following exact sequence:

$$0 \to \mathbb{Z}_2 \to H^0(\Sigma) \to H_3(M) \to H^1(M) \to H^1(\Sigma) \to H_2(M) \to W \to 0.$$

By virtue of exactness, the alternating sum of the ranks, or equivalently the dimensions of these vector spaces over $\mathbb{Z}_2$, must vanish. Now the Euler characteristic $\chi(M)$ is given by:

$$\chi(M) = \sum_{i=0}^{i=4} (-1)^i \dim H_i(M; \mathbb{Z}_2)$$

while the $\mathbb{Z}_2$ Kervaire semi-characteristic $s(\Sigma)$ is given by:

$$s(\Sigma) = \dim H^0(\Sigma; \mathbb{Z}_2) + \dim H^1(\Sigma; \mathbb{Z}_2).$$

If dimensions are taken modulo 2 we may reverse any of the signs in these expressions to obtain the relation:

$$\chi(M) - s(\Sigma) = \dim W \bmod 2.$$

So far we have not used the condition that the compact 4-manifold M is a spin manifold. To do so we consider the cup product, $\cup$ which gives a map:

$$H^2(M, \Sigma) \times H^2(M) \to H^4(M).$$

For a compact connected 4-manifold $H^4(M; \mathbb{Z}_2) \equiv \mathbb{Z}_2$ so the cup product provides a well defined $\mathbb{Z}_2$ valued bilinear form Q on the image of $H^2(M, \Sigma)$ in $H^2(M)$ under the same homomorphism as above. In other words Q is non-degenerate on the vector space W. [A symmetric bilinear form Q on a vector-space W is non-degenerate if and only if $Q(x, y) = 0 \ \forall x \in W \Rightarrow y = 0$.]

The obstruction to the existence of a spin structure, the second Stiefel-Whitney class $w_2 \in H^2(M; \mathbb{Z}_2)$, is characterized by [12]:

$$w_2 \cup x = x \cup x \quad \forall x \in H^2(M; \mathbb{Z}_2).$$

Thus if M is a spin manifold w_2 must vanish and hence

$$Q(x, x) = x \cup x = 0 \quad \forall x \in H^2(M; \mathbb{Z}_2).$$

Now over $\mathbb{Z}_2$, a symmetric bilinear form which vanishes on the diagonal is the same thing as skew-symmetric bilinear form. But a skew-symmetric bilinear form over any field must have even rank and since Q is non-degenerate this implies that the dimension of W must be even. Indeed, one may identify the dimension of W modulo two as the second Stiefel-Whitney class in this situation. We have thus established that for an orientable spin-cobordism

$$\chi(M) = s(\Sigma) \bmod 2$$

and hence:

$$u(\Sigma) = s(\Sigma) \bmod 2.$$

Thus, for example, $u(\mathbb{RP}^3) = 0$ since it is connected and $H_1(\mathbb{RP}^3; \mathbb{Z}) = \mathbb{Z}_2$. It is straightforward to check this example directly by regarding $\mathbb{RP}^3$ as the boundary of the cotangent bundle of the 2-sphere, $T^*(S^2)$. Similar remarks apply to the lens spaces $L(k, 1)$ which may be regarded as the boundary of the 2-plane bundle over S^2 with first Chern class $c_1 = k$ and which have $H_1(L(k, 1); \mathbb{Z}) = \mathbb{Z}_k$. If the integer k is even they spin-Lorentz bound and if it is odd they do not.

The properties of our invariant $u(\Sigma)$ under disjoint union and connected sum now follow straightforwardly from the behaviour of homology groups under these operations.

Generalized Spinor Structures

One way of introducing spinors on a manifold which does not admit a conventional spinor structure is to introduce a $U(1)$ gauge field with respect to which all spinorial fields are charged, the charges being chosen so that the unremovable ± 1 ambiguity in the definition of conventional spinors is precisely cancelled by the holonomy of the $U(1)$ connection [13]. In other words we pass to a $Spin^c(4) \equiv Spin(4) \times_{\mathbb{Z}_2} U(1)$ structure. For general n it is not always possible to lift the tangent bundle of an orientable manifold, with structural group $SO(n)$ to a $Spin^c(4)$ bundle because the obstruction to lifting to a $Spin(n)$, i.e. the second Stiefel-Whitney class w_2, may not be the reduction of an integral class in $H^2(M; \mathbb{Z})$. However, according to Killingback and Rees [14] (see also Whiston [15]) this cannot happen for a compact orientable 4-manifold. From a topological point of view we may clearly replace $Spin^c(4)$ by its Lorentzian analogue: $SL(2, \mathbb{C}) \times_{\mathbb{Z}_2} U(1)$. Thus from a purely mathematical point of view we could always get around the difficulty of not having a spinor structure by using the simplest generalization of a spinor structure at the cost of introducing an extra and as yet unobserved $U(1)$ gauge field. Another possibility would be to use a non-abelian gauge field as suggested by Back, Freund, and Forger [16] and discussed by Isham and Avis [17]. There is no evidence for a gauge field that is coupled in this way to all fermions. It is also not clear that one could arrange that all the anomalies that would arise from such a coupling would cancel.

Acknowledgements. We would like to thank Michael Atiyah, Nigel Hitchin, Ray Lickorish, and Graeme Segal for helpful discussions and suggestions.

References

1. Morris, M.S., Thorne, K.S., Yurtsever, U.: Phys. Rev. Lett. **61**, 1446–1449 (1988)
2. Novikov, I.D.: Zh. Eksp. Teor. Fiz. **95**, 769 (1989)
3. Frolov, V.P., Novikov, I.G.: Phys. Rev. D **42**, 1057–1065 (1990)
4. Geroch, R.P.: J. Math. Phys. **8**, 782–786 (1968)
5. Milnor, J.: L'Enseignement Math. **9**, 198–203 (1963)
6. Reinhart, B.L.: Topology **2**, 173–177 (1963)
7. Yodzis, P.: Commun. Math. Phys. **26**, 39 (1972); Gen. Relativ. Gravit. **4**, 299 (1973)
8. Sorkin, R.: Phys. Rev. D **33**, 978–982 (1982)
9. Bichteler, K.: J. Math. Phys. **6**, 813–815 (1968)
10. Geroch, R.P.: J. Math. Phys. **9**, 1739–1744 (1968); **11**, 343–347 (1970)
11. Gibbons, G.W.: Nucl. Phys. B **271**, 479 (1986); Sanchez, N., Whiting, B.: Nucl. Phys. B **283**, 605–623 (1987)
12. Kirby, R.: Topology of 4-manifolds. Lecture Notes in Mathematics. Berlin, Heidelberg, New York: Springer
13. Hawking, S.W., Pope, C.N.: Phys. Letts. **73** B, 42–44 (1978)
14. Killingback, T.P., Rees, E.G.: Class. Quantum. Grav. **2**, 433–438 (1985)
15. Whiston, G.S.: Gen. Relativ. Gravit. **6**, 463–475 (1975)
16. Back, A., Freund, P.G.O., Forger, M.: Phys. Letts. **77** B, 181–184 (1978)
17. Avis, S.J., Isham, C.J.: Commun. Math. Phys. **64**, 269–278 (1980)

Communicated by N. Yu. Reshetikhin

PHYSICAL REVIEW D   VOLUME 46, NUMBER 2   15 JULY 1992

Chronology protection conjecture

S. W. Hawking

Department of Applied Mathematics and Theoretical Physics, University of Cambridge,
Silver Street, Cambridge CB3 9EW, United Kingdom
(Received 23 September 1991)

It has been suggested that an advanced civilization might have the technology to warp spacetime so that closed timelike curves would appear, allowing travel into the past. This paper examines this possibility in the case that the causality violations appear in a finite region of spacetime without curvature singularities. There will be a Cauchy horizon that is compactly generated and that in general contains one or more closed null geodesics which will be incomplete. One can define geometrical quantities that measure the Lorentz boost and area increase on going round these closed null geodesics. If the causality violation developed from a noncompact initial surface, the averaged weak energy condition must be violated on the Cauchy horizon. This shows that one cannot create closed timelike curves with finite lengths of cosmic string. Even if violations of the weak energy condition are allowed by quantum theory, the expectation value of the energy-momentum tensor would get very large if timelike curves become almost closed. It seems the back reaction would prevent closed timelike curves from appearing. These results strongly support the chronology protection conjecture: *The laws of physics do not allow the appearance of closed timelike curves.*

PACS number(s): 04.20.Cv, 04.60.+n

I. INTRODUCTION

There have been a number of suggestions that we might be able to warp spacetime in such a way as to allow rapid intergalactic space travel or travel back in time. Of course, in the theory of relativity, time travel and faster-than-light space travel are closely connected. If you can do one, you can do the other. You just have to travel from A to B faster than light would normally take. You then travel back, again faster than light, but in a different Lorentz frame. You can arrive back before you left.

One might think that rapid space travel might be possible using the wormholes that appear in the Euclidean approach to quantum gravity. However, one would have to be able to move in the imaginary direction of time to use these wormholes. Further, it seems that Euclidean wormholes do not introduce any nonlocal effects. So they are no good for space or time travel.

Instead, I shall consider real-time, Lorentzian metrics. In these, the light-cone structure forces one to travel at less than the speed of light and forward in time in a local region. However, the global structure of spacetime may allow one to take a shortcut from one region to another or may let one travel into the past. Indeed, it has been suggested by Morris and Thorne and others [1–3] that in the future, with improved technology, we might be able to create traversable wormholes connecting distant regions of spacetime. These wormholes would allow rapid space travel and, thus, travel back in time. However, one does not need anything as exotic as wormholes. Gott [4] has pointed out that an infinite cosmic string warps spacetime in such a way that one can get ahead of a beam of light. If one has two infinite cosmic strings, moving at high velocity relative to each other, one can get from A to B and back again before one sets out. This example is

worrying, because unlike wormholes, it does not involve negative-energy densities. However, I will show that one cannot create a spacetime in which one can travel into the past if one only uses finite lengths of cosmic string.

The aim of this paper is to show that even if it is possible to produce negative-energy densities, quantum effects are likely to prevent time travel. If one tries to warp spacetime to allow travel into the past, vacuum polarization effects will cause the expectation value of the energy-momentum tensor to be large. If one fed this energy-momentum tensor back into the Einstein equations, it appears to prevent one from creating a time machine. It seems there is a chronology protection agency, which prevents the appearance of closed timelike curves and so makes the universe safe for historians.

Kim and Thorne [5] have considered the expectation value of the energy-momentum tensor in a particular model of a time machine. They find that it diverges, but argue that it might be cut off by quantum-gravitational effects. They claim that the perturbation that it would produce in the metric would be so small that it could not be measured, even with the most sensitive modern technology. Because we do not have a well-defined theory of quantum gravity, it is difficult to decide whether there will be a cutoff to quantum effects calculated on a background spacetime. However, I shall argue that even if there is a cutoff, one would not expect it to come into effect until one was a Planck distance from the region of closed timelike curves. This Planck distance should be measured in an invariant way, not the frame-dependent way that Kim and Thorne adopt. This cutoff would lead to an energy density of the Planck value, 10^{94} g/cc, and a perturbation in the metric of order 1. Even if "order 1" meant 10^{-2} in practice, such a perturbation would create a disturbance that was enormous compared with chemical binding energies of order 10^{-9} or 10^{-10}. So one could not hope to travel through such a region and back into the past. Furthermore, the sign of the energy-momentum tensor of the vacuum polarization seems to be such as to resist the warping of the light cones to produce closed timelike curves.

Morris and Thorne build their time machine out of traversable Lorentzian wormholes, that is, Lorentzian spacetimes of the form $\Sigma \times R$. Here R is the time direction and Σ is a three-dimensional surface, that is, asymptotically flat, and has a handle or wormhole connecting two mouths. Such a wormhole would tend to collapse with time, unless it were held up by the repulsive gravity of a negative-energy density. Classically, energy densities are always positive, but quantum field theory allows the energy density to be negative locally. An example is the Casimir effect. Morris and Thorne speculate that with future technology it might be possible to create such wormholes and to prevent them from collapsing.

Although the length of the throat connecting the two mouths of the wormhole will be fairly short, the two mouths can be arbitrarily far apart in the asymptotically flat space. Thus going through a wormhole would be a way of traveling large distances in a short time. As remarked above, this would lead to the possibility of travel into the past, because one could travel back to one's starting point using another wormhole whose mouths were moving with respect to the first wormhole. In fact, it would not be necessary to use two wormholes. It would be sufficient just for one mouth of a single wormhole to be moving with respect to the other mouth. Then there would be the usual special-relativistic time-dilation factor between the times as measured at the two mouths. This would mean that at some point in the wormhole's history it would be possible to go down one mouth and come out of the other mouth in the past of when you went down. In other words, closed timelike curves would appear. By traveling in a space ship on one of these closed timelike curves, one could travel into one's past. This would seem to give rise to all sorts of logical problems, if you were able to change history. For example, what would happen if you killed your parents before you were born. It might be that one could avoid such paradoxes by some modification of the concept of free will. But this will not be necessary if what I call the *chronology protection conjecture* is correct: *The laws of physics prevent closed timelike curves from appearing.*

Kim and Thorne [5,6] suggest that they do not. I will present evidence that they do.

II. CAUCHY HORIZONS

The particular time machine that Kim and Thorne [5] consider involves wormholes with nontrivial topology. But as I will show, to create a wormhole, one has to distort the spacetime metric so much that closed timelike curves appear. I shall therefore consider the appearance of closed timelike curves in general, without reference to any particular model.

I shall assume that our region of spacetime develops from a spacelike surface S without boundary. By going to a covering space if necessary [7], one can assume that

spacetime is time orientable and that no timelike curve intersects S more than once. Let us suppose that the initial surface S did not contain any wormholes: Say it was simply connected, like R^3 or S^3. But let us suppose we had the technology to warp the spacetime that developed from S, so that a later spacelike surface S' had a different topology, say, with a wormhole or handle. It seems reasonable to suppose that we would be able to warp spacetime only in a bounded region. In other words, one could find a timelike cylinder T which intersected the spacelike surfaces S and S' in compact regions S_T and S_T' of different topology. In that case the topology change would take place in the region of spacetime M_T bounded by T, S, and S'. The region M_T would not be compact if it contained a curvature singularity or if it went off to infinity. But in that case, extra unpredictable information would enter the spacetime from the singularity or from infinity. Thus one could not be sure that one's warping of spacetime would achieve the result desired if the region M_T were noncompact. It therefore seems reasonable to suppose that M_T is compact. In Sec. V, I show that this implies that M_T contains closed timelike curves. So if you try to create a wormhole to use as a time machine, you have to warp the light-cone structure of spacetime so much that closed timelike curves appear anyway. Furthermore, one can show the requirement that M_T have a Lorentz metric and a spin structure imply that wormholes cannot be created singly, but only in multiples of 2 [8]. I shall therefore just consider the appearance of closed timelike curves without there necessarily being any change in the topology of the spatial sections.

If there were a closed timelike curve through a point p to the future of S, then p would not lie in the future Cauchy development [7] $D^+(S)$. This is the set of points q such that every past-directed curve through q intersects S if continued far enough. So there would be a future Cauchy horizon $H^+(S)$ which is the future boundary of $D^+(S)$. I wish to study the creation of closed timelike curves from the warping of the spacetime metric in a bounded region. I shall therefore consider Cauchy horizons $H^+(S)$ that are what I shall call "compactly generated." That is, all the past-directed null geodesic generators of $H^+(S)$ enter and remain within a compact set C. One could generalize this definition to a situation in which there were a countable number of disjoint compact sets C, but for simplicity I shall consider only a single compact set.

What this condition means is that the generators of the Cauchy horizon do not come in from infinity or a singularity. Of course, in the presence of closed timelike curves, the Cauchy problem is not well posed in the strict mathematical sense. But one might hope to predict events beyond the Cauchy horizon if it is compactly generated, because extra information will not come in from infinity or singularities. This idea is supported by some calculations that show there is a unique solution to the wave equation on certain wormhole spacetimes that contain closed timelike curves [15]. But even if there is not a unique solution beyond the Cauchy horizon, it will not affect the conclusions of this paper because the quantum effects that I shall describe occur in the future Cauchy development $D^+(S)$, where the Cauchy problem is well posed and where there is a unique solution, given the initial data and quantum state on S.

The inner horizons of the Reissner-Norström and Kerr solutions are examples of Cauchy horizons that are not compactly generated. Beyond the Cauchy horizon, new information can come in from singularities or infinity, and so one could not predict what will happen. In this paper I will restrict my attention to compactly generated Cauchy horizons. It is, however, worth remarking that the inner horizons of black holes suffer similar quantum-mechanical divergences of the energy-momentum tensor. The quantum radiation from the outer black-hole horizon will pile up on the inner horizon, which will be at a different temperature.

By contrast, the Taub-Newman-Unti-Tamburino (NUT) universe is an example of a spacetime with a compactly generated Cauchy horizon. It is a homogeneous anisotropic closed universe, where the surfaces of homogeneity go from being spacelike to null and then timelike. The null surface is a Cauchy horizon for the spacelike surfaces of homogeneity. This Cauchy horizon will be compact and therefore will automatically be compactly generated. However, I have deliberately chosen the definition of compactly generated, so that it can apply also to Cauchy horizons that are noncompact. Indeed, if the initial surface S is noncompact, the Cauchy horizon $H^+(S)$ will be either noncompact or empty. To show this one uses the standard result, derived in Sec. V, that a manifold with a Lorentz metric admits a timelike vector field V^a. (Strictly, a Lorentz metric implies the existence of a vector field up to a sign. But one can choose a consistent sign for the vector field if the spacetime is time orientable, which I shall assume.) Then the integral curves of the vector field give a mapping of the future Cauchy horizon $H^+(S)$ into S. This mapping will be continuous and one to one onto the image of $H^+(S)$ in S. But the future Cauchy horizon $H^+(S)$ is a three-manifold without boundary. So, if S is noncompact, $H^+(S)$ must be noncompact as well. However, that need not prevent it from being compactly generated.

An example will illustrate how closed timelike curves can appear without there being any topology change. Take the spacetime manifold to be R^4 with coordinates t, r, θ, ϕ. Let the initial surface S be $t=0$ and let the spacetime metric g_{ab} be the flat Minkowski metric η_{ab} for t negative. For positive t let the metric still be the flat Minkowski metric outside a timelike cylinder, consisting of a two-sphere of radius L times the positive-time axis. Inside the cylinder let the light cones gradually tip in the ϕ direction, until the equator of the two-sphere, $r=\frac{1}{2}L$, becomes first a closed null curve γ and then a closed timelike curve. For example, the metric could be

$$ds^2 = -dt^2 + 2f\,dt\,d\phi - f\,d\phi^2 + dr^2$$
$$+ r^2(d\theta^2 + \sin^2\theta\,d\phi^2)\,,$$

$$f = r^2 t^2 \sin^4\theta \sin^2\left[\frac{\pi r}{L}\right]\,.$$

The Cauchy horizon will be generated by null geodesics that in the past direction spiral toward the closed null geodesic γ. They will all enter and remain within any compact neighborhood C of γ. Thus the Cauchy horizon will be compactly generated.

One could calculate the Einstein tensor of this metric. As I will show, it will necessarily violate the weak energy condition. But one could take the attitude that quantum field theory in curved space allows violations of the weak energy condition, as in the Casimir effect. One might hope, therefore, that in the future we might have the technology to produce an energy-momentum tensor equal to the Einstein tensor of such a spacetime. It is worth remarking that, even if we could distort the light cones in the manner of this example, it would not enable us to travel back in time to before the initial surface S. That part of the history of the universe is already fixed. Any time travel would have to be confined to the future of S.

I shall mainly be interested in the case where the initial surface S is noncompact, because that corresponds to building a time machine in a local region. However, most of the results in this paper will also apply to the cosmological case, in which S can be compact.

The Cauchy horizon is generated by null geodesic segments [7]. These may have future end points, where they intersect another generator. The future end points will form a closed set B of measure zero. On the other hand, the generators will not have past end points. If the horizon is compactly generated, the generators will enter and remain within a compact set C. One can introduce a null tetrad $l^a, n^a, m^a, \bar{m}^a$ in a neighborhood of $(H^+(S)-B)\cap C$. The vector l^a is chosen to be the future-directed tangent to the generators of the Cauchy horizon. The vector n^a is another future-directed null vector. Because I am using the signature $-+++$, rather than the $+---$ signature of Newman and Penrose, I normalize them by $l^a n_a = -1$. The complex-conjugate null vectors m^a and $\bar{m}^a$ are orthogonal to l^a and n^a and are normalized by $m^a \bar{m}_a = 1$. One can then define the Newman-Penrose quantities [9,10]

$$\epsilon = -\tfrac{1}{2}(n^a l_{a;c} l^c - \bar{m}^a m_{a;c} l^c)\,,$$

$$\kappa = -m^a l_{a;c} l^c\,,$$

$$\rho = -m^a l_{a;c} \bar{m}^c\,,$$

$$\sigma = -m^a l_{a;c} m^c\,.$$

Note that these definitions have the opposite sign to those of Newman and Penrose. This is because of the different signature of the metric.

Because the generators are null geodesics and lie in a null hypersurface, $\kappa = 0$ and $\rho = \bar{\rho}$. The convergence ρ and shear σ obey the Newman-Penrose equations along γ:

$$\frac{d\rho}{dt} = \rho^2 + \sigma\bar{\sigma} + (\epsilon + \bar{\epsilon})\rho + \tfrac{1}{2}R_{ab}l^a l^b\,,$$

$$\frac{d\sigma}{dt} = 2\rho\sigma + (3\epsilon - \bar{\epsilon})\sigma + C_{abcd}l^a m^b l^c \bar{m}^d\,,$$

where t is the parameter along the generators such that

$l^a = dx^a/dt$.

The real and imaginary parts of ϵ, respectively, measure how the vectors l^a and m^a change compared to a parallel-propagated basis. By choosing an affine parameter $\tilde{t}$ on the generators, one can rescale the tangent vector l^a so that $\epsilon + \bar{\epsilon} = 0$. The generators may be geodesically incomplete in the future direction; i.e., the affine parameter may have an upper bound. But one can adapt the lemma in Ref. [7], p. 295, to show that the generators of the horizon are complete in the past direction.

Now suppose the weak energy condition holds:

$$T_{ab} l^a l^b \geq 0 ,$$

for any null vector l^a. Then the Einstein equations (with or without cosmological constant) imply

$$R_{ab} l^a l^b \geq 0 .$$

It then follows that the convergence ρ of the generators must be non-negative everywhere on the Cauchy horizon. For suppose $\rho = \rho_1 < 0$ at a point p on a generator γ. Then one could integrate the Newman-Penrose equation for ρ in the negative $\tilde{t}$ direction along γ to show that ρ diverged at some point q within an affine distance ρ_1^{-1} to the past of p. Such a point q would be a past end point of the null geodesic segment γ in the Cauchy horizon. But this is impossible because the generators of the Cauchy horizon have no past end points. This shows that ρ must be everywhere non-negative on a compactly generated Cauchy horizon if the weak energy condition holds.

I shall now establish a contradiction in the case that the initial surface S is noncompact. The argument is similar to that in Ref. [7], p. 297. On C one can introduce a unit timelike vector field V^a. One can then define a positive definite metric by

$$\hat{g}_{ab} = g_{ab} + 2 V_a V_b .$$

In other words, $\hat{g}$ is the spacetime g with the sign of the metric in the timelike V^a direction reversed.

One can normalize the tangent vector to the generators by $g_{ab} l^a V^b = 1/\sqrt{2}$. The parameter t on the generators then measures the proper distance in the metric $\hat{g}_{ab}$. One can define a map

$$\mu_t : (H^+(S) - B) \cap C \to (H^+(S) - B) \cap C ,$$

by moving each point of the Cauchy horizon a parameter distance t to the past along the generators. The three-volume (measured with respect to the metric $\hat{g}_{ab}$) of the image of the Cauchy horizon under this map will change according to

$$\frac{d}{dt} \int_{\mu_t(H^+(S) \cap C)} dA = 2 \int_{\mu_t(H^+(S) \cap C)} \rho \, dA .$$

The change in volume cannot be positive because the Cauchy horizon is mapped into itself. If the initial surface S is noncompact, the change in volume will be strictly negative, because the Cauchy horizon will be noncompact and will not lie completely in the compact set C. This would establish a contradiction with the requirement that $\rho \geq 0$ if the weak energy condition is satisfied.

Thus a compactly generated Cauchy horizon cannot form if the weak energy condition holds and S is noncompact.

On the other hand, the example of the Taub-NUT universe shows that it is possible to have a compactly generated Cauchy horizon if S is compact. However, in that case the weak energy condition would imply that ρ and σ would have to be zero everywhere on the Cauchy horizon. This would mean that no matter or information, and in particular no observers, could cross the Cauchy horizon into the region of closed timelike curves. Moreover, as will be shown in the next section, the solution will be classically unstable in that a small matter-field perturbation would pile up on the horizon. Thus the chronology protection conjecture will hold if the weak energy condition is satisfied whether or not S is compact. In particular, this implies that if no closed timelike curves are present initially, one cannot create them by warping the metric in a local region with finite loops of cosmic string. If the weak energy condition is satisfied, closed timelike curves require either singularities (as in the Kerr solution) or a pathological behavior at infinity (as in the Godel and Gott spacetimes).

The weak energy condition is satisfied by the classical energy-momentum tensors of all physically reasonable fields. However, it is not satisfied locally by the quantum expectation value of the energy-momentum tensor in certain quantum states in flat space. In Minkowski space the weak energy condition is still satisfied if the expectation value is averaged along a null geodesic [11], but there are curved-space backgrounds where even the averaged expectation values do not satisfy the weak energy condition. The philosophy of this paper is therefore not to rely on the weak energy condition, but to look for vacuum polarization effects to enforce the chronology protection conjecture.

III. CLOSED NULL GEODESICS

The past-directed generators of the Cauchy horizon will have no past end points. If the horizon is compactly generated, they will enter and remain within a compact set C. This means they will wind round and round inside C. In Sec. V it is shown that there is a nonempty set E of generators, each of which remains in a compact set C in the future direction, as well as in the past direction.

The generators in E will be *almost* closed. That is there will be points q such that a generator in E will return infinitely often to any small neighborhood of q. But they need not actually close up. For example, if the initial surface is a three-torus, the Cauchy horizon will also be a three-torus, and the generators can be nonrational curves that do not close up on themselves. However, this kind of behavior is unstable. The least perturbation of the metric will cause the horizon to contain closed null geodesics. More precisely, the space of all metrics on the spacetime manifold M can be given a C^∞ topology. Then, if g is a metric that has a compactly generated horizon which does not contain closed null geodesics, any neighborhood of g will contain a metric g' whose Cauchy horizon *does* contain closed null geodesics.

The spacetime metric is presumably the classical limit

of an inherently quantum object and so can be defined only up to some uncertainty. Thus the only properties of the horizon that are physically significant are those that are maintained under small variations of the metric. In Sec. V it will be shown that in general the closed null geodesics in the horizon have this property. That is, if g is a metric such that the Cauchy horizon contains closed null geodesics, then there is a neighborhood U of g such that every metric g' in U has closed null geodesics in its Cauchy horizon. I shall therefore assume that in general E consists of one or more disjoint closed null geodesics. The example given above of the metric with closed timelike curves shows that the Cauchy horizon need not contain more than one.

I shall now concentrate attention on a closed null geodesic γ in the Cauchy horizon. Pick a point p on γ and parallel propagate a frame around γ and back to p. The result will be a Lorentz transformation Λ of the original frame. This Lorentz transformation will lie in the four-parameter subgroup that leaves unchanged the null direction tangent to the generator. It will be generated by an antisymmetric tensor

$$\Lambda = e^{\omega} .$$

The null vector l^a tangent to the null geodesic will be an eigenvector of ω because its direction is left unchanged by Λ:

$$l^a = h \omega^a{}_b l^b .$$

The eigenvalue h determines the change of scale, e^h, of the tangent vector after it has been parallel propagated around the closed null geodesic in the future direction. In Sec. V it is shown that if h were negative, one could move each point of γ to the past to obtain a closed timelike curve. But this curve would be in the Cauchy development of S, which is impossible, because the Cauchy development does not contain any closed timelike curves. This shows that h must be positive or zero. Clearly, $h = 0$ is a limiting case. In practice, one would expect h to be positive. This will mean that each time one goes round the closed null geodesic, the parallel-propagated tangent vector will increase in size by a factor e^h. The affine-parameter distance around will decrease by a factor e^{-h}. Thus the closed null geodesic γ will be incomplete in the future direction, although it will remain in the compact set C and so it will not end on any curvature singularity. Because $h \geq 0$, γ will be complete in the past direction.

If $h \neq 0$, there will be another null vector n^a, which is an eigenvector of $\omega^a{}_b$ with eigenvalue $-h$. The Lorentz transformation Λ will consist of a boost e^h in the timelike plane spanned by l^a and n^a and a rotation through an angle θ in the orthogonal spacelike plane.

The quantity h is rather like the surface gravity of a black hole. It measures the rate at which the null cones tip over near γ. As in the black-hole case, it gives rise to quantum effects. However, in this case, they will have imaginary temperature, corresponding to periodicity in real time, rather than in imaginary time, as in the black-hole case.

Another important geometrical quantity associated with the closed null geodesic γ in the Cauchy horizon is the change of cross-sectional area of a pencil of generators of the horizon as one goes round the closed null geodesic. Let

$$f = \ln \left| \frac{A_{n+1}}{A_n} \right| ,$$

where A_n and A_{n+1} are the areas of the pencil on successive passes of the point p in the future direction. The quantity f measures the amount the generators are diverging in the future direction. Because neighboring generators tend toward the closed null geodesic γ in the past direction, f will be greater than or equal to zero. Again, $f = 0$ is a limiting case. In general, f will be greater than zero.

The quantity f determines the classical stability of the Cauchy horizon. A small, high-frequency wave packet going round the horizon in the neighborhood of γ will have its energy blueshifted by a factor e^h each time it comes round. This increased energy will be spread across a cross section transverse to γ. On each circuit of γ, the two-dimensional area of the cross section will increase by a factor e^f. The time duration of the cross section will be reduced by a factor e^{-h}. So the local energy density will remain bounded and the Cauchy horizon will be classically stable if

$$f > 2h .$$

This is true of the wormholes that Kim and Thorne consider, provided they are moving slowly. But it seems they will still be unstable quantum mechanically.

One can relate the result of going round γ to integrals of the Newman-Penrose quantities defined in the last section:

$$\oint \rho \, dt = -\tfrac{1}{2} f ,$$

$$\oint \epsilon \, dt = -\tfrac{1}{2}(h + i\theta) ,$$

where e^h is the boost in the l^a-n^a plane and $e^{i\theta}$ is the spatial rotation in the m^a-$\overline{m}^a$ plane of a tetrad that is parallel propagated after one circuit of γ. One can also define the distortion q of an initially circular pencil of generators by

$$\oint \sigma \, dt = -\tfrac{1}{2} q .$$

One can choose the parameter t on γ so that $\epsilon + \overline{\epsilon}$ is constant and so that the parameter distance of one circuit of γ is 1. Then

$$\epsilon + \overline{\epsilon} = -h .$$

One can now integrate the Newman-Penrose equation for ρ around a circuit of γ and use the Schwarz inequality to show

$$\oint R_{ab} l^a l^b dt \leq -[hf + \tfrac{1}{2}(f^2 + q\overline{q})] \leq 0 .$$

This gives a measure of how much the weak energy condition has to be violated on γ. In particular, it cannot be satisfied unless $f = q = 0$.

IV. QUANTUM FIELDS ON THE BACKGROUND

Quantum effects in the spacetime will be determined by the propagator or two-point function

$$\langle T\phi(x)\phi(y)\rangle .$$

This will be singular when the two points x and y can be joined by a null geodesic. Thus quantum effects near γ will be dominated by closed or almost-closed null geodesics.

One can construct a simple spacetime that reproduces the Lorentz transformation Λ on going around γ, but not the area increase e^f, in the following way. One starts with Minkowski space and identifies points that are taken into each other by the Lorentz transformation Λ. For simplicity, I will just describe the case where Λ is a pure boost in the n^a-l^a plane. Consider the past light cone of the origin in two-dimensional Minkowski space. The orbits of the boost Killing vector will be spacelike. Identify a point p with its image under the boost Λ. This gives what is called Misner space [12,7] with the metric

$$ds^2 = -dt^2 + t^2 dx^2 ,$$

on a half-cylinder defined by $t < 0$ with the x coordinate identified with period h. This metric has an apparent singularity at $t=0$. However, one can extend it by introducing new coordinates

$$\tau = t^2, \quad v = \ln t + x .$$

The metric then takes the form

$$ds^2 = -dv\, d\tau + \tau\, dv^2 .$$

This can then be extended through $\tau = 0$. This corresponds to extending from the bottom quadrant into the left-hand quadrant. One then gets a metric on a cylinder. This develops from a spacelike surface S. However, at $\tau = 0$, the light cones tip over and a closed null geodesic appears. For negative τ, closed timelike curves appear. The full four-dimensional space is the product of this two-dimensional Misner space with two extra flat dimensions. One can identify these other dimensions periodically if one wants to have a spacetime in which the initial surface S and the Cauchy horizon $D^+(S)$ are compact. However, such a compactification will not change the nature of the behavior of the energy-momentum tensor on the horizon.

Misner space has a four-parameter group of isometries and is also invariant under an overall dilation. It is therefore natural to expect the quantum state of a conformally invariant field also to have these symmetries. By the conservation equations and the trace-anomaly equation, the expectation value of the energy-momentum tensor for a conformally invariant field must then have the form

$$\langle T_{ab}\rangle_0 = \text{diag}(K, 3K, -K, -K), \quad K = \frac{B}{t^4} ,$$

in an orthonormal basis along the (t,x,y,z) axes. The coefficient B will depend on the quantum state and spin of the field.

Because the space is flat, it is easy to calculate a propagator $\langle T\phi(x)\phi(y)\rangle_0$ for a particular quantum state of any free field with these symmetries. One just takes the usual Minkowski propagator and puts in image charges under Λ. One can then calculate the expectation value of the energy-momentum tensor by taking the limit of this propagator minus the usual Minkowski propagator. This has been done by Hiscock and Konkowski [13] for the case of a conformally invariant scalar field. They found that B is negative, implying that the expectation value of the energy density is negative and diverges on the Cauchy horizon.

The quantum state that the propagator $\langle T\phi(x)\phi(y)\rangle_0$ corresponds to is a particularly natural one, but is certainly not the only quantum state of the spacetime. The propagator in any other state will obey the same wave equation. Thus it can be written

$$\langle T\phi(x)\phi(y)\rangle = \langle T\phi(x)\phi(y)\rangle_0$$
$$+ \sum \tfrac{1}{2}[\psi_n(x)\bar\psi_n(y) + \text{c.c.}] ,$$

where ψ_n are solutions of the homogeneous wave equation that are nonsingular on the initial surface S. The expectation value of the energy-momentum tensor in this state will be

$$\langle T_{ab}\rangle = \langle T_{ab}\rangle_0 + \sum T_{ab}^{cl}[\psi_n] ,$$

where $T_{ab}^{cl}[\psi_n]$ is the classical energy-momentum tensor of the field ψ_n. One can think of the last term as the energy momentum of particles in modes corresponding to the solutions ψ_n.

One could ask if there was a propagator that gave an energy-momentum tensor that did not diverge on the Cauchy horizon. I have found propagators that give the expectation value of the energy momentum to be zero everywhere, but they do not satisfy the positivity conditions that are required for them to be the time-ordered expectation values of the field operators in a well-defined quantum state. I am grateful to Bernard Kay for pointing this out. One way of getting a propagator that was guaranteed to satisfy the positivity conditions would be to add particle excitations to the $\langle\ \rangle_0$ state. However, no distribution of particles would have a stress in the x direction that is 3 times the energy density. Unless the energy-momentum tensor of the particles had the same form as that of $\langle T_{ab}\rangle_0$, it would not diverge with the same power of distance away from the horizon and so could not cancel the divergence. Thus I am almost sure there is no quantum state on Misner space for which $\langle T_{ab}\rangle$ is finite on the horizon, but I do not have a rigorous proof.

In the general case in which there is a negative Ricci tensor and $f > 0$, it is difficult to calculate the expectation value of the energy-momentum tensor exactly because one does not have a closed form for the propagator. However, near the Cauchy horizon the metric and quantum state will asymptotically have the same symmetries and scale invariance as in Misner space. Thus one would still expect the same Bt^{-4} behavior, where the value of t at a point is now defined to be the least upper bound of the lengths of timelike curves from the point to the closed null geodesic γ. If $h > 0$, t will be finite on $D^+(S)$.

Again, the coefficient B will depend on the quantum state. Approximate WKB calculations by Kim and Thorne [5] for a wormhole spacetime indicate that there is a quantum state for this spacetime for which B is negative. Because the classical stability condition $f > 2h$ is satisfied, it does not seem possible to cancel the negative-energy divergence with positive-energy quanta. Thus it seems that the expectation value of the energy-momentum tensor will always diverge on the Cauchy horizon for any quantum state.

V. GLOBAL RESULTS

If there is a timelike tube T connecting surfaces S and S' of different topology, then the region M_T contains closed timelike curves.

This is a modification of a theorem of Geroch [14]. I shall describe it here because it involves constructions that will be useful later. One first puts a positive-definite metric $\tilde{g}_{ab}$ on the spacetime manifold M. (This can always be done.) Then one can define a timelike vector field V^a as an eigenvector with negative eigenvalue of the physical metric g_{ab} with respect to $\tilde{g}_{ab}$:

$$g_{ab} V^a = -\lambda \tilde{g}_{ab} V^a .$$

One can normalize V^a to have unit magnitude in the spacetime metric g_{ab}. With a bit more care, one can choose the vector field V^a so that it is tangent to the timelike tube T. One can define a mapping

$$\mu : S_T \to S_T' ,$$

by moving points along the integral curves of V^a. If each integral curve that cuts S_T were also to cut S_T', μ would be one-to-one and onto. But this would imply that S_T and S_T' have the same topology, which they do not. Therefore there must be some integral curve γ which cuts S_T but which winds round and round inside the compact set M_T and does not intersect S_T'. This implies there will be points $p \in M_T$ that are limit points of γ. Through p there will be an integral curve $\bar{\gamma}$, each point of which is a limit point of γ. But because $\bar{\gamma}$ is timelike, it would be possible to deform segments of γ to form closed timelike curves.

A compactly generated Cauchy horizon $D^+(S)$ contains a set E of generators which have no past or future end points and which are contained in the compact set C.

Let λ be a generator of the Cauchy horizon. This means that it may have a future end point (where it intersects another generator), but it can have no past point. Instead, because the horizon is compactly generated, in the past direction λ will enter and remain within a compact set C. This means that there will be points q in C which are such that every small neighborhood of q is intersected by λ an infinite numbers of times. Let B be a normal coordinate ball about a limit point q. There will be points p and r on ∂B to the future and past of q which will be limit points of where λ intersects ∂B. It is easy to see that p and r must lie on a null geodesic segment γ through q. By repeating this construction about p and r,

one can extend γ to a null geodesic without future or past end points, each point of which is a limit point for λ. Because λ enters and remains within C, γ must remain within C in both past and future directions. the set E consists of all such limit geodesics γ.

If γ is a closed null geodesic with $h < 0$, then γ can be deformed to give a closed timelike curve λ to the past of γ.

Let $l^a = dx^a / dt$ be the future-directed vector tangent to γ and let a be defined by

$$l^a_{;b} l^b = al^a .$$

Then $a = (\epsilon + \bar{\epsilon})$, and so

$$\oint a \, dt = -h .$$

Let V^a be a future-directed timelike vector field normalized so that $l^a V^b g_{ab} = -1$. Then one can find a one-parameter family of curves $\gamma(t, u)$ such that

$$\gamma(t, 0) = \gamma(t) ,$$

$$\frac{\partial x^a}{\partial t} = l^a ,$$

$$\frac{\partial x^a}{\partial u} = -x V^a ,$$

where x is a given function on γ. Then

$$\begin{aligned}
\frac{\partial}{\partial u} (l^a l^b g_{ab}) &= -2x l^a_{;c} V^c l^b g_{ab} \\
&= -2(x V^a)_{;c} l^c l^b g_{ab} \\
&= -2(x V^a l^b g_{ab})_{;c} l^c + 2x V^a l^b_{;c} l^c g_{ab} \\
&= 2 \frac{\partial x}{\partial t} - 2ax .
\end{aligned}$$

Let

$$x = \exp \left[\int_0^t a \, dt + htb^{-1} \right] ,$$

where $b = \oint dt$. Then, for sufficiently small $v > 0$, $\gamma(t, v)$ will be a closed timelike curve to the past of γ.

If the metric g is such that the Cauchy horizon $H^+(S)$ contains a closed null geodesic γ with $h > 0$ and $f - |q| \neq 0$, then the property of having a closed null geodesic is stable; i.e., g will have a neighborhood U such that for any metric $g' \in U$, there will also be a closed null geodesic in the Cauchy horizon.

Let p be a point on γ. A point q in $I^-(p)$, the chronological past of p, will lie in the Cauchy development $D^+(S)$, and $J^-(p) \cap J^+(S)$, the intersection of the causal past of p with the causal future of S, will be compact. This means that a sufficiently small variation of g will leave q in the Cauchy development of S. On the other hand, because $h > 0$, the previous result implies there is a closed timelike curve λ through a point r just to the future of p. A sufficiently small variation of the metric will leave λ a closed timelike curve and hence will leave r not in the Cauchy development. Thus the existence of a Cauchy horizon will be a stable property of the metric g. Similarly, the positions, directions, and derivatives of the generators will be continuous functions of the metric g in a neighborhood of γ.

Let W be a time like three-surface through p transverse to the Cauchy horizon. Then the generators of the horizon near γ define a map

$$\nu: W \cap D^+(S) \to W \cap D^+(S) ,$$

by mapping where they intersect W to where they intersect it again the next time round. If $f - |q| \neq 0$, the eigenvalues of $d\nu$ will be bounded away from 1. It then follows that the existence of a closed orbit is a stable property.

VI. CONCLUSIONS

As one approaches a closed null geodesic γ in the Cauchy horizon, the propagator will acquire extra singularities from null geodesics close to γ that almost return to the original point. In the Misner-space example in Sec. IV, these extra contributions came from the image charges under the boost. When one approached the Cauchy horizon, which corresponded to the past light cone of the origin in two-dimensional Minkowski space, these image charges became nearly null separated and their light cones became nearly on top of each other. It was therefore natural to find that the expectation value of the energy-momentum tensor diverged as one approached the Cauchy horizon.

If the boost h on going round γ is zero, the distance t from γ to any point to the past of γ in the Cauchy development will be infinite. This is rather like the fact that there is an infinite spatial distance to the horizon of a black hole with zero surface gravity. If the expectation value were of the form of Bt^{-4} with finite B, it would therefore be zero. Even if the energy-momentum tensor of individual fields did not have this form and still diverged on the Cauchy horizon, one might expect that the total energy-momentum tensor might vanish in a supersymmetric theory, because the contributions of bosonic and fermionic fields might have opposite signs. However, one would not expect such a cancellation unless the spacetime admitted a supersymmetry at least on the horizon. This would require that the tangent vector to the horizon corresponded to a Killing spinor, which would imply

$$\theta = \rho = \sigma = 0 ,$$

in addition to

$$h = 0 .$$

These conditions will not hold on a general horizon, but it is possible that the back reaction could drive the geometry to satisfy them, as the back reaction of black-hole evaporation can drive the surface gravity to zero in certain circumstances.

If one assumes that the expectation value of the energy-momentum tensor diverges on the horizon, one can ask what effect this would have if one fed it back into the field equations. On dimensional grounds one would expect the eigenvalues of the energy-momentum tensor to diverge as Bt^{-4}, where B is a constant that depends on the quantum state and t is the distance function to the horizon. However, because of boost and other factors, the energy density measured by an observer who crosses the Cauchy horizon on a timelike geodesic will go as $Bd^{-1}s^{-3}$, where s is proper distance along the observer's world line until the horizon and d is some typical length in the problem. In Misner space, d is the length of the spacelike geodesic from the origin orthogonal to the observer's world line.

To get the metric perturbation generated by this energy-momentum tensor, one has to integrate with respect to s twice. Thus the metric perturbation will diverge as $GBd^{-1}s^{-1}$. Kim and Thorne [5] agree that the metric perturbation diverges, but claim that quantum-gravitation effects might cut it off when the observer's proper time before crossing the Cauchy horizon, s, is the Planck time. This would give a metric perturbation of order

$$Bl_p d^{-1} .$$

If d were of order 1 m, the metric perturbation would be of order 10^{-35}. This is far less than about 10^{-19}, which is the best that can be detected with the most sensitive modern instruments.

It may be that quantum gravity introduces a cutoff at the Planck length. But one would not expect any cutoff to involve the observer-dependent time s. If there is a cutoff, one would expect it to occur when the invariant distance t from the Cauchy horizon was of order the Planck length. But t^2 is of order ds. So a cutoff in t at the Planck length would give a metric perturbation of order 1. This would completely change the spacetime and probably make it impossible to cross the Cauchy horizon. One would not therefore be able to use the region of closed timelike curves to travel back in time.

If the coefficient B is negative, the energy-momentum tensor will have a repulsive gravitational effect in the equation for the rate of change of the volume. This will tend to prevent the spacetime from developing a Cauchy horizon. The calculations that indicate B is negative therefore suggest that spacetime will resist being warped so that closed timelike curves appear. On the other hand, if B were positive, the graviational effect would be attractive, and the spacetime would develop a singularity, which would prevent one reaching a region of closed timelike curves. Either way, there seem to be theoretical reasons to believe the chronology protection conjecture: *The laws of physics prevent the appearance of closed timelike curves.*

There is also strong experimental evidence in favor of the conjecture from the fact that we have not been invaded by hordes of tourists from the future.

ACKNOWLEDGEMENTS

I am grateful to Gary Gibbons, James Grant, Bernard Kay, John Stewart, and Kip Thorne for many discussions and suggestions.

[1] M. S. Morris and K. S. Thorne, Am. J. Phys. **56**, 395 (1988).

[2] M. S. Morris, K. S. Thorne, and U. Yurtsever, Phys. Rev. Lett. **61**, 1446 (1988).

[3] V. P. Frolov and I. D. Novikov, Phys. Rev. D **42**, 1057 (1990).

[4] J. R. Gott III, Phys. Rev. Lett. **66**, 1126 (1991).

[5] S.-W. Kim and K. S. Thorne, Phys. Rev. D **43**, 3929 (1991).

[6] K. S. Thorne, Ann. N.Y. Acad. Sci. **63**, 182 (1991); see also V. P. Frolov, Phys. Rev. D **43**, 3878 (1991).

[7] S. W. Hawking and G. F. R. Ellis, *The Large Scale Structure of Space-Time* (Cambridge University Press, Cambridge, England, 1973).

[8] G. W. Gibbons and S. W. Hawking (unpublished).

[9] E. T. Newman and R. Penrose, J. Math. Phys. **3**, 566 (1962); **4**, 998(E) (1963).

[10] J. M. Stewart, *Advanced General Relativity* (Cambridge University Press, Cambridge, England, 1991).

[11] G. Klinkhammer, Phys. Rev. D **43**, 2542 (1991).

[12] C. W. Misner, in *Relativity Theory and Astrophysics I: Relativity and Cosmology*, edited by J. Ehlers, Lectures in Applied Mathematics Vol. 8 (American Mathematical Society, Providence, RI, 1967).

[13] W. A. Hiscock and D. A. Konkowski, Phys. Rev. D **26**, 1225 (1982).

[14] R. P. Geroch, J. Math. Phys. **8**, 782 (1967).

[15] J. L. Friedman and M. S. Morris, Phys. Rev. Lett. **66**, 401 (1991).